Human Body Composition

To my wife
Grace
Lumini vitae meae

Gilbert B. Forbes

Human Body Composition

Growth, Aging, Nutrition, and Activity

With 92 Figures

Springer-Verlag
New York Berlin Heidelberg
London Paris Tokyo

Gilbert B. Forbes, M.D. Professor of Pediatrics and
Radiation Biology and Biophysics, The University
of Rochester School of Medicine and Dentistry,
Rochester, New York 14642, USA

Library of Congress Cataloging in Publication Data
Forbes, Gilbert B.
Human body composition.
1. Body, Human—Composition. 2. Human growth.
3. Aging. 4. Nutrition. I. Title. [DNLM: 1. Body
Composition. QU 100 F692h]
QP514.2.F58 1987 612 86-31472

Typeset by David A. Seham Associates, Metuchen, New Jersey.

9 8 7 6 5 4 3 2 1

ISBN-13: 978-1-4612-9100-8 e-ISBN-13:978-1-4612-4654-1
DOI: 10.1007/978-1-4612-4654-1

Preface

Man has always been curious about himself, a curiosity that began centuries ago with an examination of the soul, and that extended in the period of the Renaissance to his anatomy and certain functions such as the circulation of the blood. Chemical science entered the scene in the 18th century, and burst into prominence in the 19th century. As the various chemical elements were discovered, many were found to be present in body fluids and tissues. Organic compounds were recognized; it became known that body heat was produced by the combustion of food; chemical transformations such as the production of fat from carbohydrate were recognized; and in the 1850s it was determined that young animals differed from adults in certain aspects of body composition. As methods for chemical analysis evolved, they were applied to samples of body fluids and tissues, and it became apparent that life depended on chemical normality; and most importantly it was realized that given the necessary amount of food and water the body had the ability to maintain a degree of constancy of what Claude Bernard called the *milieu intérieur,* in other words its interior chemical environment. Having arisen originally from the sea, animals had to take with them the salty medium they had lived in, and to incorporate it, albeit in somewhat changed form, into their body fluids—their *milieu intérieur*—which circulated inside to serve as a medium through which body tissues could communicate with the outside world and hence be nourished, and rid of wastes. This evolution from a primal aquatic environment to life on land required the development of appropriate physiologic mechanisms and compositional attributes over the course of many eons. Many of the mysteries of these features were solved by the multitudinous efforts of the 19th century chemists, while the unlocking of other mysteries remained for the advanced technology to come in the present century.

It is a truism that function follows form; and the "form" that I deal with here is body composition. In this "modern" era where the literature is saturated with papers on DNA, enzyme conformation, gene mapping, and the like, it may seem rather old-fashioned to deal, as I propose to do, with body composition techniques for the study of man, in essence to build on the concepts of former days when investigators first became interested in discovering what the body was made of, not just in anatomical terms but to exploit the methods of the new disciplines of chemistry and physiology. Recent years have seen the development of tools that can be applied to the living, to conduct a "bloodless dissection" of the human body so as to construct what James Gamble called a chemical anatomy; to discover what happens during growth and aging; and to learn some of the consequences of disease and of altered nutrition.

It is my aim to bring together the results obtained by these modern investigators. It was they who first saw clearly the phenomenon of dehydration as a body fluid volume defect, and who forged the tools and the concepts necessary for the development of a chemical anatomy. There are reasons to believe that body composition, particularly fat content, may have something to do with disease.

The literature in this field is vast indeed. My task has been eased by the efforts of others—notably Elsie Widdowson, Francis Moore, Joseph Brožek, and Donald Cheek, who have provided such excellent compendia of pertinent information, including their own work. There would be little point in constructing a mere litany of all of the published work in the field of human body composition; rather, I shall try to present some basic concepts, and some information on techniques and their application to the human state both in health and disease; to construct a work in which the curious may browse and the serious may delve. Hopefully, the former will be stimulated to explore the application of body composition techniques, and the latter will find enough information, known in modern jargon as "hard data," to satisfy their scientific needs.

This book provides information on gross composition, and I leave it to others to deal with the burgeoning literature on membrane transport, subcellular architecture, enzymatic processes, regulatory mechanisms, "tight junctions," and the like. The body as a whole, and its composition as deduced from measurements made *in vivo,* is my theme.

The human body is a marvelous machine, its composition gathered from its environment, in such a way as to provide a framework for its function. As W.A. Fowler (1984) so aptly put it in his Nobel lecture: ". . . all of the heavy elements from carbon to uranium have been synthesized in stars. Our bodies consist for the most part of these heavy elements . . . Thus it is possible to say that each one of us and all of us are truly and literally a little bit of stardust."

A comment on my own career may be in order. It has been my good fortune to be the recipient of a Research Career Award from the National Institute of Child Health and Human Development. This offered me a degree of scientific security (and with a minimum of "paper work"!), and made it possible for me to proceed at my own pace. True, this pace may have been tortoise-like in comparison to that of others, but my observations of scientists have convinced me that with rare exceptions new ideas spring forth from their minds only at long intervals. This means that they need time for reflection and for planning, and to be relieved of the burden of applying for research funds at frequent intervals. The NIH should seriously consider the reinstitution of these awards.

The policy of the University in keeping interdepartmental barriers low has been of great help: without the facilities of the Department of Radiation Biology and Biophysics, which were generously made available to me (I was trained in Pediatrics), much of my research could not have been done.

My thanks to those who have helped me so much in my own work over the years: John Hursh, Charles Lobeck, Sidney Breibart, Frank Schultz, Cenie Cafarelli, G.H.

Amirhakimi, Julio Reina, Jan Bruining, Michael Bryson, and Anne Harrison; to Anne Perley, Anne Lewis, Augusta McCoord, Richard Cuddihy, Eulalia Halloran, Janice Smith, Kathleen Morris, and Cheryl Porta; to my most competent secretary Marjorie Ewell; to the Atomic Energy Commission and the National Institutes of Health for supporting my research; and to my dear wife for her unswerving moral support.

Rochester, New York
June 1986

Contents

1

Historical Introduction

Creatio anima scientiae est, instrumenta corpus.

Anon.

Nature has been kind to the body compositionist, in providing the investigator with the necessary materials, and with the capacity to manufacture others, together with the required technological expertise. But the crowning achievement was nature's construction of an isotope of potassium (^{40}K) that is radioactive and present in sufficient quantity (0.1 μCi) to permit its estimation by specially designed scintillation counters. Knowing the isotopic abundance of ^{40}K (0.0118%) permits one to estimate total body potassium by this means, without the need to administer anything to or to obtain samples of any sort from the subject.

The last few decades have seen the emergence of a number of techniques for assessing certain aspects of body composition in man. The result is a better appreciation of the growth process and to some extent of the aging process. This book attempts to bring these observations and conclusions together.

Scientists are ever in debt to the past: efforts to discover various aspects of body composition began a century and a half ago. The pioneer 19th century biochemist, Justus von Liebig, found that the body contained many of the substances present in food, and that body fluids contained more sodium and less potassium than tissues. In the mid-19th century Albert von Bezold (1857), a young German student, discovered that growth in animals was accompanied by a decrease in water content and an increase in ash content. Then came the detailed analyses of muscle by Katz (1896) with results that are surprisingly close to those of modern chemists.

Lawes and Gilbert (1859) documented the change in body fat content as well as some increase in lean, as animals to be used for meat were fed for market; incidentally, they noted that body water varied inversely with fat content. Bischoff (1863) analyzed several adult human cadavers for water content. This was followed by similar analyses of fetuses and newborns by Fehling (1877) and by Camerer and Söldner (1900). By the end of the 19th century the chemical composition of the fetus as regards water, fat, nitrogen, and major minerals had been established; a full analysis of the adult was not accomplished until much later.

Meanwhile, Ludwig Pfeiffer (1887) had noted that the variation in water content of animal bodies could be reduced if the data were expressed on a fat-free basis; and in 1906 Adolph Magnus-Levy announced that in considering tissue composition *"Der sehr fettarme Muskel eines verhungerten Tiers kann nicht ohne weiteres dem fettreichen bei normaler Ernährung gegenüber gestellt werden. Zulässig ist nur der Vergleich fettfrei berechneter Organe."*[1] In a certain sense this was a confirmation of Claude Bernard's hypothesis of the constancy of the *milieu intérieur.*

So the concept of the fat-free body mass was born, a concept that underlies most methods for estimating body fat content in living subjects. In the 1930s Hastings and Eichelberger (1937) refined this concept by a series of tissue analyses. Neutral fat does not bind water, nitrogen, or electrolyte, and it is now common practice to express the results of tissue analysis on a fat-free basis.

At the turn of the century Voit (1901) and Rubner (1902) both spoke of an "active protoplasmic mass" to which certain physiological functions could be related, and Philip Shaffer (Shaffer and Coleman, 1909) used urinary creatinine excretion as an index of muscle mass.

Although early investigators had learned that body composition changed during growth, it remained for Moulton (1923) to announce the concept of "chemical maturity," namely that stage of life when body composition approaches the adult value and hence is considered "mature." Moulton concluded that mammals reach chemical maturity at about 4% of their life span. If this holds for man, the age would be about 3 years, which is probably too early, since childhood occupies a much longer portion of the life span of our species than is true for other mammals.

To complete the story of chemical composition one must include the whole-body assays carried out by Iob and Swanson (1938) in the 1930s and later by McCance and Widdowson (Widdowson and Dickerson, 1964), who expanded their analyses to include trace minerals in fetus, newborn, and adult.

Meanwhile the concept of body fluid volume(s) was being developed. Early efforts were made to determine blood volume by administration of carbon monoxide (see review by Gregerson and Rawson, 1959). Then in 1915 the report of Keith, Rowntree, and Geraghty on the determination of blood volume by dilution of dyes (Vital Red, Congo Red) stimulated much work in this area, including studies on patients with various abnormalities, and shock in wounded World War I soldiers.

Drawing on the results of blood volume determinations, Marriott (1923) announced the concept of "anhydremia" as an important feature of infantile diarrhea, and he and others showed the beneficial effects of restoring

---

[1]The fat-poor muscle of a starved animal cannot be compared directly to one normally nourished and rich in fat. It is permitted only to compare organs on a fat-free basis.

body fluid deficits in dehydrated infants and children. His work initiated the modern study of the physiology of body fluids.

The metabolic balance technique, which involves measuring intake of an element in food and drink and output in urine and stool, the algebraic sum of which indicates a gain or loss of the element in question, had been used for some time in metabolic studies. For example, Cathcart (1907) found that nitrogen was lost from the body during fasting, and Benjamin (1914) determined that infants retained nitrogen as they grew. Benedict and associates (1919) found that even a modest reduction in food intake was accompanied by losses of body nitrogen.

However, it fell to investigators such as James Gamble, Daniel Darrow, and Edmund Kerpel-Fronius to use the metabolic balance technique to detect changes in various body fluid compartments. It was already known that blood plasma contained much more Na and much less K than tissues such as muscle, and studies of patients with massive edema had shown that the fluid that accumulated in the subcutaneous tissues and in the peritoneal space resembled an ultrafiltrate of plasma; and the same was true to a large extent of cerebrospinal and intraocular fluids. These investigators reasoned that Na balance and K balance could be used to detect changes in extracellular and intracellular fluid volumes, respectively. They were able to show that states of water loss or gain primarily involved changes in extracellular fluid volume, and that intracellular volume and composition tended to be more stable.

Soon after the discovery of deuterium, George von Hevesy (1934) used this isotope to estimate total body water by isotopic dilution, and later Francis Moore (1946) introduced the concept of total exchangeable sodium and potassium. By applying Archimedes' principle Albert Behnke (1942) showed us how to estimate the relative proportion of lean and fat in the human body, and made it possible to distinguish overweight from obesity. Keys and Brozek (1953) provided us with a detailed analysis of the densitometric technique. McCance and Widdowson (1951) applied dilution techniques to study total body water and extracellular fluid volume (with urea and thiocyanate) from which they calculated cell mass and body fat in human subjects. They were able to show that the rehabilitation of undernourished individuals was accompanied by an increase in both body fat and in cell mass, and a decrease in relative extracellular fluid volume. Although radiologists had known for some time that the human body emanated γ-rays, it remained for Rudolph Sievert (1951) to discover that the body content of ^{40}K was sufficient to be detected, and quantitated, by suitable instruments; this opened the way for the use of ^{40}K assays to estimate lean and fat in a noninvasive manner (Forbes et al., 1961). More recent years have seen the use of neutron activation (Anderson et al., 1964; Cohn et al., 1971), of computerized tomography, and of other electrical techniques for this purpose; tomorrow may witness the use of nuclear magnetic resonancc.

The result of all of this activity is that we now have a large amount of data on human body composition both in health and in disease. With these new techniques at hand, investigators no longer have to labor under the methodological constraints of the metabolic balance technique in assessing long-term changes in body composition. Changes that occur during growth, and with aging, are now well defined, as are those that accompany altered nutrition.

2

Techniques for Estimating Body Composition

> Although this may seem a paradox, all exact science is dominated by the idea of approximation.
>
> *Bertrand Russell*

Body Fluid Volumes

In normal individuals from one-half to two-thirds of the body weight is water (in the very obese it can be as little as one-third). It is generally agreed that the body fluids exist in several compartments; and although these compartments have recognizable anatomic boundaries, in the physiologic sense they are interconnected, since water, amino acids, electrolytes, carbon dioxide, and oxygen pass freely from one compartment to another.

Extracellular fluid (ECF) has two components: the blood plasma and the interstitial fluid (ISF) which bathes tissue cells. The fluid contained within cells is termed intracellular fluid (ICF). Together these account for most of the body water. The remainder are some specialized fluids, to which Francis Moore has given the name "transcellular," namely, the intraocular, synovial, and cerebrospinal fluids, and the fluid within the lumen of the intestine. In addition, Neuman and Neuman (1958) have provided evidence for the existence of a hydration layer associated with the apatite crystals of bone. To a large extent these "transcellular" fluids are in communication with plasma and ISF, yet each possesses special transport mechanisms and hence special functions, or in the case of bone, where water is bound and electrolytes are adsorbed, to the crystal surface.

In tissues such as liver and muscle, the chemically determined ECF space (i.e., Cl space) corresponds in value to that which can be determined by planimetry of suitably prepared histologic sections (Truax, 1939; Barlow and Manery, 1954). In the central nervous system, on the other hand, the microscopic ECF space is much less than the Cl space, and in bone the latter may exceed the water content (Forbes et al., 1959), and so the Cl space cannot be used as an index of ECF volume in these tissues.

Body cells are connected to the outside world by means of blood plasma and ISF, and it is these avenues that provide for gas exchange in the pulmonary alveoli, absorption of nutrients and water from the gastrointestinal tract, excretion of solutes and water by the kidney, and the transfer of materials to the fetus. These transfers of materials from the environment to the interior of the body cells, and from the cells to the outside world, have been found by modern techniques to take place very rapidly and to involve considerable quantities. Orally administered radioisotopes appear in the bloodstream within a few minutes, and when given intravenously they enter the interstitial fluid very rapidly, and quickly appear in pancreatic juice, milk, sweat, and urine. In the adult the fractional turnover of body water is about 7% per day, and that of body Na is about 7% per day. Several liters of fluid are secreted into the gastrointestinal tract, and reabsorbed, each day. A portion of the transfers of water and solutes between the various body fluid compartments is by simple diffusion and by the existence of "loose junctions" between cells; the remainder is by means of active transport mechanisms provided by cell membranes. The structure of the central nervous system differs in that the junctions between cells are "tight," so that transfers are impeded—the so-called blood–brain barrier. The stability of the several body fluid volumes in health is due to an array of hormonal and circulatory functions.

Determination of Body Fluid Volumes

Early in his career every student of chemistry is taught a simple technique for determining the volume of a given solution: one adds a known quantity of solute, mixes thoroughly, and then analyzes an aliquot of the mix for the added solute. Since the amount of added solute (Q) is constant, $C_2V_2 = C_1V_1$, where C is concentration and V is volume, so that V_2, the volume in question, is equal to Q divided by its concentration in the aliquot taken for analysis:

$$V_2 = \frac{C_1V_1\ (= Q)}{C_2} \tag{1}$$

The volume to be determined has of course been increased by the added volume V_1, but in most experiments V_1 is so much smaller than V_2 that it can be neglected.

The various techniques used for estimation of body fluid volumes are, with one exception, all based on this principle. A known amount of material is given to the subject, and after an interval of equilibration a sample of blood, urine, saliva, or in some instances exhaled water vapor is obtained for analysis, whereupon the calculation is made according to Equation 1.

The one exception is the procedure used by Cohn et al. (1984) to estimate ECF volume. Assuming that body Cl is all in the ECF compartment, these investigators determined total body Cl by neutron activation, whence

$$\text{ECF (L)} = \frac{\text{total body C1 (meq)}}{\text{meq Cl/L serum ultrafiltrate}} \tag{2}$$

This equation is the same as Equation 1, when injected radiochloride is substituted in the numerator, and radiochloride concentration in serum ultrafiltrate in the denominator.

However, biological systems pose problems not encountered when one is determining the volume of fluid in a laboratory beaker. The first problem is that many administered tracers are excreted in the urine and so are lost from the body. Hence the dilution equation becomes:

$$V_2 = (Q \text{ administered} - Q \text{ excreted})/C_2 \qquad (3)$$

In the case of inulin, the constant infusion technique has been used to estimate ECF volume. Inulin is infused intravenously at a constant rate; and after a suitable interval the serum concentration stabilizes, whereupon the infusion is stopped, and the amount excreted in the urine is determined to the point of complete disappearance of the solute from the serum. This amount, then, is the amount of solute retained in the fluid phase at the time of equilibrium; division by the equilibrium concentration in serum yields a value for the volume of distribution. It is obvious that such a procedure is difficult to perform, and its accuracy depends on how well the equilibrium concentration of solute can be achieved. Poulos et al. (1956) have described what they consider to be a better method for estimating the volume of distribution of inulin, one that does not require a knowledge of the injected dose. Postinjection blood samples are obtained at 2 and 4 hr, and the urine excretion of inulin is determined in this same interval. The volume of distribution is then calculated as urine excretion of inulin between t_1 and t_2 divided by the change in plasma concentration during the same time interval.

The second problem is that the body fluid most often used for analysis, namely blood plasma or serum, contains appreciable amounts of protein, so that each milliliter contains less water than an equivalent amount of ISF. Moreover, the Donnan equilibrium operates to render solute concentrations in serum water slightly different than in ISF: in general, cation concentrations are higher in the former than in the latter whereas the reverse is true for anions. Since ISF cannot be sampled easily, a correction—really two corrections—must be made for results of serum analyses.

The concentrations of the major constituents of plasma, ICF, and cerebrospinal fluid are given in Table 2.1. Sodium is the major cation of plasma, and hence of ECF, whereas potassium is the major cation of ICF; the major anions are, respectively, chloride and phosphate. Within each compartment the sum of cations and anions is the same, so that electroneutrality is preserved. However, there is a small potential difference across the cell membrane (the resting membrane potential), the interior being about 90 mV negative to the exterior. Osmotic equilibrium between the ECF and ICF is maintained by the free diffusion of water across the cell membrane.

The large difference in the Na/K ratio between ECF and ICF is achieved, and maintained, by an active transport process, the Na-K pump. It is now known that this "pump" is a Na^+,K^+,Mg^{2+}-dependent ATPase; the hydrolysis of ATP provides energy for the active transport of Na and K across the cell membrane. Hence Na, for example, is actively transported out of cells against an electrochemical gradient.

TABLE 2.1. Body fluid constituents.

	Blood plasma (meq/liter)		Cerebrospinal fluid (meq/liter)[a]	Intracellular fluid (meq/liter cell water)[b]
	Infant	Adult		
Cations				
$Na+$	140	142	144	10
K^+	5	4	3	150
Ca^{2+}	5 (10 mg/dl)	5	2.5	2
Mg^{2+}	2 (2 mg/dl)	2	2.1	25
Anions				
Cl^-	103	100	123	
HCO_3^-	24	27	20	10
Phosphate	3 (5.5 mg/dl)	2 (3.5 mg/dl)	1	100
Sulfate	1	1	1	20
Organic acids	7	6	4	
Protein	14 (6 g/dl)	16 (7 g/dl)	20 mg/dl	60

[a]There is disagreement on precise values for CSF ion concentrations; the values here represent averages of those reported in the literature.
[b]Calculated from the Cl^- space.

To calculate interstitial fluid Na concentration from serum Na concentration, the latter must be divided by the serum water concentration and multiplied by the Donnan factor. While the most precise result can be obtained from actual measurement of serum water, for most purposes this can be assumed to be 93%; the Donnan factor for Na is 0.95. Hence

$$\text{ISF [Na]} = \frac{\text{serum [Na]}}{0.93} \times 0.95 \qquad (4)$$

(bracketed values are concentrations).

The Donnan factor being the same value but in the opposite direction, the equation for chloride is:

$$\text{ISF [Cl]} = \frac{\text{serum [Cl]}}{0.93 \times 0.95} \qquad (5)$$

The Donnan factor for Br is 0.95. In some reports in the literature this factor appears to have been improperly applied (see Bell et al., 1984).

The third problem is that most materials used to estimate body fluid volumes enter the lumen of the gastrointestinal tract (exceptions are inulin and thiosulfate) so the calculated ECF/total water ratio will be lower, and the ICF/total water ratio higher when these two substances are used. The fluid contained therein is usually not considered as an integral part of the ECF; however, such fluids are in equilibrium with true ECF. For instance, it is estimated that several liters of water enter and leave the gastrointestinal tract each day in the adult, and although the total amount contained within the lumen at any one time in a fasting adult human is not a large fraction of the ECF volume, considerable amounts are present in ruminants.

A further consideration has to do with the time needed for equilibration of the administered tracer. In studies of human subjects lasting about 2 hr Burch et al. (1947) found that the time course of blood radiosodium following intravenous injection could be characterized by an exponential function. Fifty-three percent of the injected dose left the blood stream with a half-time of 1.2 min, and 18% with a half-time of 11 min. Then in experiments of 60 days duration the remaining radio Na declined with a half-time of 13 days (Burch et al., 1948). Equilibration kinetics were different in patients with congestive heart failure, and the half-time of this third component varied inversely with salt intake. Extrapolation of the second component back to time zero indicates a volume of distribution roughly equivalent to ECF volume.

The slow decline phase represents loss of the isotope in urine, and penetration into erythrocytes, cerebrospinal fluid, and bone. It is evident that true equilibrium for the injected isotope is never completely achieved. However, in practice it is customary to use the volume of distribution at 2–4 hr postinjection as the value for "sodium space."

This picture of radiosodium kinetics would not be complete without reference to a fourth exponential component that has a very long biologic half-life. When human subjects and animals are monitored for periods of half a year or more postinjection, 0.1–0.4% of the injected dose of ^{22}Na leaves the body with a half-time that ranges from 200 to 700 days (Forbes and McCoord, 1969). At such times about 95% of the body burden is in the skeleton. Since the half-time of injected radiocalcium is of the same order of magnitude, it is likely that the ^{22}Na representing this fourth component has been incorporated into the apatite crystal of bone.

Figure 2.1 shows the time course of serum radiochloride after intravenous injection. Burch et al. (1950) identified four exponential components: 45% of the injected dose left the blood stream with a half-time of 0.3 min, 25% with one of 2.2 min, 5% with one of 18 min, and the remainder with one of about 3 days. This last component primarily represents urine excretion. Using bromide as a tracer, others have found a half-time of 14 days for this long component in man (Moore et al., 1968). A difference in salt intake may account for this discrepancy. The effect of salt intake on the turnover of radiosodium was mentioned earlier, and Moore et al. (1968) have shown an effect of water intake on the turnover of ^{3}H in humans, from a half-time of 5 days on high intakes of water to one of 13 days on normal intakes, and then to a prolonged half-time (precise value unknown) when water intake was restricted for a period of 20 hr.

It is clear that turnover rates of injected isotopes are dependent to a large extent on the intake of the companion isotope. Metabolic rate is also a factor, since small animals generally have higher turnover rates than large animals, and human infants higher rates than adults.

Coleman et al. (1972) studied the fate of injected ^{3}H in dogs, and found that the rate of approach to equilibrium in various regions in the body varied with the rate of blood flow in those regions. It was not until 100 min had elapsed that distribution equality was established.

The situation becomes more complicated when a metabolizable tracer is used. Figure 2.2 shows the time course of antipyrine concentration in blood following intravenous administration. In such situations it is nec-

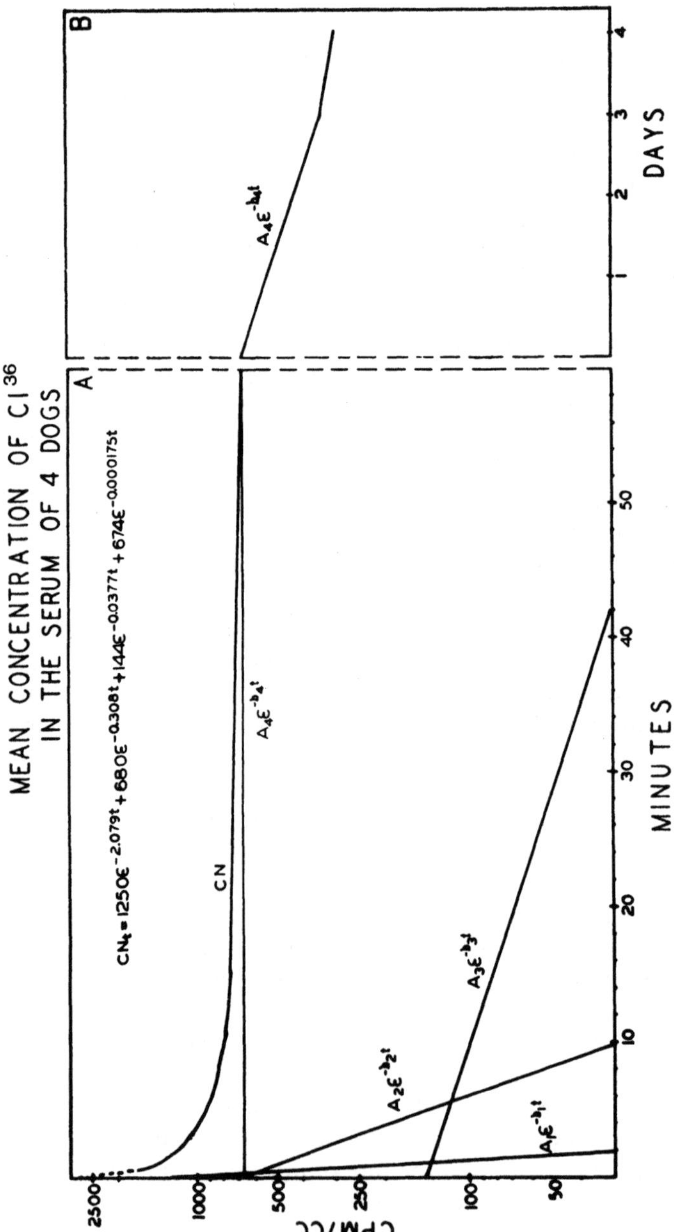

FIGURE 2.1. Time course of radiochloride (counts per minute per cc serum) after intravenous injection, resolved into four exponential functions. (From Burch GE, Threefoot SA, Ray CT: Rates of turnover and biologic decay of chloride space in the dog determined with the long-life isotope, Cl^{36}. J Lab Clin Med 35:331–347, 1950, with permission.)

FIGURE 2.2. Plasma levels of antipyrine in normal subjects at various times after intravenous injection. Extrapolation of semilog plot to zero time. (From Soberman R, Brodie BB, Levy BB, Axelrod J, Hollander V, Steel JM: The use of antipyrine in the measurement of total body water in man. J Biol Chem 179:31–42, 1949, with permission.)

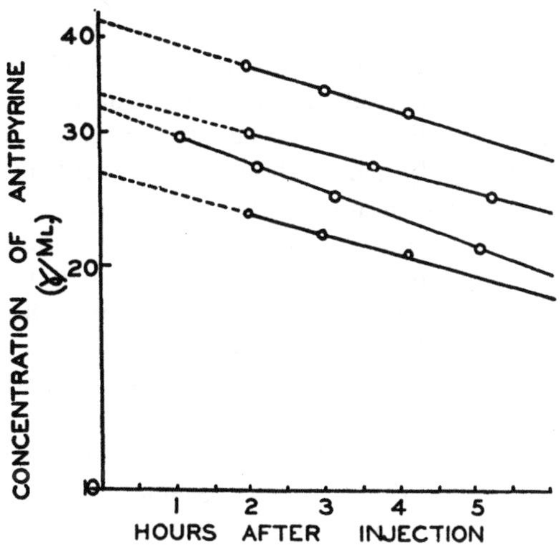

essary to take several blood samples at known intervals after injection, and provided they follow a log-linear course (which they usually but not always do) one can calculate a regression line and derive the zero time intercept. Note that the semilog slopes vary somewhat among the four subjects shown. The data shown in the paper by Soberman et al. (1949) indicate that antipyrine levels in tissue water are comparable to those in plasma water by 3 hr postinjection.

The same situation obtains when urea is used to estimate total body water, since it is rapidly excreted in urine. The rate of labelled albumin loss from the intravascular space is about 8% per day in adult man (Rossing, 1967).

It is evident that the use of these various dilution techniques involves a great deal of care. The administered dose must be accurately known, as must be the equilibrium concentration in blood; the compartment being assessed should not be undergoing changes in volume during the mixing period; the material used should not alter normal physiologic processes. It is in regard to this last point that radioactive materials have proved such a boon to the physiologist and body compositionist. For example, 50 μCi of ^{24}Na (a suitable tracer dose for an adult) contains only 5.6×10^{-12} g Na.[1] If one were to attempt to use stable Na for estimating "Na

---

[1]From the fundamental law of radioactive decay, $dN/dt = \lambda N$, one can solve for the number of atoms present (N) from a knowledge of the decay rate (dN/dt) and the decay constant (λ), which is ln 2/half-life. Hence the number of atoms present in 50 μCi of ^{24}Na ($t_{1/2}$ 14.8 hr, or 888 min; $\lambda = 7.8 \times 10^{-4}$/min) can be calculated as follows:

$$N = \frac{50\ \mu\text{Ci} \times 2.2 \times 10^{6}\ \text{dpm}/\mu\text{Ci}}{7.8 \times 10^{-4}} = 1.41 \times 10^{11}\ \text{atoms}$$

$$\text{g Na} = \frac{1.41 \times 10^{11}\ \text{atoms}}{6.03 \times 10^{23}\ \text{atoms/mol}} \times 24\ \text{g/mol} = 5.6 \times 10^{-12}\ \text{g}.$$

space," the resultant increase in serum Na would change serum osmolality and thus lead to a redistribution of body water. Because radioactive tracers can be assayed in very small quantities with reasonable accuracy they offer the distinct advantage of not adding to the *net* quantity already present in the body and so should not perturb body fluid equilibria. Furthermore, the radiation dose provided by tracer amounts of radioactive materials is not sufficient to affect cellular metabolism.

Oral administration of tracer is usually easy. The administered volume must be completely swallowed, and followed by several rinsings from the vessel. The subject should have fasted overnight, and refrain from food or drink until the test is finished. In instances where the tracer is rapidly excreted in urine, a collection should be made so that the amount of tracer lost thereby can be measured. Some investigators have failed to do so in the belief that the error is small. The results will be difficult, if not impossible, to interpret in subjects who have gastric retention, diarrhea, or polyuria. Indeed Tzamaloukas et al. (1985) found spurious values for ethanol "space" in anesthetized dogs when the material was given by gastric tube, the anesthesia having produced gastric retention.

Intravenous administration of tracer presents more room for error. The graduations on the usual glass or plastic syringes are not accurate enough for small quantities. Some investigators weigh the syringe before and after the injection, but in so doing one has to avoid contamination with the subject's blood. In the case of radioisotopes that emit γ-rays, the full syringe can be counted prior to injection and then after the injection is finished.

Figure 2.3 shows a device for accurate intravenous administration that I have found useful. With the stopcock closed the injectate is put into the upright syringe (A) by means of an accurate (Bureau of Standards calibrated) pipette. This solution is then drawn into the horizontal syringe (C), the stopcock turned, and the solution injected into the vein. This is followed by two or three washings with a suitable diluent. One can even withdraw and reinject a small amount of blood if desired. In animal studies the intraperitoneal route is frequently used.

It should be remembered that some tracers may stick to glass or plastic syringes, particularly when the specific activity (i.e., ratio of tracer to stable isotope) is extremely high.[2]

The materials used to determine body fluid volumes are listed in Figure 2.4 and Table 2.2. Not all materials used to estimate a given body fluid compartment yield the same values, and Figure 2.4 portrays these differences. For example, the volume of distribution of inulin is considerably

---

[2]Precautions are mandatory in dealing with radioisotopes. Those emanating γ-rays or strong β-rays, which penetrate glassware, must be shielded from laboratory personnel. *Never* pipette radioactive material by mouth.

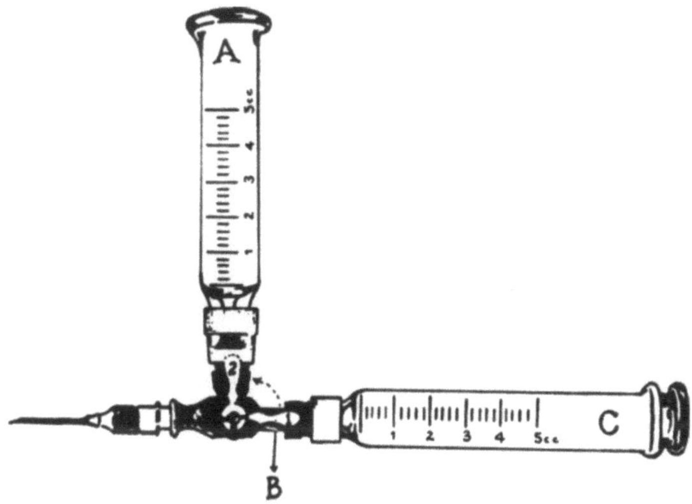

FIGURE 2.3. System for accurate intravenous injections. (From Barnett HL, Fellers FX: A simple quantitative method for intravenous injection of small volumes of fluid. Science 106:401–402, 1947, with permission.)

less than that of bromide, though both have been used as measures of "extracellular fluid" volume.

Two points are worth mentioning in connection with assays for total red cell mass. Labels such as ^{32}P, ^{51}Cr, and ^{99m}Tc can be incubated with the subject's own blood in vitro, and the labelled cells reinjected, while the radioiron label must be attached in vivo. The carrier for ^{51}Cr should be hexavalent chromium; I once had the unfortunate experience of using chromium of another valency, and so ruined an experiment because the label didn't stick to the cells.

Other labels have been tried, such as ^{42}K, and type O erythrocytes for non-type O subjects, the latter being determined in the recipient's blood by differential agglutination.

Swan et al. (1954) estimated ECF volume with several different indicators in nephrectomized dogs. Using radiochloride dilution space as 100%, the relative volumes were 61% for inulin, 71% for raffinose, 75% for sucrose, 81% for thiocyanate, and 82% for radiosulfate and mannitol. In the course of studies on the distribution of electrolytes in tendon and skin, Nichols et al. (1953) were able to confirm the theory of Manery et al. (1938) that the interstitial fluid is in actuality a connective tissue phase diluted with an ultrafiltrate of plasma.

As shown in Figure 2.5 when the inulin space is corrected for the connective tissue phase of skin and muscle—into which inulin penetrates very slowly—the inulin space corresponds to the chloride space in both tissues. In a series of in vivo and in vitro experiments Weil and Wallace (1960)

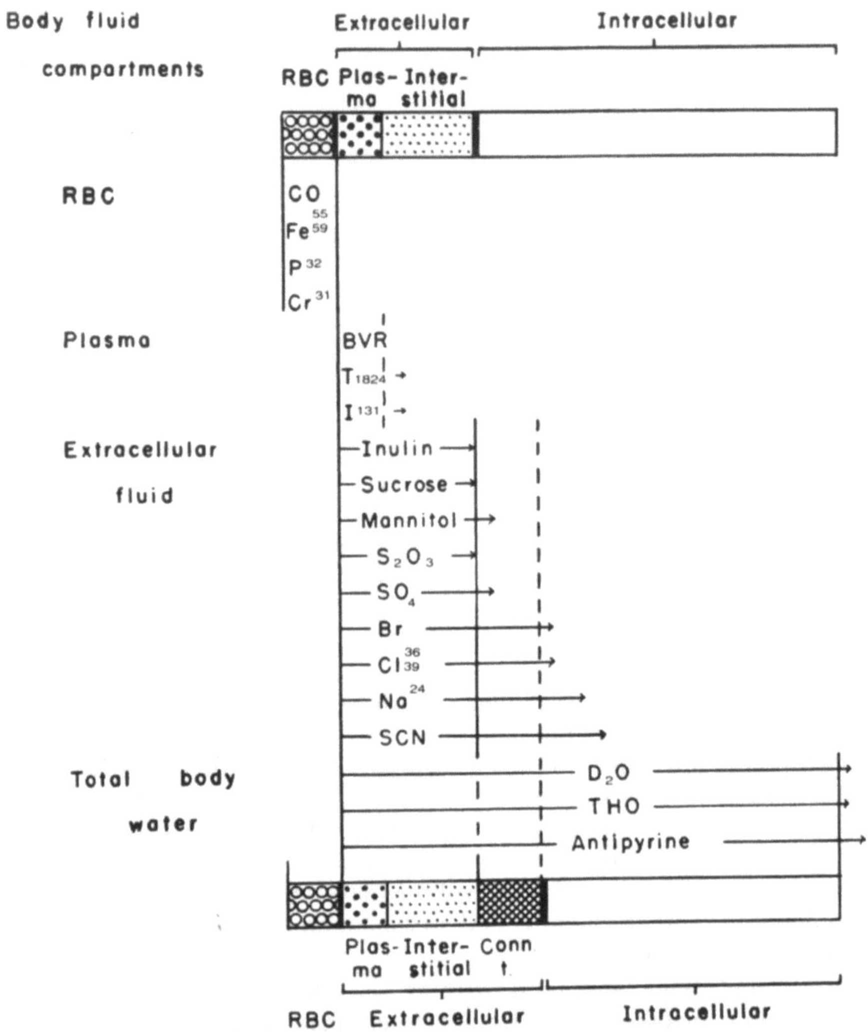

FIGURE 2.4. Materials used to estimate body fluid compartments. T_{1824}, Evans Blue dye; I^{131}, labelled serum albumin. (From Elkinton JR and Danowski TS: The Body Fluids. Baltimore, Williams & Wilkins Co., 1955, with permission.)

showed that almost all of the Na and Cl in skin and tendon were in diffusion equilibrium with plasma ultrafiltrate.

Widdowson and Dickerson (1964) estimate that of the total water in skin, connective tissue water accounts for 64% and interstitial water for about 23%. Hence a large fraction of the water considered to be extracellular in this tissue is associated with connective tissue elements. The collagen fraction of skin accounts for 80–90% of the total nitrogen content of this tissue. In the rat, collagen constitutes one fifth of total body protein (Wallace et al., 1958). Guyton (1981) pictures the ISF compartment of

TABLE 2.2. Additional materials for estimating body fluid volumes.

Volume	Remarks
Total RBC mass	
RBC labelled with ^{99m}Tc	Much lower radiation dose than radio Fe or Cr (Korubin et al., 1972); $t_{1/2}$ 6 hr
Extracellular fluid volume[a]	
^{77}Br, ^{82}Br	Stable Br analysis by fluorescent excitation (Kaufman and Wilson 1973), or by HPLC (Miller and Cappon 1984). Latter method is extremely sensitive (60μg Br/liter serum), permitting small doses. Preinjection blood sample needed.
Total body water	
$H_2{}^{18}O$	^{18}O is stable, measure in saliva as well as serum (Schoeller et al., 1980)
Urea	Rapidly excreted in urine
Alcohol	Breath analyzer (Grüner and Salmen, 1961; Loeppky et al., 1977)
n-Acetyl-4-aminoantipyrine	Not bound to plasma proteins

[a]ECF volume = Br "space" × 0.90, due to uptake by RBC.

many tissues not simply as a fluid-filled cavity, but rather as a mass of hydrated mucopolysaccharides in which are interspersed thin strands of collagen fibers and minute pockets of fluid. It behaves as a hydrated gel, and so is able to give some stability to the interstitial space as a supporting structure for the body cells. While bulk flow of fluid through the interstitium is not rapid, the results of isotopic tracer experiments show that diffusion of ions is very rapid indeed. This difference can be easily demonstrated by injecting 0.1 ml of isotonic NaCl labelled with ^{24}Na subcutaneously. The lump that has been raised thereby does not disappear until many minutes have elapsed whereas ^{24}Na can be detected in the blood stream within a minute or two. If an enzyme such as hyaluronidase is added to the solution the lump disappears much more rapidly (as does the ^{24}Na), for now the interstitial gel has been depolymerized so that bulk flow, as well as Na diffusion, are facilitated (Forbes et al., 1950). This enzyme effect has also been demonstrated in tissue preparations. Day (1952) constructed a membrane of muscle fascia and measured the transit velocity of a saline solution. When the enzyme was added, the velocity increased 10–20-fold. The volume of distribution of inulin, which under normal circumstances does not penetrate the connective tissue phase, is increased in isolated tissues by the addition of hyaluronidase (Goodford and Leach, 1962).

The gel-like character of the interstitium also acts to impede the progress of bacteria from the bloodstream to the surface of cells and so acts as one of the defense mechanisms against infection. It is a well known fact that organisms such as certain types of streptococci that produce hyaluronidase can result in lesions that spread through tissues rather rapidly, since the normal interstitial barrier has been breached.

In the presence of edema the ratio of fluid to gel in the interstitium is

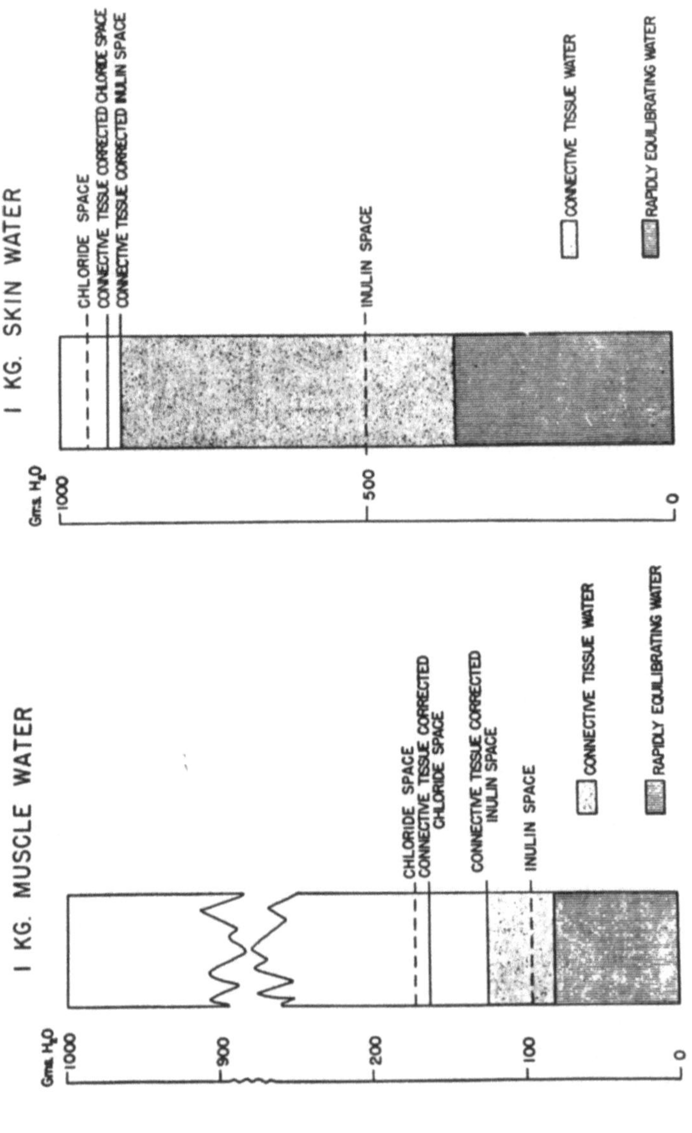

FIGURE 2.5. Inulin and chloride spaces in muscle and skin (dog). (From Nichols G, Jr, Nichols N, Weil WB, Wallace WM: The direct measurement of the extracellular phase of tissues. J Clin Invest XXXII:1299–1308, 1953, with permission.)

increased, so one would anticipate that bulk flow would be enhanced. McClure and Aldrich (1923) injected 0.2 ml of isotonic saline intradermally, and so could watch the disappearance of the bleb. In normal individuals 50–100 min elapsed before the bleb could no longer be felt, whereas in patients with edema the time was shortened to 3–34 min. Moreover the time of disappearance was inversely related to the degree of edema, and these authors proposed this technique as a test for the presence of occult edema.

In edematous individuals the subcutaneous tissues are easily indented by even moderate pressure ("pitting edema"), and are more distended over dependent portions of the body due to the influence of gravity; indeed in situations of marked edema it is possible to drain off fluid from the subcutaneous tissues with fine needles.

Under normal circumstances interstitial fluid does not behave as a simple aqueous solution. Bulk flow of fluid is impeded, the effect of gravity on body fluid distribution is minimized, and the positions of body cells within the intercellular compartment is stabilized. The gel-like property of the interstitium is most readily appreciated in tissues such as skin where the ratio of extracellular to intracellular fluid volume is relatively high. This dual function of the interstitium—providing as it does architectural stability while allowing rapid transport of materials between the blood plasma and the body cells—is of great importance to the body economy.

The interstitium also houses the lymph channels. Among other functions, they act as conduits for transporting back into the bloodstream proteins that leak from capillaries, and so help to keep the protein content of the interstitial fluids at a minimum. Such leakage of protein is evident from the fact that some materials used to estimate extracellular fluid volume (Evans Blue dye, for example) appear in lymph rather quickly following intravenous injection. Lymph flow is estimated to be about 120 ml/hr in adult man.

Intracellular water is of course the difference between total body water and ECF; its value will depend on the method used to estimate ECF.

Precision of the Various Techniques

The ultimate precision is limited by the fact that the human body is not a static system: body weight varies during the course of a day, being up to a kilogram higher in the evening than in the morning; and body weights taken in morning after voiding and prior to breakfast exhibit variations of 0.5 kg (SD) from day to day. A long drink of water can easily increase body water content by 1%, and an overnight thirst depletes body water by about 0.5%. A meal obviously stimulates the secretion of gastric and pancreatic juice and succus entericus into the lumen of the gastrointestinal tract. Diurnal variations in serum Na, Ca, K, and P have been documented, all of which suggest that body fluids ebb and flow to a certain extent. Indeed, in physicochemical terms the body is an "open" system: there is a continuous exchange of body constituents with the environment, and although such exchanges are minimized during fasting and thirsting they are never completely absent.

The careful studies of Peterson et al. (1959) on the apparent volume of distribution of sucrose indicated that this varied by as much as 8% during the course of a day. Moore Ede et al. (1975) found diurnal variations in urinary K excretion and in net ICF-ECF potassium flux even under conditions where the usual diurnal variations in posture, activity, and food intake were controlled.

Techniques for estimating body fluid volumes are also subject to technical errors: the dose of administered tracer involves measurement of the amount of solution and the tracer concentration: then analysis of blood or urine at equilibrium; hence at least three measurements must be made. In the case of a substance such as *N*-acetyl-4-aminopyrine which is rapidly excreted, several blood samples must be obtained with subsequent extrapolation to zero time.

The only way to determine the total error—biologic plus analytical—is to do repeated determinations on the same individual, under standardized conditions and in situations where body weight has not changed very much. Table 2.3 shows the results of such studies. It is apparent that body fluid volumes cannot be assayed with a precision much better than 3–5%.

Spears and associates (1974) did a series of assays on 43 adult males, and again a week later. While there was no change in average body weight during this interval, average plasma volume increased by 1.4%, erythrocyte mass fell by 1.6%, ECF volume fell by 1.3%, and total body water increased by 0.3%, and so the ratio of ECF to total body water declined by 1.6%. Only average values were presented by these authors, so one cannot determine the magnitude of the individual variations which might well have occurred.

Unless one is prepared to measure red cell and plasma volume simultaneously, the estimation of blood volume by either method alone poses yet another problem, namely that the hematocrit value (upon which the calculation is based) is known to vary with the size of blood vessels. Hahn et al. (1942) found that large vessel venous hematocrit was not representative of the vascular system as a whole in the dog. Furthermore, a certain amount of plasma, usually taken as 4%, remains trapped in the red cell layer after centrifugation. Later Nachman et al. (1950), using ^{32}P-labelled

TABLE 2.3. Reproducibility data.

Technique	Coefficient of variation	Reference
Plasma volume	1.9%, 4.7%, 6%	Haxhe (1963), von Porat (1951), Mather et al. (1947)
RBC mass	3.1%	Haxhe (1963)
Extracellular fluid volume	2.3%, 3.4%	Price et al. (1969), Haxhe (1963)
Total body water	1.8%, 2.5%	Haxhe (1963), Price et al. (1969)

red cells and T-1824 dye, reported the same phenomenon in humans, and observed the average ratio of red cell mass by ^{32}P to calculated red cell mass by T-1824 dye to be 0.80 in 38 surgical patients, with a range of 0.58–0.99. Correction of the hematocrit for trapped plasma changed the average ratio to 0.87.

Later Kirsch et al. (1971) were to find even less satisfactory results. Using ^{51}Cr-labelled red cells and ^{125}I-labelled albumin in a series of patients, they found an average value for the ratio of calculated total body hematocrit to venous hematocrit of 0.86, but the range was wide (0.70–1.20). The correspondence was particularly poor in patients with polycthemia. Studies such as these point to the fallacy of attempting to estimate total red cell mass by measuring plasma volume and venous hematocrit. Accurate estimates of total blood volume require that both components be measured by appropriate indicators.

Gibson et al. (1946a) reported this same discrepancy in both humans and dogs, and they also checked the reliability of the red cell mass determination by doing repeat studies after bleeding and transfusion. They found that changes in red cell mass as estimated by red cells labelled with radioiron corresponded very well with the amounts of erythrocytes removed or transfused. These investigators (Gibson et al. 1946b) also explored the microcirculation by means of radioiron-tagged red cells, and determined that the minute vessels of viscera, muscle, and brain together contained about 17% of the total blood volume, and further that *all* of the erythrocytes in the vascular bed were in active circulation. Contrary to the conclusions of others, they found that the spleen did not constitute a reservoir of noncirculating erythrocytes. One hastens to add, however, that the situation is different for abnormal erythrocytes, since spherocytes and sickle cells can be trapped in the spleen.

The Metabolic Balance Technique

Since chloride is largely confined to the extracellular compartment, and about 95% of total body potassium is located intracellularly, one can estimate *changes* in body fluid volumes by measuring the changes in body content of these two elements.

The metabolic balance technique consists of measuring the intake and outgo of the element in question over an interval of time. When intake exceeds outgo the balance is said to be positive, and when outgo exceeds intake the balance is negative. The procedure consists of giving a known amount of food and drink each day and collecting all excreta quantitatively. It should be remembered that tap water contains variable amounts of Na, Cl, Ca, and Mg, so it is best to provide distilled water to drink. Duplicate diets are analyzed for the element(s) in question, as are urine and feces. It is best to take 24-hr urine samples for analysis, and fecal collections for 3–6-day periods delimited by suitable stool markers. Excretion via the skin, nasal and vaginal secretions are usually neglected, if it is anticipated that the balance will be strongly positive or negative. Additional details are given in a subsequent section of this chapter.

Change in *ECF volume* is calculated from the change in Cl concentration in ECF, the external balance, and an initial ECF volume (V_1), either assumed or measured, as follows (the brackets indicate concentration):

$$V_2 = (V_1 \times [Cl_1] + \text{balance Cl})/[Cl_2] \qquad (6)$$

$$\Delta\text{ECF volume} = V_2 - V_1$$

$[Cl_2]$ and $[Cl_1]$ are calculated from their respective serum concentrations, corrected for serum water and the Donnan equilibrium factor. In situations where serum Cl concentration does not change, V_1 does not have to be known, or assumed, and Equation 6 simplifies to:

$$\Delta\text{ECF volume} = \text{Balance Cl}/[\text{Cl}] \qquad (7)$$

Change in ICF volume can be determined only from a knowledge of change in total body water and change in ECF volume. Estimation of change in total body water requires a knowledge of change in body weight, change in body solids, and the nature of the fuel that is burned during the balance period.

As described by Peters and van Slyke (1946), the procedure involves measurement of solids in excreta and ingesta; urinary nitrogen; the amounts of protein, fat, and carbohydrate metabolized, the insensible weight loss; and the change in body weight. Carbohydrate burned is taken as carbohydrate ingested; protein burned is calculated as urinary nitrogen × 6.25; fat burned is equal to insensible loss minus corrections for protein and carbohydrate burned; and insensible loss is determined from weights of intake and outgo and the change in body weight.

All of this represents a most difficult task to perform; however, investigators such as Passmore and co-workers (1963) have made use of this technique to estimate body water changes in subjects who were induced to gain or lose weight.

Lacking such an approach, one can assume that rapid changes in body weight represent changes in water content for the most part. Hence

$$\Delta\text{ICF volume} = \Delta\text{ weight} - \Delta\text{ECF volume} \qquad (8)$$

Large sweat volumes, which are most difficult to measure accurately, invalidate all such calculations.

Ill patients present an additional problem. The anions of intracellular fluids are derived in large part from polyvalent acids of organic compounds; a change in their concentrations due to disease will alter the osmotic pressure, and so distort ECF Na-ICF K relationships (Darrow and Hellerstein, 1958).

Nevertheless, Darrow et al. (1945) used the balance method to deduce shifts in intracellular Na concentration, and concluded that ICF Na diminished in states of acidosis in man. Darrow et al. (1949) also used the balance method to show that body potassium is lost in infants with diarrhea.

In summary, although the balance method offers an opportunity to investigate changes in body fluid volumes, the technique is time-consuming and demands meticulous attention to detail, a large number of chemical analyses, and a great deal of cooperation on the part of the subject and his attendants. In expert hands, the technique is capable of defining relatively small changes in body fluid volumes, changes smaller than those that can be detected by dilution techniques. The relative ease of these latter procedures has increased their popularity in recent years.

Changes in Red Cell Mass and Plasma Volume

Change in plasma volume *(PV)* can be calculated from the relative red cell volume (hematocrit, Hct) and the hemoglobin (Hb) concentration of whole blood, and an assumed or measured initial plasma volume (PV_1), as follows:

$$PV_2 = \frac{PV_1\,(1 - \mathrm{Hct}_2)\,\mathrm{Hb}_1}{(1 - \mathrm{Hct}_1)\,\mathrm{Hb}_2} \tag{9}$$

$$\Delta PV = PV_2 - PV_1$$

This calculation is based on the assumption that the total red cell mass has not changed and that the hematocrit is indicative of the entire blood volume hematocrit. The hemoglobin values correct for possible changes in erythrocyte water content.

In performing exchange transfusions it is sometimes desirable to end the procedure with a different red cell mass than that which obtained initially. Formulas have been developed for use in infants receiving exchange transfusions for the treatment of hyperbilirubinemia.

The procedure of exchange transfusions consists of drawing a syringe full of blood from the patient and then injecting an equal volume of donor blood; this is repeated many times. The percent of the recipient's own blood volume remaining after n withdrawals and replacements is:

$$\left[\frac{\text{recipient's blood volume} - \text{syringe volume}}{\text{recipient's blood volume}}\right]^n \tag{10}$$

For example, if initial blood volume of the infant is 200 ml and the syringe volume is 10 ml, after 10 exchanges 60% of the infant's own blood remains, after 20 trials 36%, after 30 trials 21%, and so forth.

The final hematocrit Hct_f achieved by exchange transfusions is calculated thus:

$$\mathrm{Hct}_f = \mathrm{Hct}_d - [\mathrm{Hct}_d - \mathrm{Hct}_i]\left[\frac{V_b - V_e}{V_b}\right]^n \tag{11}$$

where Hct_d is donor blood hematocrit, Hct_i is initial hematocrit of the patient, V_b is blood volume, and V_e is volume of each withdrawal/infusion, and n is number of withdrawals/infusions (Berman et al., 1979).

Example: let $Hct_d = 0.6$, $Hct_i = 0.3$, $V_b = 200$ ml,and $V_e = 10$ ml. For $n = 1$, $Hct_f = 0.6 - 0.3\,(200 - 10)/200 = 0.315$; for $n = 2$, $Hct_f = 0.329$, and so on.

For plasma exchange transfusions, which may be done in patients with polycythemia or to remove toxins, Hct_d is zero.

The change in hematocrit following a transfusion with blood of a hematocrit different than the recipient's hematocrit depends on assumptions regarding the induced change in blood volume. If there is no change, which is often the case some hours following the transfusion, the calculation is as follows: assume initial blood volume 70 ml/kg, initial hematocrit 30%; transfuse 10 ml blood/kg with hematocrit 70%; then initial RBC mass is $70 \times 0.30 = 21$ ml/kg, transfused RBC mass is $10 \times 0.7 = 7$ ml/kg, and the final hematocrit will be $(21 + 7)/70 = 40\%$. On the other hand, if blood volume is increased by the amount transfused, the final hematocrit will be $(21 + 7)/(70 + 10) = 35\%$.

The general formula is:

$$Hct_f = \frac{\text{blood volume} \times Hct_i + \text{transfused blood} \times Hct_{(t)}}{\text{blood volume}} \tag{12}$$

Total Exchangeable Electrolyte

This technique also makes use of the isotope dilution method. A known amount of a tracer is administered, usually intravenously, and after a period of equilibration a sample of blood or urine is obtained for analysis. Body content (Q) of the element in question is calculated as follows:

$$Q = \frac{Q^* \text{ administered} - Q^* \text{ excreted}}{Q^*/Q \text{ (serum, urine)}} \tag{13}$$

where Q^* is the tracer isotope (^{24}Na, ^{22}Na, ^{42}K, ^{82}Br, or stable Br). Bromide is used as a tracer for Cl since there are no convenient radioisotopes of Cl. The result is expressed as total exchangeable content—Na_e, K_e, Cl_e—since this procedure underestimates total body content of both Na and K in man.

Total Exchangeable Na (Na_e)

Although injected radiosodium rapidly equilibrates with extracellular fluid, and quickly enters the gastrointestinal tract and appears in the urine, and although it also readily enters the aqueous and cerebrospinal and articular fluids and to a slight extent the interior of cells, the exchange with erythrocytes and bone is incomplete. About 30% of total body Na is in skeleton, and in the adult animal only about 15% of bone cortex Na undergoes exchange with serum Na. Confirmation of the fact that Na_e is less than total body Na has recently been adduced from measurements of Na_e by isotope dilution, and of total body Na by neutron activation in human subjects (Ellis et al., 1976; Kennedy et al., 1983). These results confirm the early postulation of Forbes and Perley (1951) that only 70–75% of total body Na is exchangeable in the adult. Ellis et al. (1976) found that the amount of this "excess" Na is correlated with total body calcium (r = 0.75) and so in all likelihood is located in bone.

Radiosodium exchange is probably more complete in the young organism. This opinion is based on two considerations. First is the fact that exchange in bone occurs much more readily in young animals than in the adult: exchangeable bone Na in the young rat represents about 70% of total bone Na (Forbes et al., 1959), whereas this value drops to 40% in the older animal, and to as low as 15% in cortical bone of the rabbit and cat (Forbes, unpublished data). Second is the fact that the infant skeleton has a large component of cartilage, estimated to be 70% in the newborn infant (Forbes, 1960), and studies in my laboratory have shown that cartilage Na undergoes complete exchange with serum Na within half an hour. The opportunity was presented some years ago to dissect the tibia of a 3-year-old child; 25% of the total weight was cartilage.

The usual period of equilibration for estimating Na_e is 24 hr (Forbes and Perley, 1951) although some investigators have used shorter periods. Urine must be collected during the equilibration period since 1–5% of the injected dose will appear in the urine. ^{24}Na was used initially, but this isotope has a half-life of only 14.8 hr, which means that the counting of blood and urine samples must be timed to the nearest 10 min in order to realize suitable corrections for radioactive decay (^{24}Na decays by 1% in 13 min). Moreover, large quantities (50 mCi or more) must be on hand if the isotope is to be used for more than 2 or 3 days. This isotope has both a strong γ- (1.38 and 2.76 MeV) and β-ray (1.39 MeV), so personnel must take suitable precautions.

^{22}Na has a much longer half-life (2.6 years) and hence there need be no hurry about using it once the supply arrives in the laboratory. It, too, has an energetic γ- (1.3 MeV) and β- (0.5 MeV) ray. As mentioned earlier, a small portion of the injected dose (less than 1%) is very slowly eliminated from the body of experimental animals, and hence probably from man; this feature adds to the total radiation dose. Although this aspect makes one hesitate to use ^{22}Na in infants and children, it has been used extensively in adults.

Veall et al. (1955) have used the whole body counter to estimate total exchangeable sodium. ^{22}Na is given to the subject, and the amount retained by the body is assayed in the whole body counter to determine the numerator of Equation 13 (p. 22); the denominator is urine specific activity.[3] These authors found that Na_e determined 7 days after administration of ^{22}Na was about 10% greater than that at 24 hr.

[3]Ratio of radioisotope to stable isotope: For example, ^{24}Na disintegrations per unit time per milligram Na.

Although Na_e obviously does not provide a value for total body Na, it does provide information on that portion of body Na that exchanges fairly readily with serum Na, and so represents a quantity of some physiological importance. Rapid changes in body Na undoubtedly involve only Na_e, and so its estimation is of some interest to physiologists and clinicians.

Total Exchangeable K (K_e)

The most readily available isotope of potassium (^{42}K) also has a very short half-life, namely 12.4 hr. Hence, counting times are critical if suitable corrections for physical decay are to be made, since it decays by 1% in the short time of 10 min. It has a strong γ- and β-ray. Its short half-life means that the radiation dose from tracer amounts is very small.

Wilde (1962) has studied the dynamics of ^{42}K equilibrium in animals. Intravenously administered ^{42}K leaves the blood very rapidly; 89% of the dose leaves with a half-time of 0.3 min. Four additional exponential coefficients have been identified, with half-times of 2.6–4400 min, the longest one accounting for 1.7% of the injected dose. In the rat at 2 hr postinjection the ratio of ^{42}K/chemical K content exceeds 1.0 in kidney, lung, liver, intestine, and skin, and is <1.0 in brain and muscle. By the seventh hour the ^{42}K/chemical K ratios are close to 1.0 for all tissues save brain and erythrocytes. In the rabbit, erythrocytes had only 89% and brain only 63% of muscle specific activity at 62 hours postinjection. Figure 2.6 shows the time course of ^{42}K in plasma, urine, and erythrocytes of pigs following intravenous injection. Boddy et al. (1973) found that values for K_e determined at 44 hr were 7–12% greater than at 24 hr postinjection, showing that the isotope continues to penetrate slowly exchanging tissues.

Total exchangeable potassium has been compared to chemically determined K in animals. Talso et al. (1960) studied 12 rats ranging in body weight from 150 to 310 g, and found that K_e underestimated body K by about 2%. Corsa et al. (1950) studied rabbits in a similar manner, and found an underestimation of 4%. On the other hand, Shizgal et al. (1977) found only a 1% underestimation by their method of estimating K_e. The turnover rates of many substances are much faster in small animals than in adult man, so these results may not be applicable to humans in quantitative terms.

When K_e is compared to total body K by ^{40}K counting in human subjects it turns out that K_e underestimates body K content by 3–10% (Rundo and Sagild, 1955; Surveyor and Hughes, 1968; Jasani and Edwards, 1971; Davies and Robertson, 1973; Boddy et al., 1974; Lye et al., 1976). The short half-life of this isotope means that the equilibration time cannot be much longer than 36 hr (three half-lives) unless large doses are given.

Investigators in Dr. Francis Moore's laboratory devised an ingenious method for measuring Na_e and K_e simultaneously (McMurrey et al., 1958). Serum and urine samples are run through an ion-exchange column that neatly separates the two isotopes for counting. When combined with other

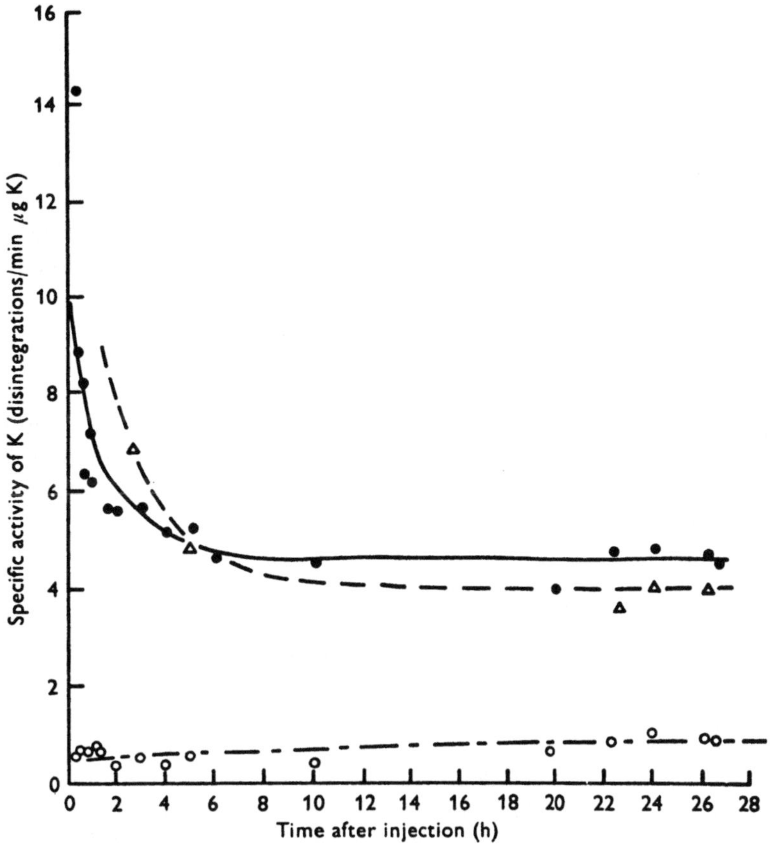

FIGURE 2.6. Time course of specific activity of potassium in plasma (•), urine (△) and erythrocytes (○) in pigs after intravenous injection of ^{42}K. (From Fuller MF et al. (1971), Br J Nutr 26:203–214, with permission.)

techniques, this group of investigators were able to simultaneously estimate total body water, ECF volume, Na_e and K_e, red cell mass, and plasma volume in their subjects. However, such a procedure demands a smoothly functioning laboratory, precise timing, and many calculations, and is obviously unsuited to routine use.

Samples of ^{42}K should be checked for possible contamination with ^{24}Na before use. Sodium can be found in many materials, and because it has a very high cross-section (i.e., activation potential) for the radiations used to prepare ^{42}K, whether these be neutron or deuteron bombardment, its presence even in trace amounts can result in ^{24}Na production. The chances of ^{24}Na contamination are therefore real, and the fact that the half-times for ^{42}K and ^{24}Na are so close means that they cannot be distinguished on the basis of decay rates.

^{43}K should be more convenient to use than ^{42}K because its physical half-life is twice as long, but this isotope is not readily available. However, there are reports of its use to estimate K_e (Davies and Robertson, 1973; Skrabal et al., 1970; 1972; Boddy et al., 1973).

Shizgal et al. (1977) have devised a method for estimating total exchangeable K (K_e) without the need to use the short-lived and inconvenient ^{42}K. The method derives from the demonstrations by Edelman et al. (1958) of a good relationship ($r = 0.989$) between total body water and the ratio of $Na_e + K_e$ to serum Na + K (serum water basis), in a variety of patients. Total body water and total exchangeable Na (Na_e) are measured (the latter with ^{22}Na) together with Na, K, and H_2O concentrations in whole blood. Then

$$K_e = \left[\frac{Na + K}{H_20}\text{(blood)} \times \text{total body } H_20\right] - Na_e. \tag{14}$$

In studies on a large number of subjects, including a variety of patients, these authors found a good correlation between K_e as estimated by the usual method and K_e as estimated by this new technique [standard error of estimate (SEE) = 141 meq K].

Total Exchangeable Chloride (Cl_e)

Unfortunately there are no convenient radioisotopes of chloride: ^{34}Cl and ^{38}Cl have half-lives of 33 and 37 min, respectively, whereas ^{36}Cl has a very long half-life (10^6 years) and hence cannot be prepared in high specific activity. Consequently, either ^{82}Br or stable Br is used as a tracer for chloride; ^{77}Br has a longer half-life, but is not readily available.

Cheek and West (1955) compared the results of stable Br dilution in a series of rats whose body fat content ranged from 1 to 19% with total body Cl as determined by carcass analysis. The average value for the latter method was 8.48 meq, and for the former it was 8.45 meq. Gamble et al. (1953) determined the volumes of distribution of intravenously administered ^{82}Br, stable Br, and ^{38}Cl in man. At 2.5 hr postinjection, the stable Br "space" was 7% greater than the Cl "space," and the ^{82}Br "space" was 2% greater.

Wallace and Brodie (1939) and Weir and Hastings (1939) studied the tissue distribution of injected bromide in animals. The former authors found identical tissue: serum ratios for Br and Cl in many organs, the one exception being brain where the ratio for Br was only one-third to one-half that for Cl. The latter authors calculated apparent ECF volume for both ions in various tissues: Br produced a higher value for stomach and a lower value for brain, and equilibration in cerebrospinal fluid was incomplete. In all of the many other tissues examined Br and Cl yielded similar results.

Edelman (1961) has derived values for the distribution of body water, sodium, chloride, and potassium from work in his own laboratory and from the literature. Table 2.4 presents these estimates for a young adult male as percent of total body content allocated to various body compo-

TABLE 2.4. Distribution of water and electrolyte (young adult man).

	Water (%)	Sodium (%)	Chloride (%)	Potassium (%)
Plasma	7.5	11.2	13.6	0.4
Interstitial fluid—lymph	20	29	37.3	1.0
Dense connective tissue—cartilage	7.5	11.7	17.0	0.4
Bone	7.5	36.5	15.2	7.6
Transcellular fluid	2.5	2.6	4.5	1.0
Total extracellular	45.0	91.0	87.6	10.4
Total intracellular	55.0	9.0	12.4	89.6
Total	100	100	100	100
(Total exchangeable content)	(100)	(70)	(94)	(90)

From Edelman (1961), with permission.
The listed value for bone K is too high: cortical bone contains 6 meq/kg (Table 2.8), so the adult bony skeleton contains about 36 meq K, or only 1% of total body K.

nents. Fortunately, in adult humans the values for the "transcellular" compartment are small. The values for bone should be accepted with some reservation, since, as will be explained later, they depend on the care which is used to rid the samples of marrow, periosteum, and cartilage prior to analysis. The cellular elements of this tissue—osteocytes, osteoblasts, and osteoclasts—make up such a minute fraction of the total that in anatomical terms bone can be considered as a part of the extracellular component of the body, and the same is true for cartilage. However, in functional terms much of the sodium in bone is associated with the apatite crystals and hence is not able to participate actively in the general extracellular fluid economy of the body.

The significance of total exchangeable sodium and total exchangeable potassium is manifested by their relationship to total body water in a variety of conditions in man. Except for situations associated with high concentrations of glucose and urea, serum sodium values correlate with serum osmolality and intracellular potassium concentrations with intracellular osmolality. The identity of intracellular and serum osmolality is maintained by the passive diffusion of water. On this basis Edelman and his associates (1958) reasoned that there should be an association between serum sodium and the ratio of Na_e and K_e to total body water. As mentioned on p. 26, they did indeed find a high correlation in a variety of patients in whom serum Na ranged from subnormal to supranormal levels.

Other Elements

Although the possibility exists for estimating total exchangeable magnesium by ^{28}Mg ($t_{½}$ 21 hr) dilution, there is no evidence in the literature to suggest that this technique has been exploited. About two-thirds of total body Mg is located in the skeleton, and like Na, bone Mg is more exchangeable in young animals than in adult animals (Breibart et al., 1960).

Estimation of Lean Body Mass and Fat

Neutral fat does not bind water or electrolytes, so the measurement of total body water or total body potassium offers a means for estimating the nonfat component of the body. This component is designated either as lean body mass (LBM) or fat-free mass (FFM). The author interprets these terms as synonymous, and prefers the former as the more delicate of the two. As used here, LBM means body mass minus ether-extractable fat, and hence includes the stroma of adipose tissue.

Behnke and Wilmore (1974) have characterized the LBM as possessing 3–5% of its weight as "essential lipid," i.e., structural as opposed to storage lipid. However, the exact amounts of such structural lipids, which are contained in cell membranes and as phospholipid in nerve and brain, present in the body have never been defined. Keys and Brozek (1953) have estimated that structural lipids in brain and nerve amount to only 200–300 g in the adult. In evaluating such postulates, it should be remembered that "structural" lipids are not removed by the usual methods for fat extraction.

Brozek et al. (1963) prefer to use an entity called the "reference man," a young adult man designated to contain about 15% of his weight as fat, to which can be added variable amounts of "obesity tissue." In their study, the composition of such "obesity tissue" was estimated from assays of individuals who had been induced to gain weight by overfeeding. The results indicated that the tissue gained was composed of extracellular fluid and protein as well as neutral fat.

The total body water and total body potassium methods for estimating LBM are based on the assumption that this component of the body has a constant composition. Hence division of the measured value for total H_2O or total K by their respective concentrations in the LBM yields a value for LBM. Body fat is then weight minus LBM.

This concept considers the body to be a simple two-component system; although obviously an oversimplification (both H_2O and K contents of fat-free tissue differ among various body tissues) it has nevertheless been found useful in many situations. Incidentally, the densitometric technique rests on the same assumption.

As will be mentioned later, the water and K contents of the LBM are not age invariant: both the newborn infant and the aged differ from the young adults in these respects.

Water Content of the LBM

Sheng and Huggins (1979) have recently reviewed the literature on this topic and have compared total body water by ^2H and/or ^3H dilution with results obtained by carcass analysis. Unfortunately, some investigators have analyzed the eviscerated carcass of their animals, and so their results cannot be taken as indicative of the situation in vivo. Moreover, in some of the species studied, especially sheep, goats, guinea pigs, and cattle, an appreciable fraction of the total body water is located within the lumen of the gut, and since administered ^3H and ^2H readily enter the gut, total body water estimated by isotopic dilution will exceed that measured by carcass analysis when gut contents are not included. This problem of excessive gut water is undoubtedly related to the high fiber content of the diet of these animals, and to their rather large gastrointestinal tract.

Fortunately, studies have shown that this problem is not a major one in man, for the gut lumen contains 1 to 2% of total body water in our species, at least in the adult (Table 2.4). There is no such information for infants.

Analyses of five adult humans show that the water content of the LBM varied between 69.4 and 73.2% (Widdowson and Dickerson, 1964). Addition of earlier data (Forbes, 1962) produces a grand average of 72.4%. Adults of other mammalian species—monkey, pig, dog, cat, rabbit, rat, and mouse—reveal values ranging from 72.0 to 78.0% (Widdowson and Dickerson, 1964).

Sheng and Huggins (1979) came to the conclusion that the water content of the LBM lay between 70 and 76% for most species analyzed.

Many of the specimens analyzed, including those from man, have not contained large amounts of fat. Since assays of total body water are used to estimate total body fat *in vivo*, the question as to whether the water content of the LBM varies with body fat content is of some importance. The careful studies of Annegers (1954) in which the entire animal (minus hair) was analyzed for water and fat do show a H_2O/LBM ratio of 73.0% ± 0.73 (SD) for male rats, 72.2% ± 0.79 for female rats, and 74.0% ± 1.4 for the female mouse over a range of 6–35% body fat.

We analyzed 41 adult female rats, some of whom were induced to gain weight by feeding a palatable diet. After removal of the brain, gut contents, and 2 g of liver, the minced carcasses were dried to constant weight and then extracted with ether; body fat content ranged from 5 to 52%. The water content of the fat-free body was 70.6% + 0.78 (SD) for 35 animals with body fat $<$40%, with no evidence of a trend; however, the average was slightly lower, 69.3% + 0.48 for the six animals with 41–52% body fat (unpublished data).

Lesser et al. (1963) estimated body fat by gas absorption (see p. 56) and total body water by tritium dilution in eight women and eight men of varying age and body fat content. For the males the calculated water content of the fat-free body was 70.8% + 2.2 (SD) and for the females it was 70.9% + 3.9. In their carcass analyses of a large number of rats and mice, whose body fat content varied from 7 to 69%, Cox et al. (1985) found a high correlation (r = 0.988) between percent body water and percent body fat. The calculated water content of the fat-free mass was 75.4%.

Gnaedinger et al. (1963) reported an excellent correlation (r = 0.97) between percent water and percent body fat in pigs; their value for the water content of the fat-free mass as determined by carcass analysis was 74.5% + 0.55.

The water content of various organs is shown in Table 2.5. Included are several mammalian species in addition to man. Although there is some variation among the organs, the range of values for any one organ is not very great. The value for bone cortex pertains to samples that have been cleaned of marrow, periosteum, and trabecular bone; the last is very difficult to clean without losing some water in the process. Unfortunately, some authors do not state whether marrow has been removed prior to analysis. This is important, because the more marrow that is included the higher the observed water content will be. Incidentally, dental enamel has only about 3% water.

The effect of marrow contamination on analytical values for bone has been studied (Forbes, 1960). The long bones were removed from adult rats, and cleaned of periosteum. The bone cortices from the right extremities were carefully cleaned of marrow while the marrow was left intact in those from the left extremities. On a fresh weight basis, water content was 26% in bones containing marrow but only 15% in those with marrow removed; ash content was 52% versus 63%, Na content 0.18 meq/g versus 0.21, Cl content 0.18 meq/g versus 0.16, and Ca content 9.0 meq/g versus 11.0. These data show that failure to remove marrow prior to analysis leads to spuriously high values for water and Cl content and lower values for Na, Ca, and ash contents. In these animals we also found that cancellous bone had a lower ash content and a higher water content then cortical bone.

The values listed in Table 2.5 pertain to adults; as will be shown in a subsequent chapter, water content of many organs does vary with age.

TABLE 2.5. Water content of various organs.[a]

Organ	Percent fat-free wet weight	Percent fresh weight
Skeletal muscle	77.4–80.9[b]	73.5–79.2
Cardiac muscle		82.7
Smooth muscle		77.4–81.0
Myometrium		79.4–81.9
Adipose tissue	81, 83, 84[c]	
Tendon	66.7, 61.8[d]	
Skin	67.1–72.8	
Liver	73.8–77.0	
Erythrocytes		66.4
Kidney		75.7–81.2
Spleen		77.5–79.0
Pancreas		77.1
Lungs		78.7–82.1
Placenta		92.9
Brain		75.8–78.4
Spinal cord		64.6–68.5
Peripheral nerve		54.7–68.9
Bone cortex[e]		12–16
Cartilage[f]		70–85

[a]From Widdowson and Dickerson (1964) unless otherwise stated.
[b]Five muscles from a single individual.
[c]Forbes (1962); Thomas (1962); Manery et al. (1938).
[d]Nichols et al. (1953); Manery et al. (1938).
[e]Forbes and McCoord (1963), rat; cancellous veal bone is 52% water (Forbes, 1960).
[f]Epiphyseal, costal (Manery, 1954).

In using ^3H or ^2H for estimating total body water it should be remembered that these isotopes undergo some exchange with nonaqueous hydrogen. Culebras et al. (1977) have computed on theoretical grounds the maximum amount of such hydrogen to be 5.2% of the total exchangeable hydrogen of the body. Special candidates for such exchange are the hydrogens on carboxyl and hydroxyl groups. Thompson (1953) made a study of the long-term fate of injected tritium in rats: after an initial fall in plasma tritium concentration (half-time 3.3 days) there remained a minor fraction (<0.1% of initial dose) that exhibited a half-time of 80–280 days among various organs. As Sheng and Huggins (1979) point out, ^3H dilution measures a volume that is 4–16% larger than that obtained by desiccation.

It is the usual practice to use the following formula for calculating LBM from THO or D_2O dilution in adult men:

$$\text{LBM} = \text{total body } H_20 \times \frac{0.95}{0.73} \qquad (15)$$

$$\text{Fat} = \text{weight} - \text{LBM}$$

When the isotopes are given intravenously 1 or 2% of the dose may be excreted in the urine and through the skin and breath during the equilibration period of 3–5 hr. In man, the half-time of an injected isotope of hydrogen is about 10 days, so the fractional loss rate from the body is about 7% per day; hence the loss in 5 hr is about 1.4%. The turnover rate in small animals such as the rat is about 21% per day, so 4.4% of the dose will be lost in 5 hr. A similar rapid rate has been found in human infants (MacLennan et al., 1983).

The error produced by assuming that the water content of the LBM is, for example, 70% in a given subject rather than 73% is about 4% for calculating LBM. However, for an individual with 14% body fat, the error in fat estimation is much larger, namely 10%. On the other hand, for an individual who is 50% fat the error in fat is the same as that for LBM.[4]

[4]The same problem obtains for all methods that carry the assumption of constancy of LBM composition. The following formula applies to all:

$$\text{Percent error in fat} = \%\text{ error in lean} \times \frac{\text{lean}}{\text{fat}}$$

where lean + fat = weight. Examples for situations in which the error for estimating lean is 4%:

(a) Subject with 60 kg LBM, 10 kg fat:

$$\text{error in fat} = 4 \times \frac{60}{10} = 24\%$$

(b) Subject with 60 kg LBM, 30 kg fat:

$$\text{error in fat} = 4 \times \frac{60}{30} = 8\%$$

This is why very thin subjects may appear to have zero percent fat, or even negative values on rare occasions. Such anomalies have also been recorded for thin subjects assayed by densitometry (Behnke and Wilmore, 1974).

TABLE 2.6. Reference man.

Component	Weight (g/kg)	Density (36° C)
Water	624	0.9937
Protein	164	1.34
Fat	153	0.9007
Bone mineral	47.7	2.982
Nonosseous minerals	10.5	3.317
Total	999.2	1.064

From Brozek et al. (1963), with permission.

Density of the LBM and of Body Fat

Keys and Brozek's (1953) extensive review explains the basis for estimating the density (weight/volume) of the LBM; and Brozek et al. (1963) have provided an update on some quantitative assumptions. Since it is impossible to measure the density of the LBM *in toto,* its density must be deduced from the densities of its components together with their relative contribution to the weight of the entire LBM.

Brozek et al. (1963) made the following compilation based on the composition of the "reference man" (Table 2.6). Omitting fat, the density of the fat-free mass becomes 1.100, and it is this difference between the density of lean and fat that permits one to calculate the relative proportions of each from a measurement of the density of the body as a whole.

Archimedes is said to have discovered this principle. What he apparently did was to determine that the specific gravity[5] of a crown that was claimed to be gold was actually less than that of pure gold, and so proved that the object had been adulterated with another metal. Although a number of attempts had been made in the past, Albert Behnke and his associates (1942) were the first to use the specific gravity technique in a systematic fashion in humans.

The formula for calculating the proportions of lean and fat from the density of the body is derived as follows: let W_f represent weight of fat, W_l the weight of lean, and W_b the weight of the body; and V_f, V_l, and V_b their respective volumes. Since density (D) = W/V, the relationship $V_b = V_f + V_l$ can be written as:

$$\frac{W_b}{D_b} = \frac{W_f}{D_f} + \frac{W_l}{D_l}$$

[5] "Specific gravity" is not the same as "density." The former is the ratio of the density of a substance to the density of water. Density is weight/volume, and its determination by underwater weighing requires a knowledge of the density of water at a given temperature, and if one wants to be very precise, also the density of air.

and since $W_l = W_b - W_f$, it follows that

$$\frac{W_b}{D_b} - \frac{W_b}{D_l} = \frac{W_f}{D_f} - \frac{W_f}{D_l}$$

whence

$$W_b\left(\frac{1}{D_b} - \frac{1}{D_l}\right) = W_f\left(\frac{1}{D_f} - \frac{1}{D_l}\right).$$

Dividing W_f by W_b yields the following (letting $\frac{1}{D_f} - \frac{1}{D_l} = \frac{1}{a}$):

$$\frac{W_f}{W_b} = \frac{a}{D_b} - \frac{a}{D_l}$$

Since D_l is known (letting $\frac{a}{D_l} = a'$):

$$\text{fraction fat } \frac{W_f}{W_b} = \frac{a}{D_b} - a', \text{ and} \qquad (16)$$

$$\text{fraction lean} = 1 - \frac{W_f}{W_b}.$$

Using the values for density of lean (1.100) and of fat (0.9007 at 36°C, Table 2.6), the formula becomes

$$\frac{F}{W} = \frac{4.971}{D_b} - 4.519 \text{ (temperature 36°C)}, \qquad (17)$$

and percent fat is $\frac{F}{W} \times 100$.

This formula is very sensitive to changes in the chosen values for D_f and D_l. For example, the density of fat at 37°C is 0.9000 (Fidanza et al., 1953), so if this temperature is chosen (on the assumption that all body tissues, including subcutaneous fat are at that temperature) the formula becomes

$$\frac{F}{W} = \frac{4.95}{D_b} - 4.50 \qquad (18)$$

The method of Brozek et al. (1963) uses the concept of the "reference man" ($D = 1.064$, Table 2.6) to which fat is added or subtracted ($D_f = 0.9007$) to yield the following:

$$\frac{F}{W} = \frac{4.570}{D_b} - 4.142 \qquad (19)$$

The difference in calculated percent fat using these three formulas is fortunately not very great. As an example, for an individual with body density 1.050, calculated body fat would be 21.5%, 21.4%, and 21.0% by these three formulas.

As is the case for the total body water, the densitometric technique involves the assumption that the density of the lean component is constant; hence the ratios of skeletal weight to water to protein are assumed not to vary. Although this may be a reasonable assumption in the normal adult, it obviously does not hold for those who are edematous or dehydrated, and as will be shown later, the density of the fat-free body mass in infants and children undoubtedly differs from the adult.

TECHNIQUE: UNDERWATER WEIGHING

This involves weighing the subject in air (W_a), and then submerged in water during maximum exhalation (W_w), whence

$$D_b = W_a/(W_a - W_w)$$

Three additions must be made to this basic formula in order to account for (1) the effect of temperature on the density of water (D_w); (2) the effect of pulmonary residual volume (RV); and (3) the effect of air in the gastrointestinal tract. Consequently, the formula for body density becomes

$$D_b = \frac{W_a}{[(W_a - W_w)/D_w] - RV - 100 \text{ ml}} \tag{20}$$

The density of air can be neglected, and since it is very difficult to determine the amount of air in the gastrointestinal tract, this is usually assumed to be 100 ml.

Pulmonary residual volume (RV) can be measured by one of several methods: oxygen rebreathing, hydrogen rebreathing, nitrogen washout, and helium dilution (Rahn et al., 1949; Brozek and Henschel, 1961). Repeat studies show that the standard deviation of the difference distribution is 59–84 ml, or about 6% of mean residual volume. Wilmore (1969) has simplified the original nitrogen washout technique by reducing the dead space of the apparatus; the measurement of RV can be done quickly because the subject is required to take only five to eight breaths from the oxygen-filled spirometer. Other investigators have estimated residual volume as 25% of vital capacity.

The importance of this measurement lies in the fact that a 200 ml error in an individual whose RV is 1200 ml will change calculated body density by 0.0025 g/cm^3. Better accuracy is achieved when RV is measured with the subject submerged, since the pressure of the water on the chest causes a reduction in RV of 200–300 ml (Craig and Ware, 1967).

Ideally, the subject should be fasting, and should not have eaten foods that promote the formation of intestinal gas during the preceding day; the bladder should be emptied. The actual procedure consists of weighing the subject nude (or in a close-fitting bathing suit) in air to the nearest 20–50 g, then weighing again while completely submerged and in maximum expiration, with the breath held long enough for the scales to be read. Bathing caps are prohibited because they trap air bubbles. Since underwater weight tends to increase with successive weighings (hence calculated density will increase), at least four successive underwater weighings should be made. The temperature of the water, which should be close to 37°C, is noted, and the submerged weight of the chair in which the subject is seated is recorded. The calculation of density proceeds according to Equation 20, with the underwater weight of the chair being subtracted from the underwater weight of subject plus chair.

This technique is obviously not suited to young children, the elderly, or those who are ill or who have pulmonary abnormalities. Success demands excellent cooperation on the part of the subject, and there will be a few otherwise normal individuals who are unable to complete the procedure satisfactorily.

Water Displacement

The procedure is the same as that for underwater weighing, except that the volume of water displaced by the submerged subject is read on a manometer. Corrections for water temperature and pulmonary residual volume are the same, as is the need for subject cooperation.

Irsigler et al. (1979) and Garrow et al. (1979) have devised a method that does not require the subject to immerse his head. The subject is immersed up to the neck, and the volume of the head and neck is determined by application of Boyle's law.

Duplicate assays on large numbers of subjects by various investigators have shown errors ranging from 0.0004 to 0.0043 g/cm^3 (SD difference distribution), with most in the range of 0.0011–0.0043 (Buskirk, 1961). Durnin and Taylor (1960) did repeated assays on several rather thin subjects over a period of several days during which their body weights varied by <1 kg. They found coefficients of variation of 0.3–1.5% for LBM, and determined that the 90% confidence limit for a single assay was ± 0.0046 g/cm^3; Keys and Brozek's (1953) value was ± 0.005 g/cm^3. These values should be viewed in the context of the total range of possible body densities, namely 0.90–1.10 g/cm^3. One subject who assayed her body density (by water displacement) on 10 occasions over a 4-month period, during which her body weight ranged from 52.5 to 53.6 kg, found a coefficient of variation for density of 0.6% (Dr. Nancy Butte, personal communication). As pointed out earlier, the resultant coefficients of variation for LBM and fat depend on the relative proportion of each; in this instance (body fat 9.4 kg, or 17.7%), the coefficient of variation for LBM was 2.3% and for fat it was 15.5%. In terms of percent fat, one standard deviation encompasses values from 15.0 to 20.5%.

To illustrate the combined biologic and technical error, let us choose two subjects each weighing 70 kg, one with 10% body fat, the other with 30% fat. For the former, LBM is 63 kg and fat 7 kg, and calculated body density is 1.0761; using an error of 0.0046 density units, LBM will lie between 61.6 and 64.4 kg and fat between 5.6 and 8.4 kg. For the latter subject, who has 49 kg LBM and 21 kg fat (body density 1.0312), calculated LBM will lie between 47.4 and 50.5 kg and fat between 19.5 and 22.5 kg. In terms of percent fat, the value for the thinner subject will vary from 8 to 12%, and for the fatter one from 27.9 to 32.2%.

Although such errors may not be of great moment in evaluating a single subject, they must be taken into account in assessing the significance of recorded changes in body composition resulting from some procedure, such as diet change, exercise, hormone administration, and so forth.

Other Methods for Estimating Body Volume

Helium dilution involves injecting a known amount of helium into a chamber in which the subject is seated and then determining the concentration of helium in the chamber. No correction is necessary for pulmonary residual volume, but gastrointestinal air must be estimated. Then density is simply weight/volume. A trial of this method in children was unsuccessful (Halliday, 1971).

With the subject in an airtight chamber, body volume could theoretically be determined by altering the volume of the chamber a known amount by means of a piston mechanism, or by adding a known amount of water, and reading the change in pressure, according to Boyle's law. It is not necessary to measure pulmonary residual volume.

While these methods are capable of estimating the volume of an inanimate object, investigators have had little success with live subjects. The mere act of breathing alters the pressure–volume relationships of the chamber, as does the temperature differential and the change in humidity provided by the subject. The method devised by Gundlach et al. (1980) in which the subject chamber is filled with polyurethane foam is said to eliminate interference from water vapor and heat production. In comparing this technique, which is based on Boyle's law, with underwater weighing, the authors report a correlation coefficient of 0.991 and a standard error of estimate of 1.4 liters. A discussion of these techniques can be found in the books edited by J.T. Reid (1968) and by Brozek and Henschel (1961).

Complicated photogrammatic techniques have been used to estimate body volume; the subject is photographed at various angles with superimposed contour lines (W.R. Pierson, 1963). This method too has not found wide usage.

A new method now undergoing trials involves an acoustic principle (Deskins et al., 1985). Since the resonance frequency of a sound wave is a function of the volume of the resonance chamber, the volume of the subject can be determined by introducing a sound of known frequency into an empty chamber, repeating the procedure with the subject in the chamber, and recording the shift in wave frequency. Trials are now being made with small animals, and it is hoped that the method may be found suitable for infants.

Total Body Potassium

Total body potassium can be estimated by isotopic dilution on or by ^{40}K counting. The former involves the administration of an artificial isotope (^{42}K), the latter involves the detection and quantitation of γ-rays that emanate from the body by virtue of its content of the naturally occurring radioisotope of potassium (^{40}K).

The procedure for estimating total exchangeable potassium (K_e) was mentioned on p. 24. The short half-life of ^{42}K, namely 12.4 hours, makes for great difficulty in using this technique except as a research tool. Shizgal's (1977) modification, in which ^{22}Na (half-life 2.6 years) and isotopic hydrogen are used, is better suited to routine use (see p. 26). However, a possible disadvantage is the long half-life of this radioisotope, and the fact that a small fraction of the injected dose of ^{22}Na is eliminated from the body with a half-life of a year or more.

Potassium Content of the LBM

This has been determined in several species of mammals (Table 2.7). These generally show a reasonably consistent K/LBM ratio for each species, although there is some interspecies variation. We recently analyzed four adult female rats for K, N, and fat; two had become quite obese on a "cafeteria diet," being members of the group mentioned on p. 29. Values for the thin rats (body fat 8%) were 81 meq K and 37 g N/kg fat-free weight, and those for the obese rats (body fat 40%) were the same, namely 80 meq K and 37.5 g N/kg fat-free weight.

Table 2.8 lists the potassium contents of various organs for several mammalian species. The results are more variable than those for water content (see Table 2.5), even when expressed on a fat-free basis. There are two possible reasons for such variation: the ratio of connective tissue to true muscle in the sample taken for analysis (the former has a much lower K content than the latter), and the amount of blood present in the sample. Some investigators have analyzed their tissue samples for hemoglobin and made a correction for the amount of blood present; others have not. If the samples are taken from exsanguinated animals, it is likely that the hypoxia so produced will cause potassium to leak from cells. In reviewing the reported results for human muscle, Bergström (1962) showed that the potassium contents of samples obtained under general anesthesia were about 10% lower, and sodium content somewhat higher than those obtained under local anesthesia. Halliday et al. (1957) found that the variation in muscle K content was less when the values were related to intracellular water rather than to fat-free dry solids. The latter method is preferred by some investigators because it avoids the problem of water loss from the sample prior to analysis; however, when dry solids is used as a reference basis, variable amounts of connective tissue, with its low K content, are included.

TABLE 2.7. Body K—LBM relationships.

Species	Body fat	meq K, LBM	Variation	Reference
Rat	1–19%	K = 0.0747 LBM (g) − 0.33	0.70 meq (SEE)	Cheek and West (1955)
Rat[a]	<15%	K = 0.316 FFDS[b] + 0.20	r = 0.989, SEE 1.18	Talso et al. (1960)
Guinea pig	2–16%	K = 67.9 LBM (kg) − 0.77	r = 0.99	Sheng et al. (1982)
Pig	18–36%	69 meq K/kg LBM	± 4 (SD)	Filer et al. (1960)
Pig	16–38%	68.6 meq K/kg LBM	± 3.6 (SD)	Fuller et al. (1971)
Ham samples	4–13%	LBM (kg) = 0.0057 (cpm ^{40}K) + 1.47	r = 0.96, SEE 0.37	Kulwick et al. (1961)
Man (3M, 1F)	18–27%	66.5, 66.6, 72.8, 66.8		Forbes and Lewis (1956)

[a]K(meq) = 2.67 N (g) − 0.3, r = 0.987, SEE 1.29.
[b]Fat-free dry solids.
[c]Percent K in LBM = 0.965 − 0.0094% H_2O (r = 0.99); hence K content of LBM tends to fall as water content increases, suggesting that the increase in hydration represents ECF.
N content is 34.4 g/kg LBM ± 2.3 (SD) in these same animals, and water content is 74.1% ± 1.4.

TABLE 2.8. Potassium content of various organs[a].

Organ	meq K/kg fat-free W	meq/kg fresh W
Skeletal muscle[b]	78–90	87–110
Heart		66
Smooth muscle		72–83
Myometrium		57–84
Adipose tissue[c]	61, 55, 32	
Tendon[d]	8, 13	
Skin	15–24	
Liver	73–86	
Erythrocytes[e]		103
Kidney		57–71
Spleen		81–99
Pancreas		81
Lungs		55–63
Placenta		24–40
Brain		76–85
Spinal cord		71–92
Peripheral nerve		31–57
Bone, cortical[f]	6 ± 1.1	
Cartilage[g]		35–68

[a]From Widdowson and Dickerson (1964) unless otherwise stated.
[b]Five muscles from a single individual.
[c]Kirton and Pearson (1963); Forbes (1962); Thomas (1962).
[d]Manery et al. (1938); Nichols et al. (1953).
[e]Considerable interspecies variation, value shown is for man.
[f]Cat bone, thoroughly cleaned (G. Forbes, unpublished).
[g]Epiphyseal, costal (Manery, 1954).

Bone poses a difficult problem, because of the likelihood of marrow contamination, and also because of the large amount of calcium present. The Ca/K ratio is so large in this tissue (500–1,000:1) that the intensity of the potassium emission in most flame photometers is altered.

The values for water, Na, K, Cl, and N that are listed in various publications under the item "skeleton" should not be taken as indicative of bone *per se*, because such specimens usually include marrow, periosteum, and cartilage. This is certainly true of the data published by ourselves for man (Forbes and Lewis, 1956), and by Sheng, Huggins, and their associates for several mammalian species.

The fact that there are so few values for adipose tissue in the literature attests to the difficulty in removing all of the fat from the sample without extracting some electrolyte along with it. Whereas mesenteric adipose tissue may contain less than 50% fat, perirenal and epididymal fat and even some subcutaneous fat samples may contain as much as 80–90%. The value of 55 meq K/kg fat-free weight in Table 2.8 derives from the author's laboratory, where we were careful to analyze the ether extract for potassium and to add this amount to the residue. Using this method, we found that the K content per unit of fat-free weight was independent of the fat content of the sample.

There is some evidence that potassium concentrations may vary within the cell. The nuclei of amphibian oocytes, for example, have a higher K content, and a lower Na content, than the cytoplasm (Century et al., 1970).

Malignant tissues show considerable variation in K content. Values for lymphatic tumors are close to that for normal muscle, whereas K concentrations in carcinomas are only one-half to two-thirds as great, and Na concentrations are two to four times higher. Such high Na/K ratios are to be expected in rapidly growing tissue (Waterhouse et al. 1955).

Total Body Potassium by ^{40}K Counting

There are three isotopes of potassium in nature, with isotopic abundances of 93.1% ^{39}K, 6.9% ^{41}K, and 0.0118% ^{40}K. It so happens that ^{40}K is radioactive; it emits a very strong γ-ray (1.46 MeV), and hence can be easily detected by external counting. It also has a very long physical half-life (1.3×10^9 years), so corrections for physical decay are not necessary.

There are two principal components of the decay scheme for this isotope: 11% of the disintegrations are in the form of γ-rays associated with electron capture, the product being stable ^{40}Ar, and 89% are strong β-rays (1.31 MeV), the product being stable ^{40}Ca. The ^{40}K/chemical K ratio in human body ash is about 2% less than the ratio in analytical grade KCl (Fenn et al., 1942), suggesting that the body discriminates against the heavier isotope to a slight extent, or conversely that some discrimination in the reverse direction occurred during the preparation of the analytical grade reagent.

^{40}K is of geologic as well as biologic interest. A considerable proportion of the heat generated in the interior of the earth comes from the decay of radioisotopes such as ^{238}U, ^{235}U, ^{232}Th, and ^{40}K; and of the total amount of heat thus generated some 15% is due to ^{40}K (Turcotte and Schubert, 1982). It is the lightest of the known naturally radioactive elements. According to Weisskopf (1977), "Naturally occurring radioactive substances are the last remaining embers of the nuclear fire of a great star explosion in which terrestrial matter was created, five or six billion years ago." From the viewpoint of the body compositionist, it is indeed lucky that ^{40}K was one of those "embers."

Under the assumption that all of the element ^{40}Ar has been formed from the decay of ^{40}K, the ratio of ^{40}K to ^{40}Ar in rocks and minerals (where Ar has been imprisoned) can be used to calculate the age of the rocks. Indeed this method has been used for dating the rocks captured from the moon, and even to date the antiquity of Lucy, the 3.5 million-year-old human skeleton found near the horn of Africa (Johanson and Edey, 1981). Estimates of the age of the earth based on potassium-argon dating correspond fairly well to those based on other methods.

Each mole (39.1 g) of potassium contains 6.03×10^{23} atoms (Avagadro's number) and 0.012%, or 7.2×10^{19}, of these are radioactive; hence each gram of K contains 1.8×10^{18} radioactive atoms. From the fundamental law of radioactive decay, the number of disintegrations per unit time *(dN/dt)* is equal to λN, where N is number of atoms and λ is $\ln 2/t_{1/2}$. In the case of ^{40}K, $t_{1/2}$ is 1.3×10^9 years, λ is 0.53×10^{-9} per year, so $dN/dt = 0.53 \times 10^{-9} \times 1.8 \times 10^{18} = 0.95 \times 10^9$ disintegrations per year. Since a year contains $60 \times 24 \times 365 = 5.26 \times 10^5$ minutes, each gram of K should emanate $9.5 \times 10^8/(5.26 \times 10^5) = 1.8 \times 10^3$ disintegrations per minute (dpm). Of these about 198 dpm are γ-rays, and 1602 dpm are β-rays.

This inherent radioactivity can pose a problem in analyzing tissue samples for radioactivity. For example, each gram of muscle contains about 0.0035 g K; this amount of K produces 0.7 γ-rays and 5.6 β-rays per minute, activities that can be detected by very sensitive counting equipment, and so lead to spurious results for analyzing samples for low-level radioactivity for other isotopes. Indeed, we use a bottle of KCl to check the performance of our whole body counter.

Since the average adult man contains about 150 g K, he produces about 30,000 γ-rays each minute and 240,000 β-rays each minute from ^{40}K decay. The average adult woman is not quite as "hot," since her body contains only about 100 g K, yielding 20,000 γ-rays and 160,000 β-rays each minute. The adult body contains about 0.1 μCi ^{40}K, or 15 mg of this isotope.

It can be calculated that ^{40}K emanations account for about 16% of the total background radiation dose received by man, i.e., from cosmic radiation, internal heavy isotopes such as radium, and heavy radioisotopes in soil and rocks. Of the total dose from these sources, amounting to about 0.1 rem/year, internal ^{40}K contributes about 16 millirem.

Assuming that cells of all organisms contain about the same amount of K, those alive 1.3×10^9 years ago, i.e., one half-life, received twice as much radiation from this isotope as do those of today, and 2.6 billion years ago they received four times as much. It is interesting to ponder the question as to whether ^{40}K could have been, by virtue of its radioactivity which is known to alter DNA, a factor in evolution.

Technique

Three general features are common to all ^{40}K techniques: (1) a heavily shielded room for the subject so as to minimize the natural background rate (cosmic rays, contaminants in concrete, ^{40}K in wood and in personnel); (2) a sensitive γ-ray detector placed in proximity to the subject; and (3) suitable instrumentation for isolating and recording the γ-rays from ^{40}K. This last feature is essential because the human body contains a number of radioactive isotopes: radium, thorium, and ^{137}Cs, the last a fallout product from above-ground atomic bomb tests that took place in the 1950s. Although these materials are present in extremely small amounts (for example, the body burden of radium is about 3×10^{-8} mg and that of ^{137}Cs is about 2 nCi as of today) so too is ^{40}K present in very small amounts; fortunately the γ-ray energies of most of these contaminants are less than those from ^{40}K, so appropriate multichannel pulse height analyzers can isolate the ^{40}K photopeak generated by sodium iodide crystal detectors. However, plastic and liquid scintillation detectors have rather poor resolution, so the photopeaks from ^{137}Cs (0.66 MeV) and ^{40}K (1.46 MeV) are spread out to a considerable degree so that they cannot be completely resolved. In the period from 1950 to 1965 the human body burden of ^{137}Cs was rather large (about 8 nCi) which posed a problem for instruments employing plastic or liquid scintillation detectors, necessitating elaborate corrections for the recorded ^{40}K activity; however, once the nuclear test ban treaty was signed in 1963, the ^{137}Cs level in man began to drop, reaching its present level of 2 nCi by the mid-1970s.

The instrument to be used should not be located near any source of intense γ radiation, such as cyclotrons, radiocobalt facilities, etc. Subjects who have been given γ-emitting radioisotopes should be carefully scrutinized.

The detection room is usually made of thick steel and lined with lead, the room being large enough to accommodate an adult in the sitting, lying, or standing position. The door must be closed, so adequate ventilation and voice communication are needed. The construction material must be as free as possible of radioactive contaminants.[6] Some investigators have gone to great lengths, using pre-World War II battleship steel, and lead obtained from Medieval church roofs where centuries of weathering have depleted its contaminants.

Figures 2.7 and 2.8 show several arrangements that have been employed for detecting the signal from the subject. In the first two arrangements the subject is partially or completely surrounded by a detector made of polystyrene or a tank containing a liquid scintillator. The detector is viewed by several photo multiplier tubes.

Another type is the so-called "shadow shield" counter. As depicted in Figure 2.8, the subject is moved slowly beneath a shielded NaI crystal so that the partially shielded body is scanned from foot to head. This arrangement has the advantage of being much lighter than a completely shielded room so that it can be located on the upper floors of a building, or even transported by truck.

Figure 2.9 shows the author's instrument, constructed in 1959 by Dr. John B. Hursh at the University of Rochester after the design of Charles Miller (International Atomic Energy Agency 1970). The room measures 130 × 183 × 216 cm (inside dimensions); the walls are 20.3 cm steel lined with 3 mm lead, and the total weight is 42 tons (the 2-foot-wide door alone

[6]We were amazed to find that the background readings in our instrument increased after a new television monitor was installed. The culprit proved to be the camera lens; apparently it is common practice to add thorium to the glass to improve the quality of the image. Incidentally, a heavy snowstorm can deposit enough radioactive particulate material to cause a measurable increase in background counts.

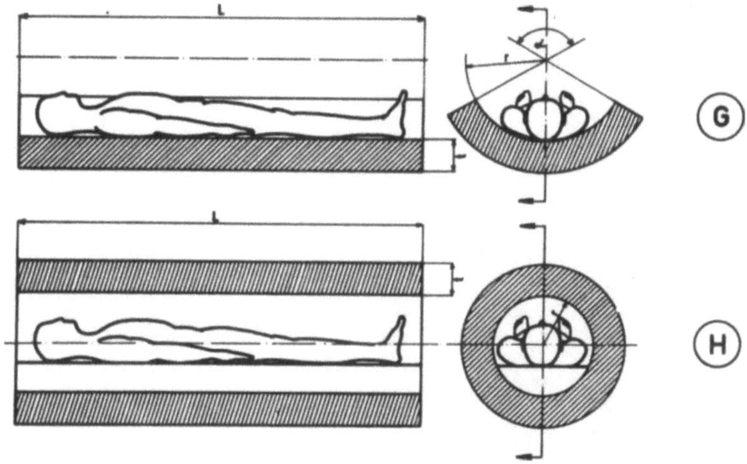

FIGURE 2.7. Subject in position for whole body count. **Top:** the detector (cross-hatched area) partly surrounds the subject (2 π geometry). **Bottom:** the detector completely surrounds the subject, except at either end (4 π geometry). (From International Atomic Energy Agency: Directory of Whole-Body Radioactivity Monitors, 1970 edition. Vienna, 1970, pp 1–25, with permission.)

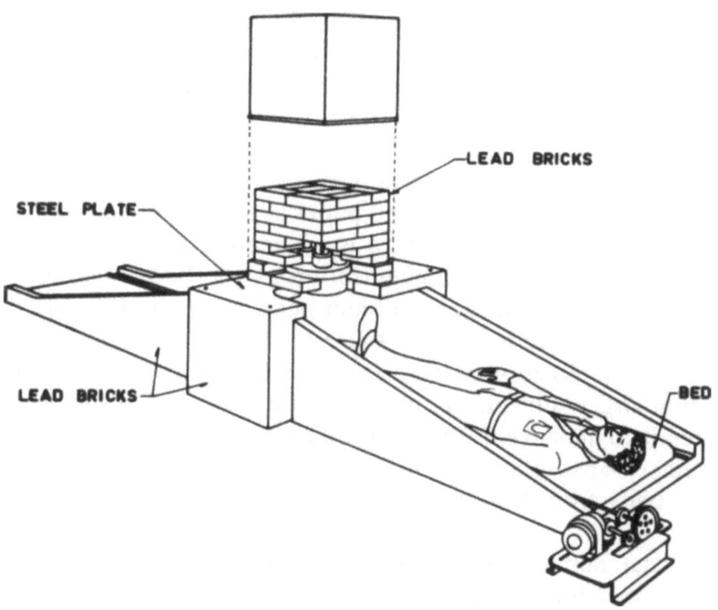

FIGURE 2.8. Subject in position to begin count in "shadow shield" counter. (From International Atomic Energy Agency: Directory of Whole-Body Radioactivity Monitors, 1970 edition. Vienna, 1970, p. US 9.2–3, with permission.)

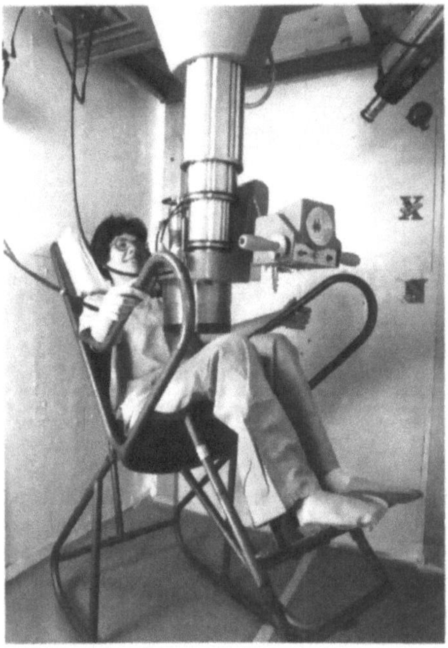

FIGURE 2.9. Rochester whole body counter.

weighs 2 tons). The room is provided with continuous ventilation, two-way voice communication, and a television monitor. A single 20.3 × 10.2 cm thallium-activated sodium iodide crystal is positioned above the subject, who sits quietly in a metal chair, in the arrangement shown in the figure.

Since clothing contains potassium, our subjects are asked to bathe and to dress only in cotton pajamas and paper slippers, and to remove jewelry before entering the counter. There is no need to fast, since many foods contain about the same K concentration as the body. (However, some animals such as sheep pose a problem because of the K content of their wool and the dirt that is trapped there.)

The crystal is positioned in relation to the chair, not the subject; this means that larger subjects are closer to the crystal than smaller subjects, which tends to improve the geometry and to partially counter-balance the increased self-absorption of ^{40}K γ-rays. This arrangement means that the subject's body from head to knees approximates a 0.5 m arc about the detector.

The crystal, and its associated "light pipe" (a nonactivated NaI crystal that tends to reduce background counting rate) is viewed by four photomultiplier tubes, and the impulses from these are fed into a 1024 channel analyzer, where they are sorted into various energy channels and stored. The counting time is 40 minutes, whereupon the counts under the ^{40}K photopeak are integrated; subtraction of the background, i.e., activity recorded from the empty room, yields a value for counts due to ^{40}K activity in the subject. Since background activity tends to vary somewhat from day to day, we generally use the average of six previous overnight background counts for this purpose.

The net counts recorded in the ^{40}K channels must then be converted to grams of body K by the use of a calibration factor. This factor was obtained in the following manner. A series of normal subjects were given known amounts of ^{42}K, which happens to have a peak γ-ray energy (1.52 MeV) close to that of ^{40}K (1.46 MeV), shortly after having had a ^{40}K count. Urine was collected for the next few hours, and the subjects counted once more. Then two bottles, one containing a known amount of KCl, the other a known amount of ^{42}K were counted individually. Hence the known quantities are the cpm/μCi ^{42}K in the subject (calculated as ^{42}K counts − ^{40}K counts corrected for urinary losses), cpm/μCi ^{42}K in the bottle, grams of K in the bottle containing KCl, and the observed ^{40}K counts in the subject; the one unknown is grams of K in the subject's body, which is then calculated as follows:

$$\text{g K (subject)} = {}^{40}\text{K count (subject)} \times \frac{\text{g K (bottle)}}{{}^{40}\text{K count (bottle)}} \times \frac{{}^{42}\text{K count (bottle)}}{{}^{42}\text{K count (subject)}}$$

The calibration factor for the particular subject is thus ^{40}K cpm/g K. This calibration factor is then related to some function of body size, in such a manner as to take into consideration two fundamental phenomena, the inverse square law of radiation intensity and the self-absorption of radiation by body tissues.

To account for subjects of widely differing body size we then did simultaneous assays of ^{40}K and total body water (by D_2O dilution) in a series of normal subjects who varied in weight from 18 to 120 kg. Under the assumption that the total K/total body water ratio is constant in normal people, we further refined our calibration procedure (Forbes et al., 1968).

The technical error of the counting procedure is due in part to the random nature of radioactive emissions, which includes the background counts; it is calculated as follows:

$$\text{SD (cpm)} = \left[\frac{\text{cpm (subject + bkg)}}{\text{counting time}} + \frac{\text{cpm (bkg)}}{\text{counting time}}\right]^{1/2}$$

In our instrument the background rate is 50 cpm, counting time 833 minutes, and the gross count rate for the average adult is 150 cpm, counting time 40 min. Inserting these values in the above equation,

$$\left[\frac{150}{40} + \frac{50}{833}\right]^{1/2} = 1.95 \text{ cpm (SD)}.$$

When divided by the net count rate (100 cpm), the coefficient of variation is 2%, and the 95% confidence limits for a single assay is 3.8%. This represents the irreducible error based on counting statistics which of course will be larger for smaller subjects; and as mentioned earlier the phenomenon of biological variation must also be considered.

The coefficient of variation for three such subjects each assayed 5–10 times was 2–3%.

Lykken et al. (1983) have raised the possibility of additional errors in making assays on athletes. Radon gas is present in the atmosphere, so people engaged in exercise will inhale more radon by virtue of their increased minute volume. This isotope decays to ^{214}Bi, which has a γ-ray energy that overlaps with that of ^{40}K, and that is retained by the body for short intervals. These authors did find a modest increase in apparent ^{40}K activity in individuals who had just finished a bout of exercise, and that it took about 30 min for this excess activity to disappear. We did a similar study, and in counting subjects within 10 min of completing a session of vigorous exercise we failed to find an increase in ^{40}K activity (Forbes and Hursh, 1984). One can only surmise that this discrepancy in results could be due to the higher radon content of the air in North Dakota (where Lykkens' laboratory is located) than in Rochester, New York.

Comparison of 4π and Single-Crystal Counters

These considerations involve several factors: (1) efficiency of detecting ^{40}K γ radiation, i.e., the percent of γ-rays emanating from the subject that are detected; (2) the resolution of the detector, i.e., the width of the energy band in the ^{40}K region of the spectrum—the smaller the width the better the resolution and hence the less the chance of interference from contaminating isotopes; (3) the time required for a satisfactory count; (4) the influence of body size and adiposity on the count rate; (5) the background counting rate; and (6) stability of the detection system.

Figures 2.10 and 2.11 show the spectra of radioactivity emanating from a normal adult. It will be noted that the ^{40}K counts in the single crystal detector occur in a much narrower band width than in the 4π counter.

A comparison of several types of counters, including the shadow shield instrument (a single crystal detector less well shielded) is given in Table 2.9. Since the 4π counter is so much more efficient, the counting time for the subject can be relatively short. On the other hand, it has a much higher background rate, and thus a lower signal-to-noise ratio.

There are a number of counter designs. The one designed by Stanton Cohn and his associates at the Brookhaven National Laboratories employs 27 NaI crystals arrayed above the subject and an equal number below the subject (Cohn and Dombrowski 1970). The 2π counter in Landstuhl, Germany is positioned so that the subject can stand erect while being counted. Descriptions and operating characteristics of these and the many other whole body counters in the U.S.A. and abroad can be found in the Directory of Whole Body Radioactivity Monitors (International Atomic Energy Agency 1970).

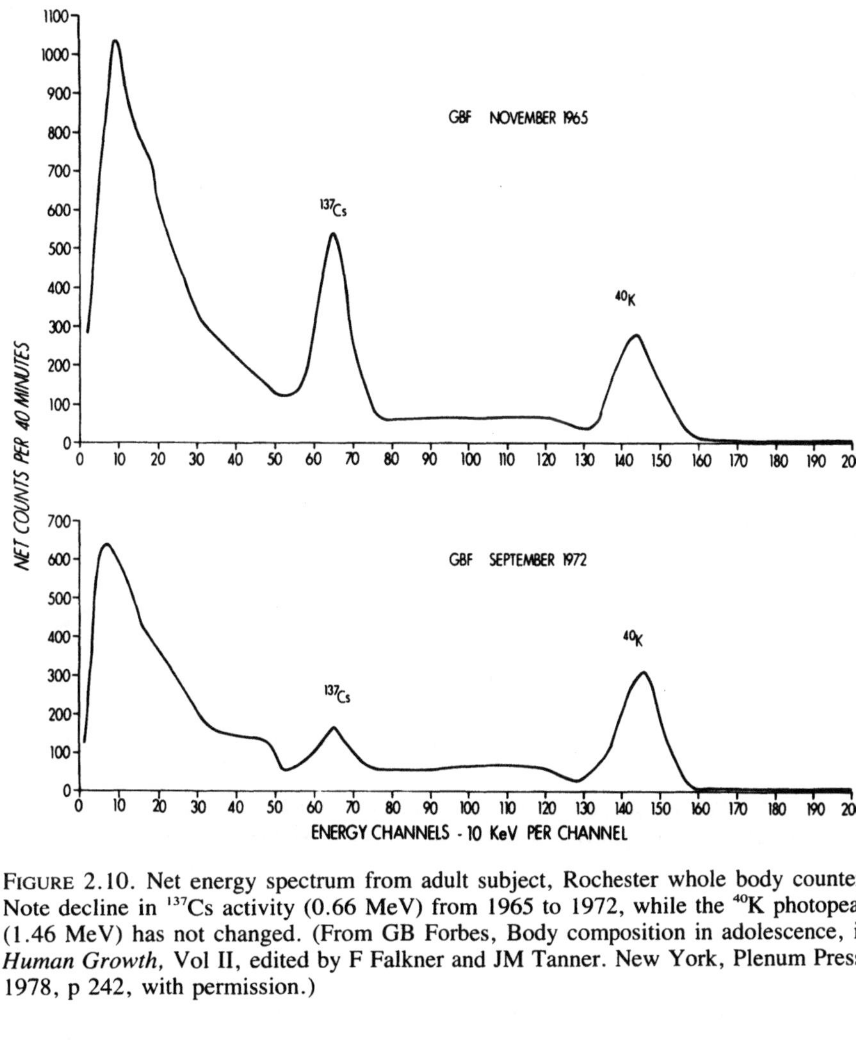

FIGURE 2.10. Net energy spectrum from adult subject, Rochester whole body counter. Note decline in ^{137}Cs activity (0.66 MeV) from 1965 to 1972, while the ^{40}K photopeak (1.46 MeV) has not changed. (From GB Forbes, Body composition in adolescence, in *Human Growth,* Vol II, edited by F Falkner and JM Tanner. New York, Plenum Press, 1978, p 242, with permission.)

Because their body potassium content is so small, infants and very young children are difficult to measure accurately. In a 4π plastic scintillator designed for infants the net count rate was 260 cpm/5 g K (the amount in a newborn infant) against a background rate of 8200 cpm (Forbes, 1968). The signal-to-noise ratio for the 4π liquid scintillators used by Garrow (see International Atomic Energy Agency 1970) and Graham et al. (1969) is only 0.03 in the case of infants. The latter instrument was designed so that the scintillation fluid tanks could be arranged to fit subjects of various size. The imprecision of ^{40}K assays for infants is evident from the fact that the coefficient of variation of body K was greater than that for body weight in the infants studied by Rutledge et al. (1976), who used the instrument designed by Graham (1969). However, by averaging the results

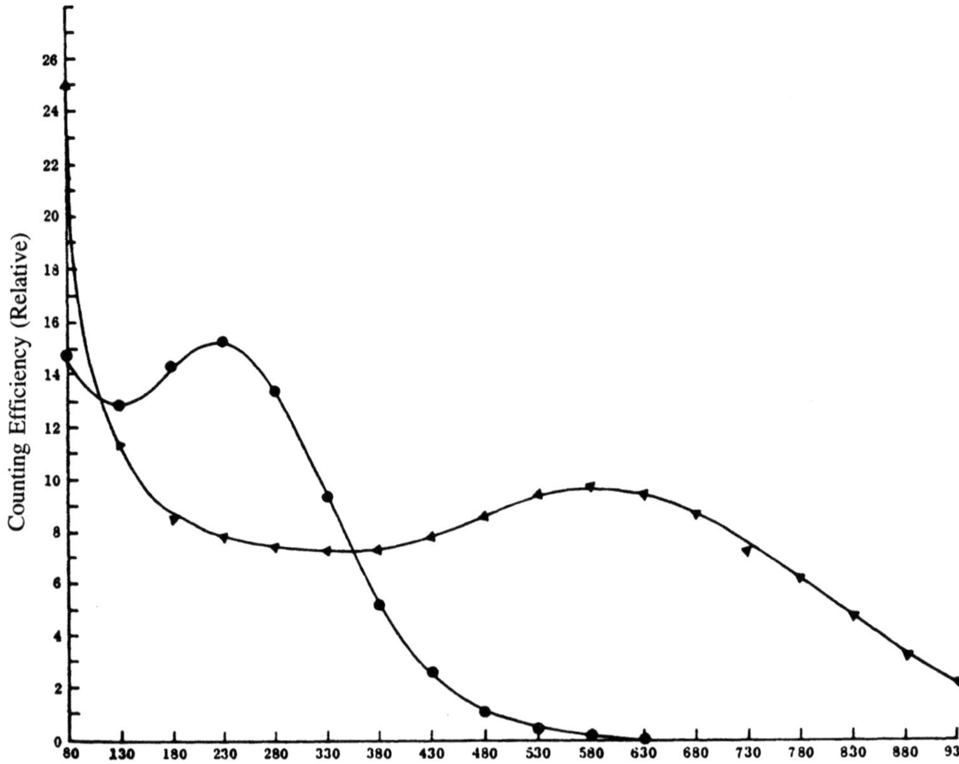

FIGURE 2.11. Net energy spectra for ^{137}Cs (●) and for ^{40}K (◄), 4π counter. [From Graham ER (1969), Internat J Appl Radiat Isotop 20:249–254, with permission.]

TABLE 2.9. Performance of various whole body counters.

	4π Liquid[a]	3π Polystyrene[e]	Single crystal NaI[b]	Multiple crystal NaI[c]	Shadow shield NaI[d]
Detector volume (dm^3)	5800	700	3.3	480	6.8
Number of phototubes	30	8	4	54	4
^{40}K band width (MeV)	1.1	0.9	0.2	0.2	0.2
Background rate (c/min)	15,000	7,750	50	690	71
Adult subject (125 g K) efficiency (cpm/g K)	41	44	0.76	4.6	0.50
Gross ^{40}K counts/bkg	1.3/1	1.7/1	3/1	1.8/1	2/1
Net ^{40}K counts/bkg	0.34/1	0.7/1	2.2/1	0.94/1	0.9/1
Usual counting time	5 min	2 min	40 min	15 min	10 min

[a]International Atomic Energy Agency: Directory Whole Body Radioactivity Monitors (1970)—U.S. 7.1.
[b]Ibid., U.S. 10.1. (author's instrument)
[c]Ibid., U.S. 5.3.
[d]Ibid., U.S. 9.2.
[e]Sköldborn et al. (1972).

of several infants studied at each of several specific ages Rutledge et al. (1976) and Novak et al. (1973) were able to show a progressive increase in body K during the first 12 months of life.

The ^{40}K technique is noninvasive and nontraumatic (the only hazard is claustrophobia) and requires a minimum of cooperation on the part of the subject. This means that repeat assays are easy to perform, and in so doing precision is improved. For example, two assays of a subject both before and after some change in body K content is anticipated reduces the 95% confidence limits from 3.8% (p. 45) to 2.8%, and so increases the chance of detecting a difference that might have resulted from the particular maneuver, or conversely, the chance of finding no difference. Figure 2.2 shows the change in body K observed to occur during nutritional rehabilitation of a patient with anorexia nervosa. The existence of inherent technical error is readily appreciated, and it is evident that the difference between any two consecutive assay points is not always representative of the overall trend as defined by the calculated regression line.

Body Cell Mass (BCM)

This concept was proposed by Francis Moore and his associates (1963). The BCM is defined as "the working, energy-metabolizing portion of the human body in relation to its supporting structures." It consists of the cellular components of muscle, viscera, blood, and brain. In designing a formula for estimating BCM they chose total exchangeable potassium (K_e)

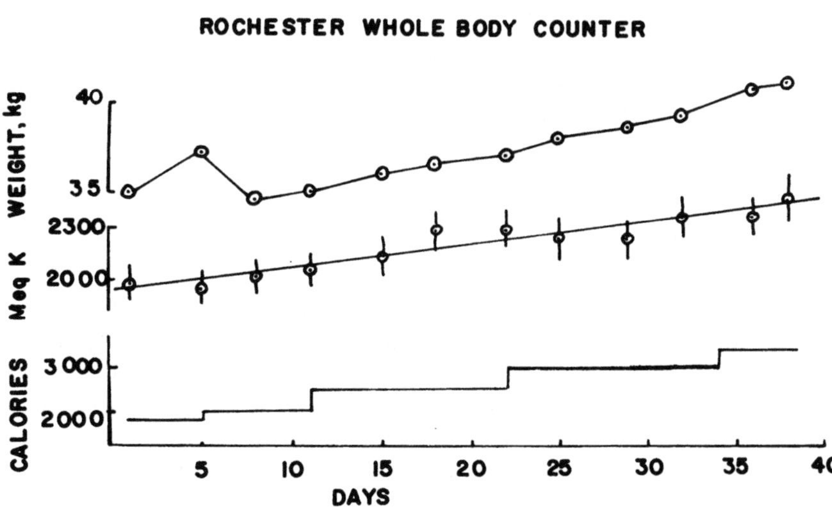

FIGURE 2.12. Progression of body weight and body K (by ^{40}K counting) during recovery from anorexia nervosa. Vertical bars define 95% confidence limits for single ^{40}K assay. Regression line calculated by method of least squares; average gain is 13 meq K/day. Lowest line shows energy intake.

as a basis: assuming an average potassium/nitrogen ratio of 3 meq/g, and a nitrogen content of 0.04 g/g wet tissue (= g N × 25),

$$\text{BCM (g)} = \text{K}_e \text{ (meq)} \times \frac{25}{3},$$

$$\text{BCM (kg)} = \text{K}_e \text{ (meq)} \times 0.00833. \tag{21}$$

ECF and bone (but not bone marrow) are neglected because these contain so little potassium.

This formula implies that the intracellular K concentration is 1/0.00833 = 120 meq/kg; indeed it defines the BCM as consisting of tissue having this ICF potassium concentration. In actuality, Moore et al. (1963) found an average value of 150 meq/K/L ICF, so if cell water is 66% of cell weight the factor becomes 0.010 rather than 0.00833. However, individual body tissues vary widely in terms of their K/N ratios, from a high of 5 meq K/g N in brain to a low of 0.45 in skin, and the ratio for the entire body is 2.0 (Widdowson and Dickerson, 1964). Burmeister and Bingert (1967) have chosen a value of 92.5 meq K/L ICF, so that the coefficient of Equation 21 becomes 0.0108. In defense of the position taken by Moore et al. (1963), they admit that their chosen value of 0.00833 is not precise; and in actuality any reasonable value would be satisfactory because the estimation of body cell mass depends entirely on the measurement of K_e or of ^{40}K, and BCM is simply body K multiplied by a factor. However, custom has enshrined the factor as 0.00833. Since LBM is considered to contain 68.1 meq K/kg and BCM to contain 120 meq K/kg, BCM is 57% of LBM.

The concept of "body cell mass" is a useful one. It encompasses those tissues most likely to be affected by nutrition or disease or physical activity over relatively short intervals of time—days or weeks. In practical terms it excludes those tissues such as bone and connective tissue which have slower turnover rates. Muscle, viscera, and brain account for most of the oxygen consumption of the body, and for most of its content of potassium.

Neutron Activation

Many elements can be made radioactive by bombardment with neutrons from the atomic pile or with deuterons or protons from the cyclotron. Indeed elements such as ^{22}Na and ^{14}C, and a number of others, are continuously being formed (in very small amounts) in the atmosphere by the action of cosmic rays. The effect of such bombardment is to make the atomic nucleus unstable, whereupon it emits particulate or electromagnetic radiation of characteristic energy, and it is this property together with the decay rate that permits identification of the particular isotope so produced. A large variety of radioisotopes are now being produced for investigative and therapeutic purposes. Indeed, many of the advances in modern biochemistry could not have been made without the use of radioisotopes.

In the Los Alamos accidents of 1945 and 1946, in which several individuals were exposed to large doses of neutrons (two of them died), it was discovered that their blood contained radioactive sodium (^{24}Na) and phosphorus (^{32}P); gold dental inlays were also radioactive (Hempelmann et al., 1952). In fact the absorbed dose of radiation received by these individuals could be calculated from the amount of ^{24}Na present.

Techniques were developed for the analysis of tissue samples by neutron activation. Extremely small amounts of several elements can be quantitated in this manner; for example, Bergström (1962) was able to do electrolyte analyses on biopsy samples of human muscle weighing a mere 20–40 mg. Then in 1964 Anderson and co-workers reported that the body content of Na, Cl, and Ca could be assayed in human subjects by irradiating the entire body with neutrons. Subsequently techniques were also developed for estimating total body nitrogen and phosphorus in laboratories both in the USA and abroad.

The subject is irradiated with neutrons of known energy, and the induced radioactivity is assayed in a whole body counter. The apparatus is calibrated by making measurements of suitable phantoms. Surprisingly, the radiation dose is rather small (~ 30 mrad). The subject's head is usually not included in the radiation field.

Table 2.10 shows the nuclear reactions employed, and Figure 2.13 shows the sort of radiation spectrum that results. Since the half-lives of most of the induced radioisotopes are very short, the counting procedure has to be done very quickly. The character of the spectrum depends on the energy of the incident neutrons.

Total body nitrogen can also be assayed by prompt γ emission: the high-energy, very short-lived emissions from ^{15}N created by neutron bombardment are counted simultaneously.

The apparatus for carrying out total body neutron activation is very expensive, and the calibration is difficult: not only must the whole body counter used to detect the induced radioactivity be calibrated for the several radioisotopes but the total dose of neutrons delivered to the body must be known. Consequently, there are very few installations in operation

TABLE 2.10. Nuclear reactions.

Substance		Activation product half-life
^{48}Ca (0.18%)[a]	(n, γ)[b] ^{49}Ca	8.8 min
^{23}Na (100%)	(n, γ) ^{24}Na	15 hr
^{37}Cl (25%)	(n, γ) ^{38}Cl	37 min
^{31}P (100%)	(n, α) ^{28}Al	2.3 min
^{14}N (99.6%)[c]	$(n, 2n)$ ^{13}N	10 min
^{26}Mg (11%)	(n, γ) ^{27}Mg	9.5 min

[a]Isotopic abundance.
[b]Nuclear reaction.
[c]also prompt γ emission $(n, \gamma)^{15}N$.

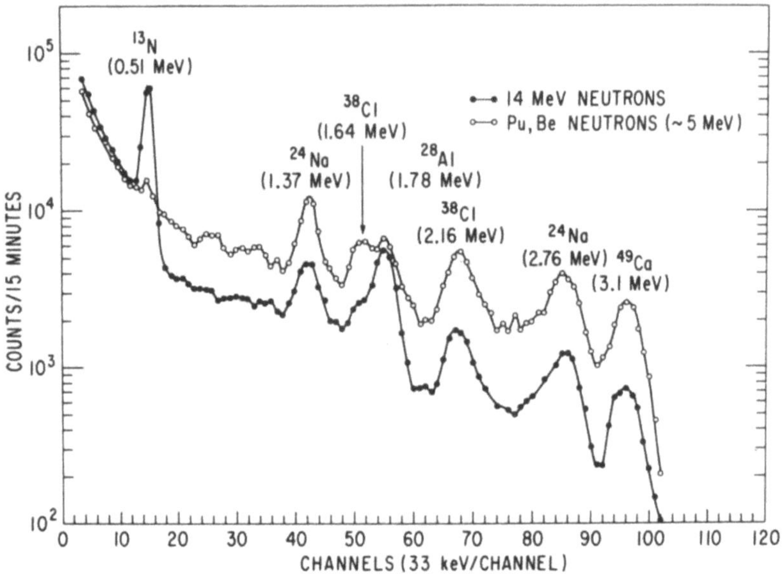

FIGURE 2.13. γ-Ray spectra for human subjects irradiated with 14 MeV neutrons and neutrons from plutonium-beryllium sources. (From Cohn SH (ed): Non-Invasive Measurements of Bone Mass and Their Clinical Application. Boca Raton, Florida, CRC Press, 1981, with permission.)

here and abroad. Although a large number of adults have been assayed by this technique, there are only two reports of its use in children (McNeill and Harrison, 1982; Archibald et al, 1983). Beddoe and Hill's (1985) recent review should be consulted for additional details.

In addition to the isotopes listed in Table 2.10, attempts have also been made to assay total body hydrogen and oxygen, liver cadmium and iron [the last by a magnetic technique (Brittenham et al., 1982)].

There are two reports showing that neutron activation analysis produces results in accord with those obtained by chemical analysis of the entire animal. Preston et al. (1985) found this to be true for total body nitrogen in the rat, and Sheng et al. (1981) for total body calcium, sodium, and chloride in the guinea pig.

Burkinshaw et al. (1979) and Cohn et al. (1980) have used data on total body K (by ^{40}K counting) and total body N (by neutron activation) to estimate the relative amounts of protein in the muscle and nonmuscle components of the body and the mass of each component. This method derives from the observation that the K/N ratio for muscle (3.03 meq/g) differs from that of the average of the nonmuscle tissues (1.33 meq/g). Their mathematical approach is as follows, letting m stand for muscle and n for nonmuscle:

let $r = K/N$,

$$r_m = K_m/N_m,$$

and
$$r_n = K_n/N_n.$$

Then

$$rN = r_m N_m + r_n N_n;$$

rearranging,

$$N_m = N (r - r_n/r_m - r_n)$$

$$N_m = K - r_n N/r_m - r_n.$$

Using square brackets to denote concentration, M_m to indicate total muscle mass, and M_n to indicate nonmuscle mass,

$$M_m = K - r_n N/[N_m] (r_m - r_n), \tag{22}$$

and

$$M_n = r_m N - K/[N_n] (r_m - r_n). \tag{23}$$

Using values of 30 g N/kg and 91 meq K/kg muscle, and 36 g N/kg and 48 meq K/kg of nonmuscle,

$$N_m = K - 1.33 N/1.70; N_n = 3.03 N - K/1.70, \tag{24}$$

and

$$M_m = K - 1.33 N/51.0; M_n = 3.03 N - K/61.2. \tag{25}$$

In this manner, the total protein (N × 6.25) in both compartments can be calculated, as well as the mass of both.

The results obtained by Burkinshaw et al. (1979) in a series of adults show an average lean muscle mass of 20 kg and average lean nonmuscle mass of 36 kg in males weighing 72 kg; respective protein contents were 3.83 kg and 8.07 kg. In women weighing 53 kg (average) lean muscle mass was 6.7 kg and lean nonmuscle mass 31 kg; respective protein contents were 1.26 kg and 6.95 kg. Using the same technique Cohn et al. (1980) report an average muscle mass of 11.0 kg and nonmuscle mass of 30 kg in women weighing an average of 64 kg; for men weighing an average of 78 kg these values were 21.1 and 36.9 kg, respectively.

Although one can admire the mathematical ingenuity of this approach as a means of compartmentalization of the LBM, certain reservations are in order. First is the fact that the K/N ratio of individual nonmuscle tissues of the body is not constant, varying from a high of 5 meq K/g N in brain to a low of 0.45 meq K/g N in skin and blood plasma. Second, the effect of the combined assay errors for body K (4%) and body N (6%) results in a coefficient of variation for muscle mass of 17% and for nonmuscle mass of 19%. Burkinshaw et al. (1979) admit that the errors of estimation are too large to permit conclusions to be drawn from a single measurement of an individual, although comparisons of mean values for groups of individuals would be possible. Indeed one of the 14 women studied by them had a negative value (−2.35 kg) for calculated muscle mass. Finally, the estimates of muscle mass reported by Burkinshaw et al. (1979) for both men and women are smaller than those that can be derived from urinary creatinine excretion.

Note should be taken of occasional aberrant results in studies of total nitrogen by neutron activation. One of Archibald and co-workers' (1983) subjects is said to have gained body N during weight loss on a low-energy diet, and consideration of the large standard deviation for N loss in the patients studied by Vaswani et al. (1983) suggests that some of them must have gained body N as they lost weight. The obvious impossibility of such a result suggests that technical error is at fault.

Combined Techniques

The calculation of lean weight from density, total body K, N, and H_2O carries the assumption that the composition of the lean body mass does not vary among normal individuals, and that the body is a two-compartment system: lean and fat. Sheng and Huggins (1979) have shown that the water content of the LBM does vary somewhat as determined by various investigators, and the data listed in Table 2.8 show a variation in K content. Bone has a low water content and a low K content but a high density in comparison to soft tissues, so in carrying out body composition assays one must assume that the bone/soft tissue ratio does not vary very much. While the two compartment concept is obviously an over simplification it may not be too far off the mark for those who are interested in estimating the amount of body fat. In the healthy young adults the ratio of total body K and total body N, which are indices of soft tissue mass, to total body Ca (an index of skeletal mass) did not vary a great deal except at the extremes of stature. The predicted K/Ca ratio was 0.10 (g/g) for a 153 cm woman and 0.13 for a 188 cm man (Ellis and Cohn, 1975a; personal communication).

However, when one is dealing with children or the elderly, it is likely that the ratio of soft tissue mass to skeletal mass will depart from the value that is characteristic of the young adult; and the same can be said for the ratio of ECF volume to total body water. In an attempt to deal with this problem some investigators have used two or more techniques simultaneously. Siri (1961) assumed that the ratio of protein to mineral was constant and not altered in states of abnormal hydration. His method involves the determination of body water and body density; and using the usual constants, the following formula is derived:

$$\text{fraction fat} = \frac{2.118}{D} - 0.78\,\frac{H_2O}{W} - 1.354 \tag{26}$$

Siri states that this method should yield better estimates of body fat in dehydrated individuals and in those with edema and serous effusions than either method alone. However, he says nothing about the problems in carrying out density assays under such circumstances.

Cohn et al. (1980, 1984) have used a combination of techniques to estimate LBM: one method employs total body N and total body Ca by neutron activation together with total body water. Multiplication of N by 6.25 yields a value for total protein, and division of total Ca by 0.34 yields a value for bone mineral. LBM is then the sum of protein, water, and bone mineral:

$$\text{LBM (kg)} = 6.25\ \text{N (kg)} + \text{H}_2\text{O (l)} + \frac{\text{Ca (kg)}}{0.34} \tag{27}$$

A second scheme makes use of assays of total body K, extracellular fluid volume, and total body Ca. Body cell mass (BCM) is calculated from total body K, and total extracellular solids are considered to be proportional to total body Ca; then LBM is the sum of BCM, ECF volume and extracellular solids:

$$\text{LBM (kg)} = \text{BCM (kg)} + \text{ECF (l)} + \frac{\text{Ca (kg)}}{0.117} \tag{28}$$

Both methods are designed to account for variations in body mineral/protein ratio. The glycogen content of the body is neglected. The LBM value obtained by the second method will depend on the technique chosen for assay of ECF volume. The disadvantage of both techniques is the need for a neutron activation facility. Those fortunate enough to possess such a facility are now engaged in studying patients with a variety of illnesses.

Beddoe et al. (1984) used assays of total body water (tritium dilution), total body N (neutron activation), and estimated mineral mass plus estimated glycogen to estimate LBM.

The ultimate precision of most techniques is limited by biological variability; body weight exhibits diurnal and day-to-day variations, and it is safe to assume that certain features of body composition also exhibit some variability. An instructive exercise is to calculate the effect of adding or subtracting 500 ml H_2O from an adult subject: this will change the density, and the K and N and H_2O contents of the LBM slightly, as well as the ratio of metabolic element balance to weight change.

Another consideration stems from the fact that the ratio ECF/total water varies somewhat with age, being higher in neonates than in young adults and then rising again in the elderly. Hence neither the K, N, or H_2O contents of the LBM, nor its density can be considered to be constant throughout the age span of man *(vide infra)*. In dealing with diseased subjects or those who are losing or gaining weight, it must be remembered that the K/N ratio varies among body tissues, from a high of 5 meq K/g N in brain, 2.8 in skeletal muscle, about 2.5 in viscera, and 2.1 in erythrocytes to a low of 0.45 in plasma and skin (Widdowson and Dickerson, 1964). Hence the ratio of change in body K to change in body N resulting from a change in weight will depend on the particular tissue components involved. In states of under- or overhydration the density of the LBM will be altered from its usual value of 1.100 g/ml, and in states of K deficiency the K content of the LBM will be subnormal.

A number of investigators have compared two or more techniques for estimating LBM and fat in the same subjects. Some report a reasonable correspondence between densitometric, body water and ^{40}K techniques (Krzywicki et al., 1974; Cohn, 1981); and Cohn et al. (1980) found rather small coefficients of variation for body N/K and body K/H_2O ratios in normal adults. A recalculation of the data published by Womersley et al. (1972) shows a good correspondence between LBM values derived from density and ^{40}K counts in both men and women; and in a subsequent paper (Womersley et al., 1976) the same was true for muscular and obese individuals of both sexes. Krzywicki et al. (1974) found a good correspondence between density and ^{40}K, and for density and body water, for women but not for men. The results of ^{40}K and densitometric assays were comparable in a series of women with varying degrees of obesity (Halliday et al., 1979); and body K was closely related to body N in a group of adult men and women (Morgan and Burkinshaw, 1983). In a group of adult men Lukaski et al. (1981a,b) found a good correspondence among all three techniques.

Average values for LBM were 74.5 kg by densitometry, 72.7 kg by ^{40}K counting, 74.0 kg tritium dilution, and 73.4 kg by deuterium dilution in the 38 adult men studied by Clark et al. (1978). The correlation between the ^{40}K and densitometric method was 0.89; and between tritium dilution and densitometry it was 0.93.

However, it must be said that the correspondence between the density and ^{40}K methods for estimating LBM is not as good for the elderly as it is for young and middle-aged adults. Both muscle and bone mass decline with age, the rates of decline vary with sex, and the decline in muscle mass is greater than that for viscera (Tzankoff and Norris, 1977, 1978). It is to be expected, therefore, that both the density and potassium contents of the LBM will differ in the elderly from the values that have been chosen for young and middle-aged adults. Unfortunately the magnitude of these changes is not known. However, the observed body K/N ratio tends to fall with advancing age (Cohn et al., 1983), which is in keeping with the observation that the ratio of connective tissue to skeletal muscle fibers is relatively high in the aged (Inokuchi et al., 1975).

Uptake of Fat-Soluble Gases

Cyclopropane, krypton, and xenon are much more soluble in fat than in lean tissues, the solubility ratios being respectively 34, 7.6, and 24 to 1. ^{85}Kr has also been used. A subject placed in a chamber containing a known amount of such a gas will progressively take up the gas into his body, and the rate of uptake will depend almost entirely on the amount of fat contained therein. This method has been used successfully in estimating total body fat in adult humans and in laboratory animals (Lesser et al. 1960, 1971; Hytten et al., 1966), and there is one report of its use in infants (Mettau, 1977, 1978). It has the advantage of being the only technique available for the *direct* measurement of body fat, lean weight being determined by difference.

Elaborate apparatus is needed, and the time for even partial equilibration in adults is long. The adult subjects studied by Lesser et al. (1960, 1971, 1963) wore an airtight headpiece for 6–9 hr; since equilibration was not yet complete by this time, a complicated mathematical treatment was necessary. In one experiment in which cylcopropane and krypton were used simultaneously, the results in some individual subjects differed appreciably. When the mean of the two was used there was a good correlation between the values obtained by this method and by total body water.

Mettau et al. (1977, 1978) placed their infant subjects in a chamber, and measured both the absorption phase and desorption phase of xenon; the total time was 270 min. These investigators found that the fat soluble gas method yielded results for total body fat in small animals (rats, guinea pigs) that were close to those obtained by carcass analysis.

The need for elaborate instrumentation, complicated mathematical treatment, numerous correction factors, and long measurement times have served to restrict greatly the use of this method.

Urinary Creatinine Excretion

This compound is formed in muscle from phosphocreatine, and provided that the diet does not contain excessive amounts of preformed creatinine (Cr), the quantity that appears in the urine is generally considered to be an index of muscle mass (Cheek, 1968). This assumption is supported by the work of Schutte et al. (1981) in dogs: they found a good relationship between total plasma creatinine (i.e., plasma Cr $\times$ plasma volume) and dissectable muscle mass in dogs, the equation being total plasma Cr (mg) $= 1.04$ muscle (kg) $+ 0.8$ over a range of 5–10 kg muscle mass. They also found that total plasma Cr was correlated ($r = 0.82$) with urinary Cr excretion in man. Each milligram of total plasma Cr corresponded to 1880 mg creatinine excreted in 24 hr; they conclude that each gram of urinary Cr/24 hr corresponds to 17.9 kg of muscle. This is an average value; the dog data show that the muscle mass/Cr ratio varies with the amount of total plasma Cr.

In a study of children and adults of widely varying body size we found an excellent correlation between LBM and urinary creatinine excretion, when two ^{40}K counts and three consecutive 24-hr collections of urine were made (Forbes and Bruining, 1976). Figure 2.14 shows this relationship; the equation is:

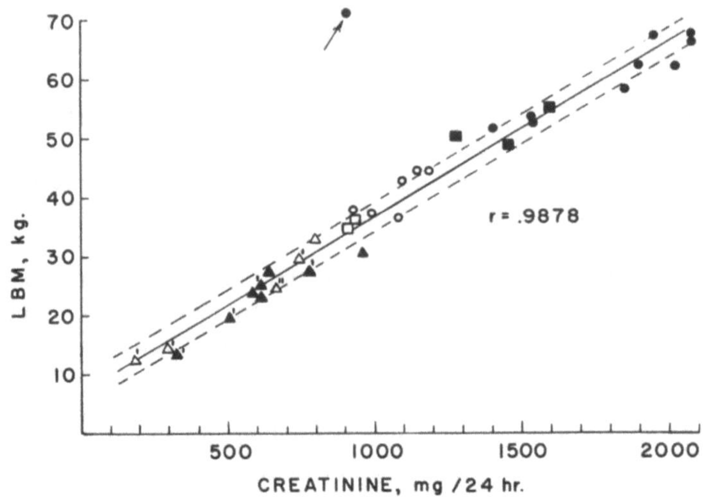

FIGURE 2.14. Plot of average LBM against average urinary creatinine excretion (n = 34). Closed symbols represent males; open, females. Dotted lines indicate one standard deviation from regression. The one aberrant point (arrow) was not included in the regression. (From Forbes and Bruining, 1976, with permission.)

$$\text{LBM (kg)} = 0.0291 \text{ Cr (mg/day)} + 7.38 \quad r^2 = 0.97 \tag{29}$$

These subjects ate *ad libitum,* so it is likely that they consumed some preformed Cr.

We have also had the opportunity of studying a group of 57 subjects who had been on a meat-free diet for several days prior to the urine collection. This group comprised 14 men and 41 women whose body weights ranged from 38 to 146 kg. For this group the equation is:

$$\text{LBM (kg)} = 0.0241 \text{ (Cr (mg/day)} + 20.7 \quad r^2 = 0.91 \tag{30}$$

(G. Forbes and M. Brown, unpublished data).

Based on carcass analyses of animals and adult humans, fat-free skeletal muscle makes up, on average, 49% of total fat-free body weight (data compiled by Forbes and Bruining, 1976). Clarys et al. (1985), using gross dissection of cadavers, found an average value of 54% ± 4.6 for muscle as a percentage of adipose tissue-free weight for men, and 48.1% ± 3.8 for women. Hence, one can use Equations 29 and 30 to estimate the ratio of muscle mass *(MM)* to urinary Cr excretion (g/day):

$$MM/\text{Cr} = 14.4 + 3.6/\text{Cr} \quad (\textit{ad libitum} \text{ diet})$$

$$MM/\text{Cr} = 11.8 + 10.1/\text{Cr} \quad (\text{meat-free diet})$$

Since equations 29 and 30 have an intercept on the *y*-axis, the ratio of LBM or of muscle mass to urinary Cr varies with Cr excretion. For example, a urinary Cr excretion of 1 g/day corresponds to a muscle mass of 18.0 kg, and one of 2 g/day to a muscle mass of 16.2 kg on *ad libitum* diet; for subjects on a meat-free diet the respective values are 22 and 16.8 kg muscle mass. A survey of the literature showed that most investigators have also found a *y*-axis intercept for the LBM/urine Cr relationship (Forbes and Bruining, 1976); hence it is not possible to speak of a constant relationship between LBM and urine Cr or between muscle mass and urinary Cr.

In performing this assay, it must be recognized that despite Shaffer's (1908) early postulate, Cr excretion does vary somewhat from day to day even in the face of a constant diet or during fasting. As an example we found a coefficient of variation of 4.8% in one subject maintained on a constant meat-free liquid diet over a period of 30 days. The urine collection period must be timed accurately, since an error as small as 15 min represents 1% of a 24-hr collection period. It is advisable to make three consecutive 24-hr collections.

Table 2.11 lists the expected urinary Cr excretion for normal men and women according to stature. As will be shown later LBM is related to stature at all ages examined thus far, so one would expect that the same would be true for Cr excretion. The table shows that stature has a significant effect: the tallest individual of either sex would be expected to excrete 15% more creatinine than the shortest. The sex difference in excretion rate is even greater when one compares men and women of the same height. Bistrian and Blackburn (1983), who compiled this table, offer it as a means for judging the nutritional status of patients who are suspected of being undernourished.

TABLE 2.11. Expected 24-hr urine Cr excretion of normal adult men and women of different heights.

Men			Women		
Height (cm)	Ideal weight (kg)	24-hr urine creatinine (g)	Height (cm)	Ideal weight (kg)	24-hr urine creatinine (g)
157.5	56.0	1.29	147.5	46.0	0.78
160.0	57.6	1.32	150.0	47.2	0.80
162.5	59.0	1.36	152.0	48.6	0.83
165.0	60.3	1.39	155.0	49.9	0.85
167.5	62.0	1.43	157.5	51.3	0.87
170.0	63.8	1.47	160.0	52.6	0.89
172.5	65.8	1.51	162.5	54.3	0.92
175.0	67.6	1.55	165.0	55.9	0.95
178.0	69.4	1.60	167.5	57.8	0.98
180.0	71.4	1.64	170.0	59.6	1.01
182.9	73.5	1.69	172.5	61.5	1.04
185.0	75.6	1.74	175.0	63.3	1.08
188.0	77.6	1.78	178.0	65.1	1.11
190.5	79.6	1.83	180.5	66.9	1.14
193.0	82.2	1.89	183.0	68.7	1.17

From Bistrian BR: Expected 24-hr urine creatinine excretion of normal adult men and women of different heights. Copyright The American Dietetic Association. Reprinted by permission from JOURNAL OF THE AMERICAN DIETETIC ASSOCIATION, *Vol. 71:393, 1977.*

Kreisberg et al. (1970) estimated muscle mass in adults by a complicated technique involving the administration of [^{14}C]creatine: the time course of the urinary excretion of labeled creatine is determined together with urinary Cr excretion and the creatine concentration in a sample of muscle.

Picou et al. (1976) have used creatine labelled with ^{15}N (a stable isotope) for this purpose in children. A tracer dose is given intravenously, and after equilibration (about 4 days) a muscle biopsy is taken for analysis of creatine. Meanwhile urine is collected for analysis of [^{15}N]Cr which was found to be excreted in exponential fashion after the first 3 days, so that a turnover rate for muscle creatine could be calculated. Then muscle mass (kg) is equal to

$$\frac{\text{urine creatinine (g/day)} - \text{creatine turnover (\%/day)}}{\text{muscle creatine } (\mu\text{g/mg})} \times 100 \quad (31)$$

This is multiplied by 1.159 to adjust for the difference in molecular weight between creatine and Cr. The assumptions are that muscle creatine is the sole precursor of urinary creatinine, that isotopic equilibrium occurs within 3 days, and that the fractional rate of Cr formation and excretion is constant. Such elegant and intricate techniques are obviously not suited to routine investigations.

Urinary Excretion of 3-Methylhistidine

This nonmetabolizable amino acid is formed in muscle where it is linked to actin and myosin. There is also some contribution from the gastrointestinal tract and other tissues. In human subjects Elia et al. (1979) report values of 3.31 mmol/kg fat-free dry weight for muscle, 1.5–2.0 for intestine, and 0.2–1.9 for other organs; so it is likely that a portion of the 3-methylhistidine that is excreted in the urine is from sources other than muscle. Indeed, in an ingenious series of experiments in which several body segments of the rat were isolated and then perfused, Wassner and Li (1982) found that while muscle accounted for 90% of the body content of 3-methylhistidine and gastrointestinal tract for 3.8%, the turnover rate in the latter organ was 24% per day compared to 1.4% per day for muscle. They estimate that 41% of the urinary 3-methylhistidine originates in the gastrointestinal tract.

Nevertheless, Lukaski and Mendez (1980) found a correlation of 0.89 between 3-methylhistidine excretion and LBM as estimated by densitometry in a group of adults. The correlation between creatinine and 3-methylhistidine excretions was poor ($r = 0.67$) in one group of subjects, and rather better ($r = 0.87$) in another group (Lukaski et al., 1981).

These investigators also made use of the Brookhaven neutron activation facility to estimate muscle and nonmuscle components of the body (see pp. 52–54). Whereas the correlation between 3-methylhistidine excretion and total fat-free mass (by density) was 0.81, that for muscle mass was 0.91. The correlation with nonmuscle mass was poor ($r = 0.33$).

Miller et al. (1982) studied a small group of normal children together with patients with cystic fibrosis. The ratio of 3-methylhistidine to creatinine excretion (molar basis) was higher in the latter (22.0) than in the former (14.6), which suggested to the authors that muscle protein catabolism was increased in patients with cystic fibrosis. The correlation between 3-methylhistidine excretion and LBM was higher for the normal children ($r = 0.94$) than for the patients with cystic fibrosis ($r = 0.76$).

One problem in using 3-methylhistidine excretion is the necessity for the subjects to be on a meat-free diet for at least 3 days prior to the urine collection. The question of day-to-day variability in the excretion rate has not been studied extensively. Lukaski et al. (1981) present average excretion rates for days 3–7 of meat-free diet for their 14 subjects: the mean values varied relatively little during this period, the range being 221 to 230 μmol/day; but the individual values are not given.

We have determined 3-methylhistidine and Cr excretion twice in each of 24 adult subjects who were fed a meat-free diet on the metabolic ward. The ratio of the 24-hr excretion on day 5 or 6 to that on day 4 was 1.09 ± 0.27 (SD) for 3-methylhistidine and 1.00 ± 0.09 for Cr; hence the day-to-day variability for the former is much greater than the latter (G.B. Forbes, unpublished data).

Based on the data available to date, one hesitates to recommend urinary 3-methylhistidine excretion as a satisfactory estimate of LBM.

Electrical Techniques

Hoffer et al. (1969) have measured the *impedance* of a weak electrical current (100 μA, 50 kHz) passed between the left ankle and the right wrist. When a correction is made for body height (squared) these authors found a good correlation ($r = 0.92$) with total body water.

A new technique is based on the change in *electrical conductivity* when the subject is placed in an electomagnetic field (5–10 MHz, 7 mW/cm^2). This change is proportional to the electrolyte content of the body, and hence should theoretically reflect the amount of lean tissue present. The instrument is variously known as EMME (electronic meat measuring equipment) or TOBEC (total body electrical conductivity). In a model designed for infants, Klish et al. (1984) found a good correlation with chemically determined LBM in rabbits, and in one designed for adults, Presta et al. (1983a,b) found good correlations with both LBM (by density) and total body water in human subjects; however the standard errors of estimate were rather high. Brocca et al. (1984) also found excellent correlations ($r = 0.95$–0.98) between TOBEC readings and total water, protein, and fat-free weight in rats; however, their animals did not vary much in body fat content.

Neither technique is hazardous, the microwave dose being far below safe limits, and neither requires much cooperation on the part of the subject.

Segal et al. (1985) have evaluated the conductivity and impedance techniques in a series of adult subjects whose body fat content (by densitometry) ranged from 5 to 55%. The results were compared to those obtained by densitometry, total body water, and ^{40}K counting. Ten conductivity measurements (TOBEC) were made in rapid succession for each subject; the coefficient of variation is said to be <2%, but no data are given for individuals measured on more than one occasion. The raw scores were transformed by taking the square root and then multiplying by the subjects' height or by height squared. Bioelectrical impedance was measured between electrodes placed on the right hand and right foot. The resistance reading (R) was transformed by multiplying $1/R$ by height squared, or by multiplying R by weight over height squared. The latter factor is of course body mass index. For this second method the transformed measurement was converted to body density by applying two constants, whence lean and fat could be calculated.

Conductivity estimates of LBM were very close to those estimated by densitometry ($r = -0.96$, SEE = 3 kg); the correlation between the latter and bioelectrical impedance was not as good ($r = 0.91$, SEE = 4.4 kg for W/H^2 transform, and $r = 0.94$, SEE = 3.4 kg for H^2/R transform). There was a tendency to overestimate LBM in subjects with a large body fat content. Correlations with total body water and total K tended to be lower ($r = 0.70$–0.94).

Lukaski et al. (1986) report an excellent correlation between H^2/resistance (by impedance) and fat-free mass (by densitometry) in a series of adult men and women whose body fat content ranged from 4 to 40%. The correlation coefficient was 0.98 and the standard error of estimate was 2.3 kg fat-free mass in this group of 114 subjects.

The inclusion of height or height squared in the transforms noted above would be expected to improve the conductivity and impedance estimates of LBM, since LBM is known to be related to stature (Forbes, 1974). Further studies of these two techniques will be awaited with interest.

Computerized Tomography

This technique is capable of distinguishing fat from other tissues; the resolution of present-day instruments is 5 mm or less. An illustration of the ability of this instrument to identify and quantitate the various tissues of the human forearm is provided by the work of Heymsfield et al. (1979) and Maughan et al. (1984), and of the thigh by Häggmark et al. (1978). Borkan et al. (1982) have used this technique to determine the ratio of intraabdominal to subcutaneous fat in human subjects. The size of liver, spleen, and kidneys can be determined. There is one report of an attempt to estimate total body fat by making a number of CT "cuts" in the extremities, trunk, and head (Tokunaga et al., 1983) and assuming cylindrical shapes for each of the regions. In five obese men (85–123 kg) and four obese women (75–98 kg) they determined that body fat made up 30–63% of total body volume. Measurements at the level of the umbilicus in 18 control and 19 obese subjects showed ratios of visceral fat to subcutaneous fat of 0.47 in the former and 0.37 in the latter. The radiation dose to the abdomen is 0.6–0.7 rad.

Sjöström et al. (1986) studied 12 women ranging in body weight from 46 to 124 kg by CT scans, ^{40}K counting, and tritium dilution. Multiple CT "cuts" were made of the extremities, trunk, and head in a manner similar to that used by Tokunaga et al. (1983), and assuming the fat content of adipose tissue to be 85% they calculated the amount of total body fat, and thus fat-free weight. As expected, body fat was greater in the heavier women, the calculated regression equation being fat (kg) = 0.826 W (kg) − 25.7 (r^2 = 0.95) over a range of 10–68 kg body fat. The correlation between body fat by CT scan and that estimated from ^{40}K counting was very good (r^2 = 0.99), while that estimated from total body water was lower (r^2 = 0.87). Based on the CT scan measurements these authors estimate that the potassium content of the fat-free mass is 62 meq/kg in women, a value not too far removed from the one of 64.2 meq K/kg suggested by Forbes et al. (1968).

Figure 2.15 shows a CT image of the mid-arm region of two individuals who happened to have about the same lean weight. The one on the left is a thin man weighing 62 kg; the right hand one is an obese woman weighing 104 kg. Respective values for body fat were 7 and 44 kg, for mid-arm circumference 29.5 and 43 cm, and for triceps skinfold thickness 3.2 and 39.5 mm. While the areas occupied by muscle are about the same (as would have been expected from their similar LBM values), the difference in subcutaneous fat is striking indeed. Heymsfield et al. (1979) have also

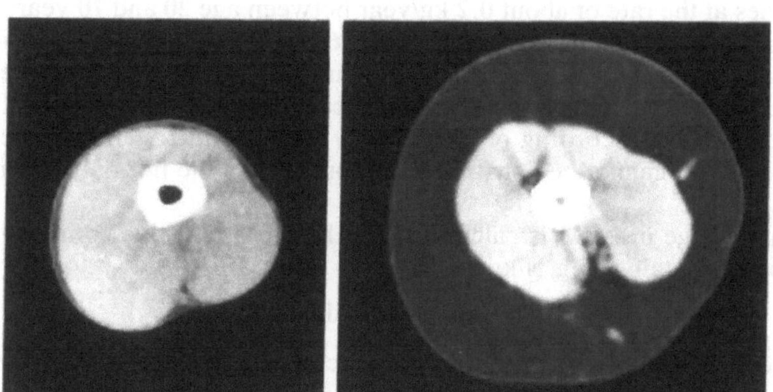

FIGURE 2.15. CAT scan of mid-arm region of a thin individual **(left)** and an obese person **(right)**.

done CT scans of the arms of malnourished individuals, and report that their muscle has a mottled appearance in addition to a smaller volume.

There is a preliminary report of the use of *infrared interactance* (Conway et al., 1984). Using five sites on the extremities, the absorption and reflection of near infrared rays (700–1100 nm) is measured; the distinction between fat and lean is that the former absorbs at a shorter wavelength than the latter. In a study of 17 adult subjects the authors claim correlations of 0.84 for men and 0.95 for women between infrared estimates and percent body fat as estimated by D_2O dilution. The authors state that the degree of tissue penetration by the infrared rays is about 1 cm.

Metabolic Balance Technique

While this technique cannot provide an answer for total body content, it can detect *changes* in body content of a number of elements, and thus help in assessing the effects of treatment (as in dietary rehabilitation) or other interventions of one sort or another. Since nitrogen, like potassium, is restricted in distribution to lean tissue, the nitrogen balance technique can yield estimates of the change in LBM, and by difference the change in body fat.

The principle is as follows: one measures N intake and N excretion, the algebraic difference being N balance. When intake exceeds outgo, the subject is said to be in positive balance; when the converse obtains, the subject is in negative balance. In the former situation, there is an apparent gain of body N; in the latter, a loss of body N. Healthy young adults enjoying constant body weight would be expected to be in zero balance. Healthy infants and children are obviously in positive N balance whereas adults, by virtue of their gradual decline in LBM, should be in slight negative balance. Indeed, as will be shown in a subsequent chapter, LBM declines at the rate of about 0.2 kg/year between age 20 and 70 years; this is equivalent to a body nitrogen loss of 6.6 g/year, or 18 mg/day. Although a negative nitrogen balance of this magnitude cannot be quantitated by the existing balance techniques, theoretical considerations point to its existence, so the recording of a distinctly positive nitrogen balance in a healthy adult cannot be construed as representing the normal state of affairs.

The change in LBM is calculated on the basis that this component of the body contains 33 g N/kg:

$$\Delta \text{LBM (kg)} = \text{N balance (g)}/33 \tag{32}$$

In the case of the newborn infant this factor is 24 g N/kg LBM based on carcass analysis; estimates for children lie between these two values (see Chapter 4). Since on average body proteins contain 16% nitrogen, the change in body protein content is 6.25 times the nitrogen balance.

The metabolic balance technique has a distinct advantage over other methods for estimating changes in LBM, in being able to detect much smaller changes. As stated earlier, the densitometric, total body water and total body K methods have inherent errors of 2–3%, so that, for example, in an adult man (LBM 60 kg) one cannot be certain that a change has occurred as a result of some maneuver unless the recorded change is at least 1.8 kg. The balance method, on the other hand, can easily detect a change in body N content of 10 g, corresponding to an LBM change of 0.3 kg.

The procedure entails the provision of a diet of known composition and the *quantitative collection* of urine and feces. The latter are analyzed for nitrogen as are duplicate diets. The balance period should be at least 6 days in duration, since urinary N excretion exhibits day-to-day fluctuations, and the intestinal transit time is 24 hr or greater; during this period the dietary N intake must be constant. A nonabsorbable marker (Brilliant Blue dye) is given at the start of the balance period,[7] and again at the end, and stool collections made between the appearance of the two markers. Some investigators prefer to administer a nonabsorbable stool marker, such as chromic oxide, cuprous thiocyanate, or polyethylene glycol PE 6400, throughout the balance period, the stools being analyzed for the marker. All offered food must be consumed, and any vomitus collected and analyzed.

Figure 2.16 shows one scheme for charting the results. The data are drawn from an actual study. Nitrogen intake is plotted upwards from the zero line, and urine and fecal losses downward from intake. Positive balance is shown by a clear area above the zero line, negative balance by the shaded area below the zero line. The variability from one balance period to another in the face of constant intake is clearly evident. Figure 2.17 shows that the daily variation in urinary N excretion is such that balance periods of 2–3 days duration may not represent the true state of affairs.

The balance technique demands meticulous attention to detail, and a considerable degree of cooperation on the part of the subjects and the attendants. Although it is possible to conduct a metabolic balance on "free-living" subjects the chances for error are great; it is far preferable to have the subject housed on a metabolic ward where trained nursing personnel are available, where dietary intake can be monitored, and where there is some assurance that excreta will be completely collected. A moment's reflection will show that the spillage of only 10 ml of urine will produce a 1% error in that day's urine nitrogen excretion.

There are several important considerations to be kept in mind.

1. Since any change in nitrogen intake is apt to be followed by a change in nitrogen balance, and in the same direction, the balance period should

---

[7]Carmine red was formerly used as a marker, but the discovery of bacterial contamination several years ago led investigators to discontinue its use.

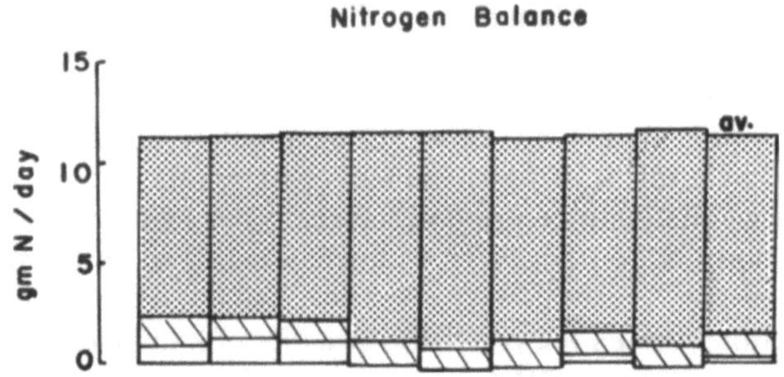

FIGURE 2.16. Nitrogen balance for eight consecutive 7-day periods in an individual on constant dietary intake. Intake is plotted up from the zero line; urine N excretion (dotted portion) and fecal N excretion (cross-hatched portion) are plotted downwards from intake. Nitrogen balance varies from +1.23 g/day to -0.28 g/day; overall average is +0.40 g/day. [From Johnston and McMillan (1952).]

be preceded by a fore-period during which the new intake is given. Adjustments in N excretion to the new intake usually take time, so that initially the subject may be in either positive or negative balance as the case may be; as time goes on the daily balance gradually returns to zero. An illustrative, though rather extreme, example is provided by Müller's (1911) study of an adult man whose nitrogen intake was abruptly increased from 12 g/day to 32–42 g/day (equivalent to 200–262 g protein) for a period of 28 days. Nitrogen balance immediately became positive and remained so for a full 24 days before a new equilibrium was finally established. The total N balance was +198 g in the face of a 4-kg gain in weight.

A more recent example is provided by Oddoye and Morgan (1979), who did careful nitrogen balances on male subjects whose diets were abruptly shifted from 12 g to 36 g N per day. Nitrogen balance immediately became positive; after some days it became less positive but failed to drop to zero by the end of the 60-day experimental period. Then when N intake was abruptly decreased, N balance became negative, and then gradually returned to zero over a period of about 50 days. Although the fact that the subjects gained weight while on the higher N diet and lost weight on the lower N diet complicates the interpretation of these results, this study

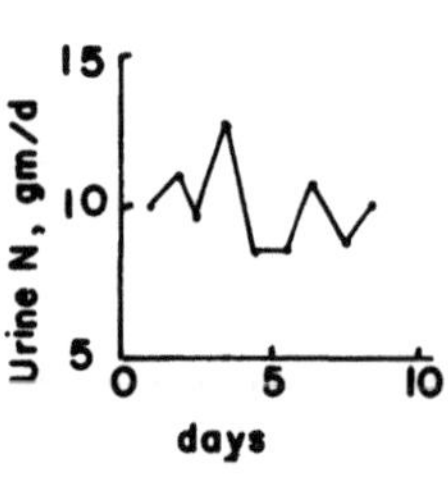

FIGURE 2.17. Daily urinary nitrogen excretion in an individual on constant diet. Mean 9.94 g/day, c.v. 14%. (Author's data.)

clearly demonstrates the existence of an adjustment period whenever N intake is changed.

2. Whereas the errors inherent in analyzing the diet and excreta for N are random in nature, the errors in assessing food intake and in collecting excreta are not. Any error in food intake is always in the direction of less than prescribed, never more; and any error in collection of excreta is always in the direction of less than recorded, never more. A simple illustration is as follows: suppose calculated N intake is 100 g over a given balance period, during which 90 g were determined to be excreted in urine and feces. The calculated balance is +10 g N. Suppose also that dietary intake actually was 1% less (plate wastage, food dribbled on chin, etc.) and that urine plus fecal volume was also 1% less (a bit of urine spilled, feces left on toilet paper). Then the actual balance was 100×0.99 minus $90 \times (1/0.99) = +8.1$ g N. Thus a 1% error in intake and outgo leads in this instance to a $(10 - 8.1)/8.1 = 23\%$ error in balance. This error will of course diminish as the difference between intake and outgo of nitrogen increases. For example, when calculated intake is 100 g N and outgo 80 g N, a 1% error in each leads to a +18.2 g balance, which is only a $(20 - 18.2)/18.2 = 9\%$ error in balance.

The end result of these nonrandom errors is that positive balances tend to be overestimated and negative balances underestimated.

3. Nitrogen is lost through the skin, nasal and vaginal secretions, and sputum as well as via urinary and fecal routes. Calloway and her associates (1971) have carefully studied such losses and conclude that they can amount to as much as 500 mg/day. They have also found that cutaneous losses are directly related to the level of blood urea nitrogen and hence to nitrogen intake. Such losses are increased in individuals who sweat, and this may be a problem in judging N balance in some obese subjects. Since these losses are most difficult to measure, most investigators add 5 or 8 mg N/kg/day to the observed urinary and fecal losses. This can add up to an appreciable amount of nitrogen in the case of a large individual; for example, the correction is 0.8 g N/day for a 100 kg person, or 5.6 g N/week. If such an individual were to ostensibly lose 20 g N in 1 week, this correction would increase the observed loss by 28%. Such corrections therefore form an appreciable fraction of balance N when gains or losses of body N are not large.

Cutaneous losses of zinc are also a function of zinc intake. Milne et al. (1983) found that such losses reduced apparent zinc balance by 12–84% in their subjects.

Nitrogen is also lost in menstrual flow. Johnston and McMillan (1952) report a range of 0.30–1.2 g N per menstrual period, and Calloway and Kurzer (1982) found a loss of 43 ± 24 mg/day during the menstrual period. Blood samples taken for analysis will contain about 3.5 g N/100 ml. Changes in blood urea nitrogen during the balance period must also be considered: a change in BUN of 10 mg/100 ml in a subject whose body contains 40 liters of water amounts to a gain or loss of 4 g nitrogen.

There is also an insensible loss of Na, K, and Cl, i.e., in the absence of sweating. These have actually been measured by a few industrious investigators both in infants and adults. The method used was to dress the subject in electrolyte-free bedclothes, and after 24 hr to wash the subject and bedclothes with distilled water and to analyze the bath water and wash water for electrolytes. Expressed as milliequivalents per square meter of surface area per 24 hr the average result for adults was 3.9 meq K/m^2 (Keutmann et al., 1939) and for young infants it was 3.4 meq K/m^2 (Swanson and Iob, 1933). The insensible losses of Na and Cl were 0.17–0.95 meq/day and 0.22–1.5 meq/day, respectively, for the infants, and 3–10 meq Na and 2–8 meq Cl in the adults. Visible sweating serves to greatly increase cutaneous losses of all three electrolytes, and for that matter elements such as Fe and Ca.

We studied two adult men who were sequentially fed diets containing different amounts of potassium but the same number of calories, so that they were in positive, negative, or zero balance, each for 2 weeks (Forbes, 1983). Repeated ^{40}K counts were made, together with analyses of diet, urine, and stool for potassium. In this way we could compare the changes in body potassium content as estimated by these two methods while body weight remained relatively constant. We found a consistent difference for all three diets: in periods of positive balance the balance method overestimated the increase in body K as estimated by ^{40}K counting; in periods of negative balance, the former method underestimated the loss of body K; and in those periods where there was no change in ^{40}K counts the potassium balance was slightly positive. The average difference between the two methods was 4.9 meq K/m^2/day. Of interest was the finding that this difference was a function of the potassium intake, the slope of the regression line being 0.057 meq K/m^2/meq K intake. For example, at an intake of 50 meq K/day, the predicted difference would be 2.9 meq/m^2/day; at one of 150 meq K/day it would be 8.6 meq/m^2/day. At first glance, such unmeasured losses constitute but a small fraction of the usual dietary intake of K (about 100 meq/day) and so might be considered inconsequential; but if one is to use K balance to estimate changes in LBM, failure to take unmeasured losses into account can seriously prejudice the results. For example, an adult (SA 1.7 m^2) will accrue, on average, an unmeasured loss of 8 meq K each day, which will amount to 1 kg of calculated LBM in 8 days. If K balance were to have been recorded as zero over that interval, such a subject would not have had a stable body composition, but would actually have lost about a kilogram of lean weight.

To my knowledge similar experiments comparing change in body N, Ca, and P by neutron activation analysis and by metabolic balance have not been done in man.

This tendency to overestimate retentions was a cause for concern to early investigators, such as Benjamin (1914), who found that the calculated increase in body protein content in growing infants was greater than expected, and particularly so in those fed cow's milk. Taken at face value his results, as well as those of many subsequent investigators, would imply that growing infants would eventually turn into solid nitrogen! Later, Mitchell (1962) spoke of "nitrogen growth" in adults since they appeared to be in positive N balance (by about 1 g/day) over long periods of time. Mitchell also found an apparent positive phosphorus balance of 170 mg/day. Nitrogen accretion during normal pregnancy similarly has been grossly overestimated by the balance technique, leading to absurdly high recommendations for protein intake. For example, Macy and Hunscher (1934) report an average N retention of 515 g. Some investigators have been led to invoke the phenomenon of "nitrogen storage" to explain the amounts of nitrogen apparently retained by growing infants and children, and by pregnant women, whereas in actuality there is no storage place for excess nitrogen in the human body.

What may be happening in those whose nitrogen balance becomes positive following an increase in food intake is an augmentation of the amount of "labile protein" in the body. Digestive enzymes would be expected to increase as would those in liver and kidney, in order to cope with the excess energy and protein load. As will be noted in a subsequent chapter, obese individuals tend to have larger viscera than the nonobese. In the opposite situation, it is well known that urinary nitrogen excretion is greater during the first few days after a sudden reduction in food intake than in the days that follow. It is known, for example, that liver size rapidly diminishes when animals are starved, and that hepatic enzyme induction follows refeeding. Consequently one could speculate that "labile protein" in the body bears some relationship to food intake, being greater in those who eat more and less in those who eat less. The fact that nitrogen balance becomes positive when excess nonprotein foods are given suggests that the responsible factor is total energy intake rather than protein intake *per se*.

It is most unfortunate that quasi-official agencies have in the past uncritically accepted the results of balance studies in devising recommendations for certain nutrient requirements for growth of infants and children. Part of the problem is that apparent nitrogen retention increases with nitrogen intake; other factors were the lack of appreciation of the nonrandom errors in assessing intake and outgo of the element in question, and the neglect of cutaneous losses. When one considers the expected nitrogen retention of a growing infant, it turns out to be a rather small amount: assuming a weight gain of 20 g/day and a body N content of 2% this infant would be expected to retain only 400 mg N each day, an amount too small to be determined with any degree of precision.

Hence the balance method cannot be expected to provide a precise value for nutrient requirements for growth. However, in situations involving large retentions of a particular element, as in recovery from malnutrition, one could anticipate better precision.

Based on animal experiments Duncan (1958) sounded a warning: "The physiological approach [to nutritional requirements] is usually made in one of two ways, each concerned with finding what changes occur in the normal animal in a given period. The first method [involves] . . . measuring the difference in body composition. The second method is that of the balance study. The results obtained by these two experimental methods should be identical. It is surprising to find how few workers seem ever to have used one to check the other . . . It has recently been done in studies of nitrogen metabolism, with disquieting results."

Doris Calloway and her associates have refined their balance technique to a high art, a recent report showing an apparent retention of only 0.72 ± 0.9 g N/day in adult women (King, 1981). In studying pregnant women they subtracted 0.72 g N from the observed retention of 1.74 g N, and thus found a value of 1.0 g/day, which is close to the expected value for the latter phase of pregnancy.

4. Finally, energy intake must be adequate, but not excessive. It was shown many years ago that nitrogen balance cannot be maintained in the face of a subnormal energy intake regardless of the amount of protein fed (Calloway and Spector, 1954). Recent investigations have shown that even modest reductions in energy intake lead to negative nitrogen balance in the face of a constant N intake (Garrel and Calloway, 1984), and that modest increases in nonprotein energy intake lead to positive balance (Butterfield and Calloway, 1984). Furthermore, it doesn't seem to matter whether the change in energy balance is achieved by alteration in energy intake or by a change in physical activity. Göranzon and Forsum (1985) determined nitrogen balance in normal adults in whom negative energy balance was achieved either by decreasing nonprotein calories by 20% while maintaining physical activity, or by increasing energy expenditure by 20% at constant energy intake; protein intake was held at the same level throughout the experiment. Both procedures produced about the same degree of negative nitrogen balance. This experiment shows that nitrogen balance is affected by alterations in energy balance regardless of how the latter is achieved.

The high degree of sensitivity of this technique, as compared to other methods for estimating changes in body composition, thus extorts a price: control of those variables known to affect balance must be as tight as possible if the effect of the procedure under investigation is to be realized.

In summary, while the metabolic balance technique is capable of detecting small changes in LBM, and (by subtraction) body fat, it demands meticulous attention to detail, a cooperative subject, an estimated correction for cutaneous losses, and is actually rather expensive when one considers the cost of operating a metabolic ward and supporting laboratory facilities.

It is, of course, the only technique available for determining the change in body content of many elements—Fe, Pb, Zn, Se, Mn, Mg, to name but a few. Although it is theoretically possible to estimate total body magnesium by neutron activation, there are formidable problems in applying this technique; and the isotopic dilution method presents problems in interpretation of the results.

Finally, it should be recognized that individual variability in metabolic balance data is the rule rather than the exception. Even under the most carefully standardized and controlled conditions nitrogen balance values exhibit sizeable coefficients of variation, so this technique shares with others the dual phenomena of technical error and biological variability.

Energy Balance

Garrow (1981) has put forth an ingenious method for estimating the composition of the tissue lost during weight reduction. Energy expenditure is estimated by means of an activity diary: the time occupied by sleeping, sitting, walking, etc. is recorded; BMR is measured; and so the total energy expended during a 24-h period can be calculated. Energy intake is measured. From the energy equivalents of fat and lean, taken as 9 and 1 k cal/g, respectively, the energy balance and the change in weight, the amount of fat lost is given as follows:

$$\Delta \text{fat (kg)} = \frac{\Delta \text{E (M cal)} - \Delta W \text{ (kg)}}{8} \tag{33}$$

The problem is the measurement of energy expenditure, a tedious and time-consuming job requiring meticulous cooperation by the subject. Nor can the possession of a whole-body calorimeter provide a solution except for the rare subject willing to be confined in such an apparatus for several days. Nevertheless, Garrow and co-workers were able to show a reasonable correlation between this method and nitrogen balance for estimating body fat loss in a group of obese women kept on low-energy diets for about 3 weeks.[8]

Some General Remarks

The application of the various techniques discussed thus far have made it possible to better define a number of aspects of human body composition, and as will be shown subsequently, they have increased our knowledge of the effects of disease, physical activity, and nutrition on the human body. However, each has its problems and its limitations, and, as with the use of any analytical technique, be it chemical, immunological, phys-

---

[8]If one were to test this technique for evaluating the composition of tissue gained during overfeeding, the energy equivalents would be different, namely 12 kcal/g for fat and 1.8 kcal/g for lean (Spady et al., 1976).

ical, or even psychological, these errors and limitations must be kept in mind.

The basic and ever present problem of what I have called "biologic variability" is best exemplified by body weight and metabolic balance. Even under the best of conditions body weight exhibits both diurnal and day-to-day fluctuations. Urinary excretion of nitrogen and electrolytes also shows daily variations even in the face of a constant intake. Energy-induced changes in body weight of as little as a kilogram or two may be associated with a change in nitrogen balance in the same direction despite a constant N intake; and the same would hold true for Na, K, and P, and perhaps other elements as well. It is absolutely necessary, therefore, to evaluate changes in metabolic balance in the light of changes in body weight.

Isotope dilution methods, densitometry, ^{40}K counting, and neutron activation techniques all are subject to technical error; all must be carefully calibrated and thoroughly tested before being put to use in conducting experiments. The magnitude of the technical error will determine the degree of change in body composition that can be detected with confidence during the course of an experimental investigation. These relatively new and wonderful techniques should be subject in operation to the same rigor that one has been used to applying in the analytical chemistry laboratory. (Indeed, their precision exceeds that of some commonly performed procedures, such as blood hormone assays.)

Strictly speaking, validation in man of the various techniques is difficult to achieve. One has to be satisfied with analogies drawn from the few cadaver analyses that have been done and from animal experiments. Comparisons of one technique, say ^{40}K counting, with another such as densitometry in the same individual or group of individuals cannot provide a complete answer since both techniques operate under the same assumption, namely that the LBM enjoys a constant composition, at least in young adults. Hence one hesitates to accept as proven, published data that purport to designate a value for the K content of the LBM as estimated by densitometry or anthropometry, or a value for "hydration" of the LBM as similarly estimated. One would be equally justified in generating a value for the density of the LBM from a measurement of body water or of total body K. Aside from the fact that the density measurement has a somewhat better precision than either the ^{40}K or total body water assays, there is no *a priori* reason for choosing it as the standard to which the others should be compared.

Anthropometry

The techniques discussed thus far are intricate, expensive, and not suited to large-scale studies or to work in the field. The anthropometric techniques to be described are simple and easy to perform; however, they suffer from a lack of precision.

Thickness of Skin Plus Subcutaneous Tissue

This is variously called skinfold thickness and fat fold thickness. Since human skin is only 0.5–2 mm thick (Edwards, 1950) the subcutaneous fat layer contributes the bulk of the measured value. The two most commonly used sites are over the triceps midway between elbow and shoulder and at the lower tip of the scapula. Others are the midbiceps region, at the lower rib margin in the anterior axillary line, the periumbilical region, over the iliac crest, and the anterior thigh.

The measurement is made by grasping the subcutaneous tissue between thumb and forefinger, shaking it gently to (hopefully) exclude underlying muscle, and stretching it just far enough to permit the jaws of the spring-actuated caliper[9] to impinge on the tissue. Since the jaws of the caliper compress the tissue, the caliper reading diminishes for a few seconds and then the dial is read. In subjects with moderately firm subcutaneous tissue the measurement is easy to make, but those with flabby, easily compressible tissue and those with very firm tissue not easily deformable present somewhat of a problem. In these latter two situations it may be difficult to achieve consistent readings. Figure 2.18 shows a caliper commonly used.

The assumptions underlying the use of this method for estimating body fat content are two in number: first, that the thickness of the subcutaneous fat mantle reflects the total amount of fat in the body, and second, the sites chosen for the measurement, either singly or in combination, represent the average thickness of the entire mantle. Neither assumption has been proven true.

There are two analyses of humans that bear on the first point: a full-term neonate in whom subcutaneous fat represented 42% of total body fat (Forbes, 1962) and an adult woman for whom the value was 32% (Moore et al., 1968). Pitts and Bullard (1968) analyzed the bodies of a number of mammalian species and found a wide variation, from 4 to 43%. Cuthbertson (1978) found a ratio of subcutaneous fat to total body fat of 0.36 in cattle, 0.46 in sheep, and 0.71 in pigs.

Southgate and Hey (1976) analyzed a number of fetuses ranging in gestational age from 180 to 290 days (only a few were at term) and found that subcutaneous fat represented 70–80% of total body fat. Although such values may be representative of premature infants, the results are pre-

[9]Three are recommended: the Harpenden Caliper, H.E. Morse Co., Holland, Michigan: the Holtain-Harpenden Caliper, Holtain Lt., Brynberian, Crymmych, Pembrokeshire, Wales; and the Lange caliper, Cambridge Scientific Industries, Inc., Cambridge, Maryland.

FIGURE 2.18. Photograph of Harpenden caliper. The opposing jaw faces each has an area of 97 mm^2, and the spring exerts a pressure of 10 g/mm^2.

sented in very brief form, and one must await the full publication before drawing conclusions.

With regard to the second point, a visit to the beach in summer time is sufficient to convince one of the regional variation in subcutaneous fat thickness, and of the individual variations in the distribution of this tissue. Some overweight individuals appear to have a centripetal distribution of subcutaneous fat, with rather thin forearms and lower legs, a configuration similar to that seen in Cushing's syndrome, whereas others have a more uniform mantle of exterior fat. The latter is particularly true of infants.

Body Circumferences and Diameters

The common sites for circumference measurements are the arm midway between elbow and shoulder, the abdomen at the level of the umbilicus, and the buttocks at its greatest perimeter. Some prefer the forearm site just distal to the antecubital crease, since this region usually contains less subcutaneous fat than the arm.

Behnke and Wilmore (1974) present a detailed, and rather complicated, analysis of the relationship of various body diameters to lean weight. The diameters of sites such as ankle, wrist, biacromial, and bitrochanteric are measured, and each is divided by a specific constant determined from measurements of a "reference" person. An average is taken of these quotients, and then squared, and finally multiplied by the subject's height.

Application of this technique to groups of adult males and females led to high correlations with LBM ($r = 0.88$–0.92) as estimated by densitometry. However, when it was applied to another group of subjects the correlations were not as high ($r = 0.73$–0.82).

Elbow breadth is now suggested as an indicator of body frame size in judging ideal body weight from the tables published by the Metropolitan Life Insurance Company. The ranges indicative of medium frame size are 6.3–7.3 cm for short men (<163 cm) to 7.3–8.3 cm for tall men (193 cm), and 5.7–6.3 for women <160 cm tall to 6.3–7.0 for women 183 cm tall. Elbow breadths less than these indicate small body frame, and those larger a large body frame. Measurements are made with the elbow flexed at 90° and the arm in supination.

Arm Circumference and Cross-Sectional Area

The muscle–bone component of the mid-arm region can be calculated from arm circumference and skinfold thickness. Although a number of investigators have used only the triceps skinfold thickness for this purpose, it would seem to be more appropriate to use the average of the biceps (B) and triceps (T) skinfold as follows:

$$\text{M + B area} = \frac{1}{4\pi}\left[\text{circumference} - \frac{\pi}{20}(\text{T + B SF})\right]^2 \qquad (34)$$

(arm circumference in cm; skinfolds in mm).

This is because the biceps and triceps skinfold thicknesses are rarely the same; an average of the two provides a better estimate of the average skinfold thickness over the circumference of the arm.

The equation for muscle + bone circumference of the arm is

$$\text{M + B circumference} = \text{arm circumference} - \frac{\pi}{20}(\text{T + B SF}). \qquad (35)$$

Some investigators have called this a "corrected" arm circumference. The area occupied by subcutaneous fat is simply:

$$(\text{arm circumference})^2/4\pi - \text{M + B area} \qquad (36)$$

A nomogram from which arm muscle and bone area and/or circumference can be read is shown in Figure 2.19. Simply place a straightedge so as to connect values for total arm circumference and skinfold thickness; the average of biceps and triceps skinfold thickness can be entered on the line marked "triceps."

It should be said here that although there are suggestions of a good correspondence between caliper and ordinary radiographic measurements of arm subcutaneous fat, correlations with measurements made by CAT scans are poor. Heymsfield et al. (1979) found arm M + B area was overestimated and hence arm fat area underestimated (and to a variable degree) by anthropometry.

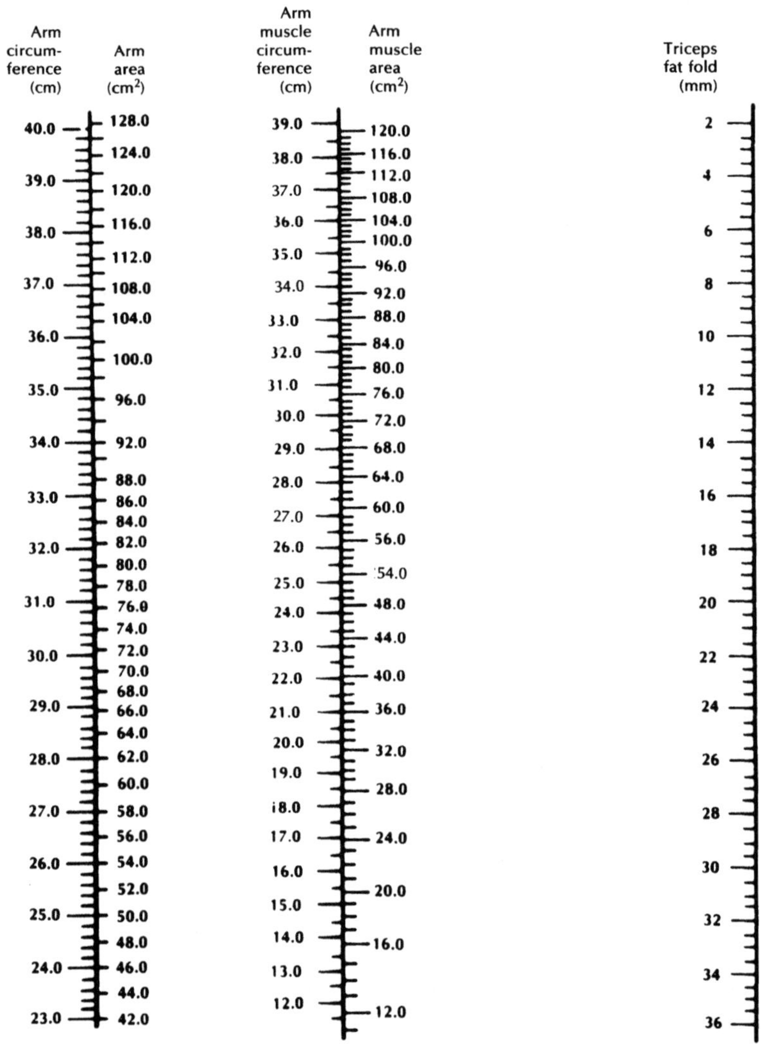

To obtain muscle circumference:

1. Lay ruler between value of arm circumference and fatfold
2. Read off muscle circumference on middle line

To obtain tissue areas:

1. The arm area and muscle area are alongside their respective circumferences
2. Fat area = arm area − muscle area

FIGURE 2.19. Nomogram for calculating arm muscle circumference and arm muscle area from triceps skinfold (or average of triceps and biceps) and arm circumference taken at a point midway from elbow to shoulder. [From Gurney and Jelliffe (1973), Am J Clin Nutr 26:912, with permission.]

These relatively simple and inexpensive techniques are not without their problems. Skinfold thickness measurements are very difficult in the obese: it is often impossible to raise a fold of subcutaneous tissue between the examiner's thumb and forefinger, and if such is possible the thickness may exceed the maximum jaw width (4 cm) of the usual calipers. In individuals who have recently lost weight the subcutaneous tissues are often flabby and easily deformed by the pressure exerted by the caliper jaws; and the same situation obtains for many elderly subjects, so that the skinfold thickness is underestimated. Hence in judging the change in body composition during weight loss, the skinfold technique may lead to an overestimation of the actual loss of body fat. Obviously dehydrated or edematous individuals are not suitable candidates for these measurements. The problem with the suprailiac and abdominal sites is the likelihood of including some muscle tissue along with subcutaneous fat in the measurement. The arms of many obese individuals taper sharply in diameter from shoulder to elbow, so that no one region is typical of the whole. Hytten and Chamberlain (1980) point out that about a third of the increase in skinfold thickness that occurs during pregnancy is rapidly lost after delivery; such a rapid change suggests that the decrease is due to loss of water or a change in skin vascularity rather than to a loss of subcutaneous fat.

Finally, in some subjects it is difficult to get the same reading twice; hence it is good practice to take two caliper readings a minute or two apart, and average the two.

Combined Measurements

Various combinations of skinfold thickness and body circumferences have been used to estimate body density and/or body fat content by numerous investigators (Brozek and Keys, 1951; Steinkamp et al., 1965; Durnin and Womersley, 1974; Behnke and Wilmore, 1974; Parizkova, 1977; Pollock et al., 1984) to mention but a few. Recently Lohman (1981) and Johnston (1982) have provided reviews of the subject. Some investigators have chosen to provide formulas for estimating body density rather than body fat or LBM *per se,* thus placing a burden on the reader who now has to translate density values into terms of body fat or percent fat by means of a formula. Body fat content is a hyperbolic function of density, not a linear one (see page 33, Equations 16 to 19).

While studies by many investigators have established relationships between various anthropometric measurements and LBM or body fat, the correlations are not very high (r^2 in the range of 0.4–0.8); however, the simplicity of the techniques means that they can be readily applied to large numbers of individuals in the field. It is obvious that such measurements as skinfold thickness and abdominal and buttocks circumferences will bear some relationship to body fat, and that biacromial, wrist, and knee diameters will vary with LBM. The quantitative relationships between anthropometric measurements and body composition vary somewhat by age and sex, and indeed among various investigators; furthermore, there have been problems in applying the results to individuals other than those from the group on whom the observations were made. Unfortunately, most authors are content merely to record correlation coefficients, and omit the standard errors of estimate. For example, Clemente et al. (1973) report a good correlation (r = 0.92) between body fat as estimated by ^{40}K counting and by an equation that includes body weight, age, and skinfold thickness; however, the standard error of estimate was rather high, 4.2 kg fat for men, and 3.3 kg fat for women.

Gnaedinger et al. (1963) found a poor correlation (r = 0.62) between back fat thickness and percent body fat as determined by carcass analysis in pigs.

However, Slaughter et al. (1978) found that the inclusion of body weight with skinfold thickness and body circumferences in a multiple regression formula produced a very good correlation with LBM (r^2 = 0.91) in adolescent boys. Hume (1966) found that LBM could be estimated just as well from height and weight (r^2 = 0.92 in men, 0.69 in women) as from skinfold thickness. Earlier Allen et al. (1956) had found a good relationship between adiposity and a weight–height cubed function, though it produced spurious results in professional football players.

Since LBM comprises 70–90% of body weight in normal children and adults, it is obvious that LBM and weight will be related; and in subjects of widely varying fat content one can anticipate that body fat and weight will also be related. As will be noted later, LBM is a function of stature at all ages (Forbes, 1974). It was the failure of weight–height functions for individuals whose excess weight consisted of muscle which led Albert Behnke to develop the specific gravity technique.

There has been a recent flurry of interest in using the ratio of abdominal circumference to buttocks circumference as an index of health. In a large group of obese adults Krotkiewski et al. (1983) found an average ratio of 0.987 for men and 0.833 for women. Those with waist/hip ratios above the average had higher values for fasting serum insulin and glucose, and higher integrated values for insulin and glucose during an oral glucose tolerance test than those with waist/hip ratios below the mean. Thus it would appear that it is the distribution of body fat as well as the total amount, which is important in determining some of the metabolic abnormalities in human obesity.

The Gothenburg longitudinal study of adult men provides additional data which bear on this question (Larsson et al., 1984). Seven hundred and ninety-two men were evaluated at age 54 years, and again 13 years later (the follow-up rate was 100%!). Evaluations at the earlier age included body mass index, skinfold measurements, abdominal and buttocks circumference, blood pressure, serum cholesterol, and smoking habits. The average waist/hip ratio was 0.927 ± 0.054 (SD); the range was 0.75 to 1.10, and the distribution of values was symmetrical (Gaussian). During this 13-year period a number of subjects developed cerebrovascular and ischemic heart disease, and some died.

The mean waist/hip ratio was slightly higher in those who suffered a stroke than in those who did not (0.958 vs. 0.925) and the same was true of those who had a heart attack (0.938 vs. 0.925); both differences were statistically significant. However, the difference in the ratio for those who died and those who lived (0.935 vs. 0.925) was not. The authors conclude that the distribution of body fat between abdominal and hip sites is a better predictor of poor health than any of the other indices of obesity which they used, even the degree of total adiposity. The same conclusion was reached by Lapidus et al. (1984), who did a similar study of 1404 women in Gothenburg; these subjects were 50–72 years old at the end of the 12-year follow-up. Their waist/hip ratios ranged from 0.59 to 1.00. These authors also report a distinct tendency for blood pressure to vary in accordance with the waist/hip ratio. However, they are careful to point out that such factors as serum cholesterol, smoking, and hypertension are even better predictors of ill health.

Because fatty acids from abdominal adipocytes drain into the portal vein, these authors offer the speculation that the exposure of the liver to higher concentrations of fatty acids could contribute to the observed effects.

With regard to estimations of body fat, the equations of Durnin and Womersley (1974) and those of Steinkamp et al. (1965) and Pollock et al. (1984) appear to have enjoyed the widest use. The first group studied 209 men and 272 women 16–72 years of age; fat content as determined by densitometry ranged from 5 to 61%. They were able to show a linear relationship between density and the logarithm of the sum of four skinfolds—biceps, triceps, subscapular, and suprailiac. The regression equation for the entire group of males is

$$\text{density (g/cm}^3) = 1.1765 - 0.0744 \log \text{SF (SEE 0.0103)} \qquad (37)$$

and for the females it is

$$\text{density (g/cm}^3) = 1.1567 - 0.0717 \log \text{SF (SEE 0.0116)} \qquad (38)$$

The coefficients of these equations, and the standard errors of estimate varied among the various age groups: intercepts ranged from 1.1333 to 1.1715, slopes from 0.0612 to 0.0779, and SEE from 0.0073 to 0.0125.

Although these relationships have been used by a number of investigators in estimating body fat, it should be noted that the distribution of the individual data points about the calculated regression line is somewhat irregular [see graph of density against sum of skinfolds in the Durnin and Womersley (1974) paper]. Hence the residual errors are not distributed symmetrically. Another problem, which was noted by the authors themselves, is the rather high estimates of percent body fat for subjects with low values for skinfold thickness (see Table 2.12). Caution should be used in applying these equations to small groups of subjects.

Table 2.12 lists estimated percent body fat based on skinfold measurements by subject age and sex. These data have been widely used in assessing individuals in the age ranges covered. The authors state that the error for percent body fat is ±3.5% for women and ±5% for men in two-thirds of instances.

Steinkamp et al. (1965) studied subjects in the age range of 25–44 years. Using skinfolds and various body circumferences they developed a series of multiple regression equations for predicting body fat content. The coefficients of the equations varied somewhat depending on the age and sex of the subjects. The correlation coefficients were all 0.90 or greater, and the standard errors of estimate were in the range of 2.2–3.8 kg fat.

Hill et al. (1978) determined total body nitrogen by neutron activation in a group of adults, and compared the results with LBM as estimated from skinfold thickness. While the correlation was good (r = 0.95) for normal subjects, it was considerably reduced (r = 0.79) in patients who had recently had a surgical procedure.

Pollock et al. (1984) present data on 308 men and 249 women aged 18–61 years, whose body fat content by densitometry ranged from 4 to 44%. Correlations between body density and sum of skinfolds from seven different sites were 0.85 for women and 0.88 for men, the respective standard errors of estimate being 0.008 and 0.009 g/cm^3. However, the standard errors were no higher when the sum of only three skinfolds was used.

Table 2.13 lists values for estimated percent body fat based on skinfold measurements, condensed from the table published by Pollock et al. (1984).

Parizkova (1977) presents a number of regression equations and nomograms for estimating percent body fat (by density) from various skinfold

TABLE 2.12. Percent body fat for the sum of four skinfolds.

Skinfolds[a] (mm)	Males				Females			
	17–29 yr	30–39 yr	40–49 yr	50+ yr	16–29 yr	30–39 yr	40–49 yr	50 + yr
20	8.1	12.2	12.2	12.6	14.1	17.0	19.8	21.4
30	12.9	16.2	17.7	18.6	19.5	21.8	24.5	26.6
40	16.4	19.2	21.4	22.9	23.4	25.5	28.2	30.3
50	19.0	21.5	24.6	26.5	26.5	28.2	31.0	33.4
60	21.2	23.5	27.1	29.2	29.1	30.6	33.2	35.7
70	23.1	25.1	29.3	31.6	31.2	32.5	35.0	37.7
80	24.8	26.6	31.2	33.8	33.1	34.3	36.7	39.6
100	27.6	29.0	34.4	37.4	36.4	37.2	39.7	42.6
120	30.0	31.1	37.0	40.4	39.0	39.6	42.0	45.1
140	32.0	32.7	39.2	43.0	41.3	41.6	44.0	47.2
160	33.7	34.3	41.2	45.1	43.3	43.6	45.8	49.2
180	35.3	—	—	—	—	45.2	47.4	50.8
200	—	—	—	—	—	46.5	48.8	52.4

From Durnin and Womersley (1974), with permission.
[a]Sum of biceps, triceps, subscapular, suprailiac.

TABLE 2.13. Percent body fat for sum of skinfolds.[a]

Sum of skinfolds (mm)	Males				Women			
	18–21 yr	28–32 yr	43–47 yr	58–61 yr	18–21 yr	28–32 yr	43–47 yr	53 + yr
9	1.3	2.3	3.9	5.5	—	—	—	—
24	6.1	7.2	8.8	10.5	9.7	10.2	10.9	11.4
36	9.8	10.9	12.6	14.3	14.8	15.3	16.0	16.5
48	13.4	14.5	16.2	17.9	19.5	20.0	20.7	21.2
60	16.9	17.9	19.7	21.4	23.7	24.2	25.0	25.5
72	20.1	21.2	23.0	24.7	27.5	28.0	28.8	29.3
84	23.2	24.4	26.1	27.9	30.9	31.4	32.2	32.7
99	26.9	28.0	29.8	31.6	34.6	35.1	35.8	36.3
111	29.6	30.8	32.6	34.4	37.0	37.5	38.2	38.7
126	32.7	33.9	35.8	37.6	39.4	39.9	40.6	41.1

[a]Triceps, iliac, thigh for women; chest, abdomen, thigh for men.
From Pollock et al. (1984), with permission.

measurements. Of interest is the fact that for boys and girls aged 8–12 years the correlation coefficients for the log triceps + subscapular sites ($r = 0.859$) are almost as good as those derived from log of 11 skinfold sites ($r = 0.875$); however, the standard errors of estimate were rather high (4.94 and 4.66 respectively). Correlations were much lower in adult men 17–40 years ($r = 0.57$–0.66); and in older adults Parizkova (1977, p. 50) states that the "percentage of body fat cannot be derived so far from skinfolds."

Beddoe et al. (1984) found a very good correlation between fat-free weight as estimated from skinfolds and total body water plus total body nitrogen in 21 men and 20 women ($r = 0.97$, c.v. 4.7%). They use hydrogen as an internal standard for their nitrogen activation procedure; this could possibly serve to increase the predictability of their method, since a sizeable portion of body hydrogen is associated with protein.

We have tested the relationship between arm muscle + bone (M + B) area and LBM by ^{40}K counting in a group of normal subjects aged 8 to 70 years. Figure 2.20 is a scattergram of arm M + B area plotted against LBM for the normal men and boys, and Figure 2.21 shows a similar plot for normal women and girls. The correlation coefficients belie the evident scatter in the data.

However, when we selected subjects who had anorexia nervosa and those with obesity the slope of the regression line was less steep (0.53) and the y-axis intercept higher (14.1 kg LBM) than those for the equations describing the normal subjects in (Figures 2.20 and 2.21).

Buttocks circumference is also related to body fat. Noppa et al (1979) report a correlation coefficient of 0.86 in a group of 227 women 44–66 years old. Interestingly enough, for these subjects the correlation between body fat and body weight was even better ($r = 0.90$). Indeed weight and buttocks circumference were each better predictors of body fat than subscapular or triceps skinfolds or the sum of these two skinfolds. For this group of women, whose body fat content ranged from 8 to 47 kg, the

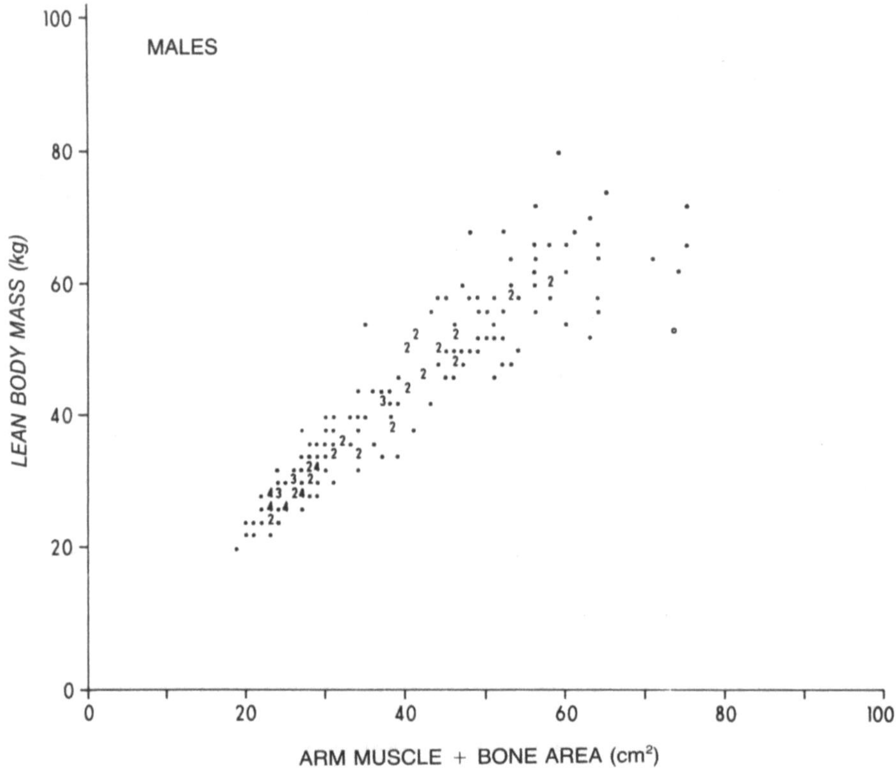

FIGURE 2.20. Plot of LBM against arm muscle-bone area for 178 males aged 8–71 years: $y = 0.953x + 5.68$ ($r = 0.93$). Datum represented by the open circle not included in the regression analysis. (Author's data.)

equation for the combined measurements, using sum of triceps and subscapular skinfolds, weight, and buttocks circumference is

$$\text{Fat (kg)} = 0.37\,W + 0.13\ \text{butt circumference (cm)} + 0.10\ \text{SF (mm)} - 21.1 \qquad (r = 0.92,\ \text{SEE } 2.96). \tag{39}$$

Equations such as these are more suitable for women than for men, for four reasons: first is the fact that the variability in LBM tends to be less than that in men; second, the regression of LBM on height has a less steep slope (see Chapter 4), and third, a larger fraction of body weight is composed of fat in women. Finally, women tend to accumulate more fat around the hips. The end result of all these factors is that variations in female body weight are more apt to reflect variation in body fat than variations in lean.

Almond et al. (1984) studied a group of ill patients by means of skinfold

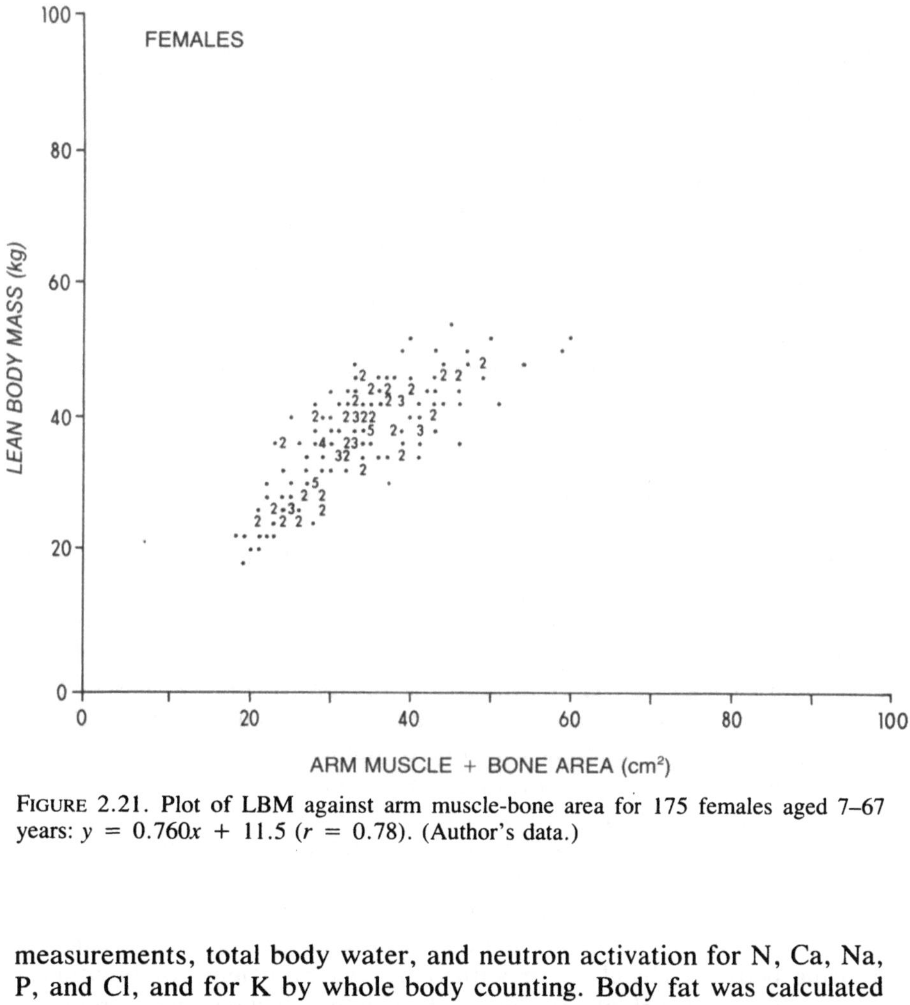

FIGURE 2.21. Plot of LBM against arm muscle-bone area for 175 females aged 7–67 years: $y = 0.760x + 11.5$ ($r = 0.78$). (Author's data.)

measurements, total body water, and neutron activation for N, Ca, Na, P, and Cl, and for K by whole body counting. Body fat was calculated as weight minus the sum of protein, water and minerals, the result being an average fat content of 9.1 kg. Calculations from skinfold thickness gave a larger value, namely 12.1 kg.

In a study of undernourished men, Spurr et al. (1981) found that the Durnin-Womersley formula (Table 2.12) underestimated fat weight, and further that the prediction worsened as the men recovered. In the hands of Forse and Shizgal (1980), arm circumference and triceps skinfold did not satisfactorily predict the degree of undernutrition in patients who had surgical illnesses or who were obese. Using anthropometric techniques to judge the composition of the weight loss in obese individuals treated with a low-energy diet, van Gaal et al. (1985) concluded that body fat represented all of the 14.4 kg weight loss. However, the fact that their subjects exhibited a decline in urinary creatinine excretion and a negative nitrogen balance leads one to conclude that they suffered some loss of LBM as well as fat. Hence it is likely that the skinfold technique overestimated the loss of body fat in these subjects. Collins et al. (1979) compared measurements of arm muscle and bone circumference and area to total body nitrogen as estimated by neutron activation in a group of surgical patients. They concluded that the anthropometric method provided such a poor prediction of total body nitrogen as to be valueless.

These examples are offered as indications that anthropometric techniques for estimating body composition may not provide correct answers in subjects who are ill or who have an altered nutritional status.

Fuchs et al. (1978) used an interesting approach in their study of a large group of adult men, whose body fat content ranged from 6 to 36%. They used the circumference of the arm with the biceps muscle *flexed*. Their equation is

$$\text{LBM (kg)} = 0.514 \text{ height (cm)} + 0.0178 \text{ arm circumference (cm)}^2 - 49.7. \quad (40)$$

The standard error of estimate is 2.95 kg, which is about 4.6% of the mean LBM. When applied to a second group of adult men, who had not been used in constructing the original relationship, this equation yielded an average LBM value of 64.4 kg, compared to one of 63.9 kg by total body water, and 63.7 kg by ^{40}K counting. The reason for the success of this two-component equation is obvious: flexed arm circumference reflects muscle mass; and LBM is known to be a function of stature.

In considering the various anthropometric techniques discussed in the preceding pages, it is fair to ask why their ability to predict body fat content, or LBM, has not been better. By using multiple regression techniques a few investigators have achieved correlations a bit above 0.9, thus explaining perhaps 85% of the variance, but many have fallen far short of this goal, with results which are not a great deal better than those achieved by using body weight alone or with body weight–height combinations. One problem is the lack of precision of the skinfold thickness measurement; another is the fact that the standards to which the anthropometric techniques are compared, such as densitometry, total body water, or ^{40}K counting, also involve errors. Attempts to improve the reliability of skinfold measurements by using corrections for compressibility of the subcutaneous tissue have not been entirely successful. The need to devise prediction formulas which differ among subjects according to age and sex is disquieting. The best results are obtained when the range of the independent variable is large.

In assessing these techniques in patients who were ill, Forse and Shizgal (1980) concluded that predictability was not nearly as good as that for normal individuals. Others have commented on the poor results in the elderly.

The subcutaneous tissue mantle is far from uniform in thickness, and some sites, such as the inner thighs which in some individuals appear to be replete with fat, are not easily accessible. As stated earlier, the ratio of subcutaneous fat to total body fat undoubtedly varies.

A recent study by Kvist et al. (1986) illustrates the problem. They did serial CAT scans at 22 locations on the trunk, extremities, and head and neck of eight women of varying body fat content. Of the total volume of subcutaneous fat, the legs and trunk regions accounted for no less than 90%, leaving only 8% for the arms and 2% for the head and neck regions. The upper legs contained the largest areas of subcutaneous fat in the thinnest women (body fat 25 kg or less), whereas these were in the pelvic and trunk regions of those who had about 60 kg of body fat. Hence, the favored site for measuring skinfold thickness, namely the triceps region, is far from ideal. These investigators also estimated "visceral fat" as being 10% of total body fat, but it is unclear whether this included such internal sites as the pelvic and retroperitoneal regions, and nothing was said about intermuscular fat.

In their studies of male monkeys, Jen et al. (1985) found that abdominal girth and abdominal skinfold thickness showed a much better correlation with body fat than either triceps or subscapular skinfolds.

While anthropometric techniques are useful in the field, and for comparing groups of individuals without the need for elaborate equipment, their relative lack of precision is a drawback in deriving accurate estimates of body composition for single individuals. Indeed, in a recent review, Johnston (1982) came to the conclusion that "accurate estimates of whole body composition from anthropometry are not possible," and he suggests that anthropometric data be presented as such without translation into terms of actual body composition. In the minds of those who are dedicated to the anthropometric technique this must represent a rather extreme view. Certainly, a subject with a large triceps skinfold thickness is undoubtedly fatter than one with a thin skinfold, and the same could be said for a large difference in buttocks circumference. Johnston's point, which is based on a review of published data, is that anthropometry lacks sufficient precision for quantitative judgments on body composition.

Functions of Height and Weight

There is now a considerable amount of evidence to show that *stature* is an important variable in assessing the results of body composition assays. LBM is a function of height at all ages thus far examined (Forbes, 1974), and as will be shown later the same is true for several types of athletes. Blood volume and urinary creatinine excretion are greater in tall individuals than in their shorter age and sex peers. The weight of the human diaphragm muscle is correlated with stature (Arara and Rochester, 1982), as is total body nitrogen (Williams et al., 1982). Mention has already been made of the influence of stature on total body calcium, and the cadaver data reported by Borisov and Morei (1974) show the same is true for skeletal weight.

While it may only represent a mathematical happenstance, note should be made of the observation that in normal adults LBM turns out to be related to the cube of height (Forbes, 1974). The same is true for blood volume (Allen et al., 1956) and for skeletal mass as estimated by neutron activation (McNeill and Harrison, 1982),[10] and for total body nitrogen (Williams et al., 1982). However, as will be shown later (Chapter 8) the exponent for the LBM-height function is larger for many types of athletes.

Although the correlations between these aspects of body composition and stature are not high enough to permit the latter to be a good predictor of the former, they are sufficient to make it necessary to consider stature as an important variable in comparing body composition data between individuals or groups of individuals. Regardless of other characteristics or circumstances, tall individuals will, on average, have a larger LBM, a larger blood volume, and a larger bone mass than their shorter age and sex peers. For example, the fact that many athletes are taller than average is one reason why they tend to have a larger LBM.

Body Mass Index (BMI)

This is body weight divided by a power of height, usually $(\text{height})^2$, which is said to be independent of stature at least in the adult (Keys et al., 1972). Calculations based on values for ideal body weight (Metropolitan Life Insurance Co., 1983) suggest that BMI for normal men and women should be in the range of 19–27 W/H^2 (weight in kg, height in m). Indeed, this range roughly corresponds to 25th–75th percentile values recorded for adult individuals who participated in the 1971–74 National Health and Nutrition Survey (Cronk and Roche, 1982). In the case of infants and children, average BMI values change with age (van Wieringen, 1972; Rolland-Cachera et al., 1982), beginning at 13 kg/m^2 at birth, reaching a peak of 18 at about 1 year, and then a nadir of 15 at about age 6 years, to be followed by a rise to adult values during adolescence. Individuals with high indices are classified as overweight, even obese, and those with subnormal indices as undernourished. While such classifications cannot be applied, for instance, to short muscular men, or to tall asthenic women, the BMI has found usefulness in evaluating groups of individuals.

However, the BMI is not always an accurate index of body composition. This is evident from the fact that the average index is about the same for both sexes during the adolescent and young adult years (Cronk and Roche, 1982; Rolland-Cachera et al., 1982) despite the obvious difference in body

---

[10]The formulas developed by Cohn (1981, p. 203) for predicting total body K from height and weight and for predicting total body Ca from total body K and height lead to much the same results. These relationships are that body $K = (f)W^{1/2}H^2$ and body $Ca = (f)KH$. Using the formula for the Body Mass Index, namely $W = (f)H^2$, and substituting in the first equation we have $K = (f)H^3$, and in the second $Ca = (f)H^{2.5}$.

fat content. Indeed, in adolescent males Haschke (1983) found a poor correlation between BMI and fatness.

However, Revicki and Israel (1986) found a somewhat better (though modest, $r = 0.71$) correlation between BMI and percent fat (by densitometry) in a large group of adult men. The standard error of estimate was 5.3% fat, which is not too far below the standard deviation of 7.3% fat for the entire sample. As expected, there was no correlation of BMI with height. While there can be no doubt as to relationship of BMI to adiposity, for the simple reason that body fat contributes an ever larger share of body weight as weight increases, BMI cannot be considered an accurate estimate.

Figure 2.22 is a nomogram from which BMI can be read.

BMI appears to have superseded the ponderal index ($W^{1/3}/H$) used in the past.

Physiologists commonly use weight $^{0.75}$ as an index of "active metabolic mass" in evaluating data from animal experiments. Indeed both Brody (1945) and Kleiber (1975) have shown that interspecies BMR is a linear function of body weight to about the 3/4 power in mammals ranging in size from the mouse to the elephant, a weight range of at least five orders of magnitude.

Brody's formula is kcal/day = 70.5 $W^{0.734}$; Kleiber's is kcal/day = 67.6$W^{0.756}$. The equation for birds is 89 $W^{0.64}$ over a weight range of 15 g–16 kg. However, these relationships pertain only to the adult members of the various species; the slopes of the regression lines were somewhat different for members of a single species.

BMR values are commonly reported in terms of surface area (SA) in adults, since early observations showed that the BMR/SA ratio had less variability than BMR/W or BMR/H. The formula developed by DuBois and DuBois (1916), namely

$$\text{SA (cm}^2) = 71.8 \times W_t^{0.425} \times H^{0.725},$$

is widely used to calculate surface area. Recently Martin et al. (1984) made measurements of skin surface in a number of adult cadavers, and found that the DuBois formula gave an accurate estimate of surface area, and that power functions of weight alone also did well in this regard.[11]

THE MELLITS–CHEEK EQUATION

Some years ago Drs. Mellits and Cheek (1970) published an equation which related total body water as a function of weight and height for children and adolescents. The original formulation by Cheek (1968) was later mod-

[11]In this connection, the DuBois formula reduces to a power function of weight if one assumes that height is proportional to $W^{1/3}$: thus $H^{0.725} = W^{0.241}$, and SA = $W^{0.425+0.241} = W^{0.666}$.

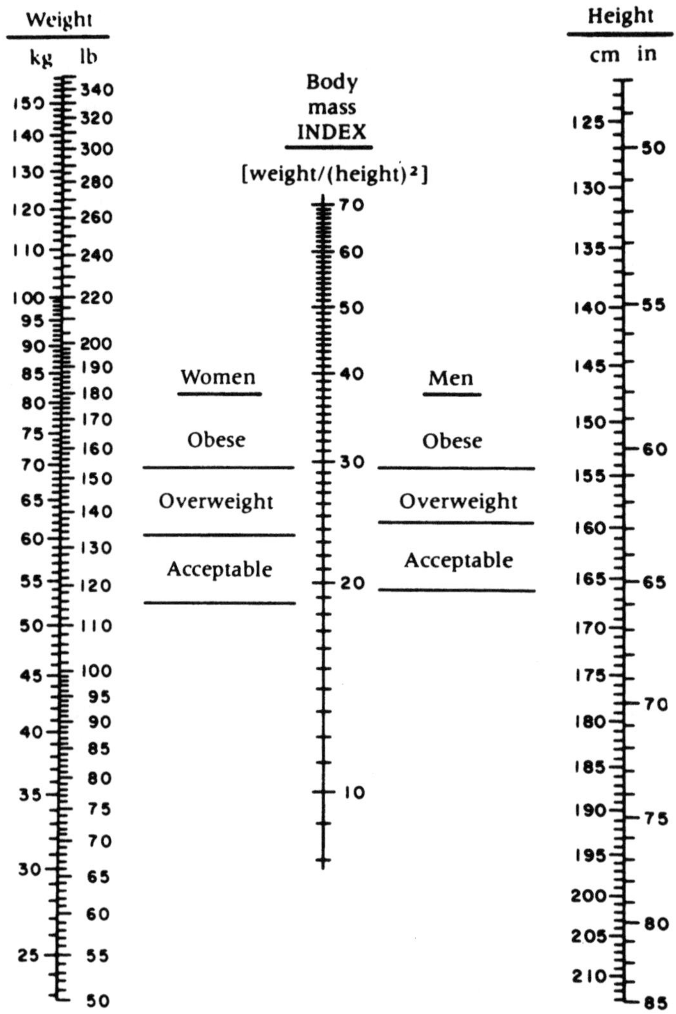

FIGURE 2.22. Nomogram for calculating BMI (W/H^2) from weight and height. [From Bistrian BR and Blackburn GL: Assessment of protein-calorie malnutrition in the hospitalized patient, in Nutritional Support of Medical Practice, 2nd edition (HA Schneider, CE Anderson, DB Coursin, eds), 1983. New York, Harper & Row, p 135, with permission.]

ified by inclusion of data from the literature to cover an age range from one month to 34 years. These equations are, for females:

$$H_2O\ (l) = 0.076 + 0.507W + 0.013H, \text{ SEE } 0.50 \text{ liters. (for } H \leqslant 110.8 \text{ cm)} \tag{41}$$

$$H_2O\ (l) = -10.313 + 0.252W + 0.154H, \text{ SEE } 1.65 \text{ liters. (for } H \geqslant 110.8 \text{ cm).}$$

and for men:

$$H_2O\ (l) = -1.927 + 0.465W + 0.045H, \text{ SEE } 0.60 \text{ liters.} \quad (\text{for } H \leqslant 132.7 \text{ cm}) \tag{42}$$

$$H_2O\ (l) = -21.993 + 0.406W + 0.209H, \text{ SEE } 2.52 \text{ liters.} \quad (\text{for } H \geqslant 132.7 \text{ cm})$$

Deuterium dilution was used as an estimate of total body water, whence LBM and fat could be calculated.

These equations have been used by a number of investigators, notably Frisch and her associates (see Chapter 4) to estimate LBM and fat. Since they lack a term for age, all individuals of a given sex, height, and weight within the stated height limits are allotted the same value for total body water, and assuming that the LBM is 72% water, the same values for LBM and fat; hence biological variation among individuals of similar size is not admitted. One is entitled to ask, therefore: how well is body water estimated by these equations?

An analysis was made of the results obtained by the equation for girls over 110.8 cm tall. The observed values for total body water for the 44 girls (I was unable to locate the 45th) on which Cheek's equation was based were compared to values predicted by the equation, and the difference between the two expressed as a percentage of the observed value. These percentages were averaged, and the standard deviation was calculated in the usual way. The results of this analysis are shown in the top row of Table 2.14. Although the mean difference is close to zero, the standard deviation is rather large, and the range indicates a high degree of uncertainty. Looked at in this manner, the predictive power of equation

TABLE 2.14. Predicted body composition (Mellits–Cheek equation).

Investigator	n	Age, yr	Predicted − observed H_2O percent difference: Mean	SD	Range	Remarks[b]
Mellits and Cheek (1970)[a]	44	6–31	+0.38	7.6	−12, +28	Females
Cheek et al. (1970)	13	9–17	+4.26	14.1	−15, +34	Obese females
G.B. Forbes (unpublished)[c]	113	8–18	−1.68	8.6	−26, +21	^{40}K counting, females
Brook (1971)	8	5–11	+11.84	4.6	−6, +20	Females
C. Young et al. (1968)	98	9–17	−2.91	11.2	−26, +28	Tritium dilution, females
Ljunggren et al. (1961)	31	18–29	+1.52	7.4	−11, +15	Anorexic, normal, obese females
Schutte (1980)	172	10–18	+2.95	6.7	−14, +29	Black males
Loucks et al. (1984)	32	19–28	−13.5[d]			Female athletes

[a]Sources listed in their publication.
[b]Normal subjects unless otherwise indicated; body H_2O assayed by deuterium dilution unless otherwise indicated.
[c]H_2O = LBM × 0.73.
[d]Calculated from % fat values; Mellits–Cheek equation overestimates percent fat by 9.8% (range −16.8 to +1.1%).

41 is far less than one would infer from the stated correlation coefficient of 0.973 and the standard error estimate of 1.65 liters H_2O.

A similar analysis was made for other groups of girls and women in whom assays of total body H_2O were actually made or were calculated from ^{40}K counting. The results of such analyses are given in Table 2.14. These are seen to be as bad, and in one instance actually worse. Indeed, the predicted value for LBM ($H_2O/0.72$) actually exceeded body weight in one of Ljunggren's subjects. The values for Schutte's male subjects were calculated from the means of eight age groups. The results for these are seen to be equally poor, despite the stated correlation coefficient of 0.97. The data shown in Table 2.14 do not inspire confidence in the ability of the Mellits-Cheek equation to generate precise estimates of body composition.

External Radiation Techniques

Bone

Barnett and Nordin (1960) were the first to suggest this simple technique for estimating skeletal size. Roentgenograms of the hand are taken, in a

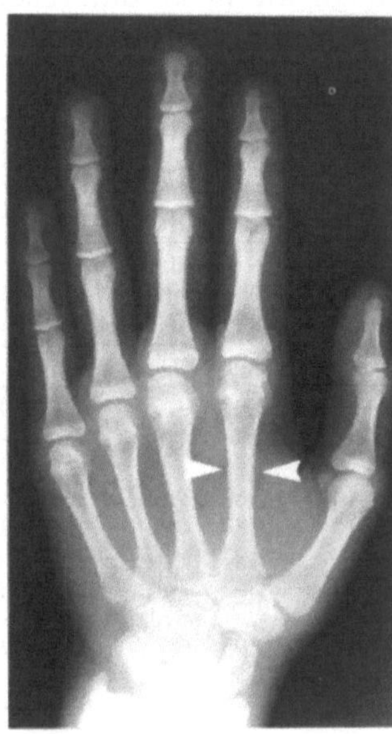

FIGURE 2.23. Hand roentgenogram of 14-year-old girl. Arrows point to mid-shaft of second metacarpal at which point caliper measurements are usually made. Cross-sectional area is calculated by Equation 43.

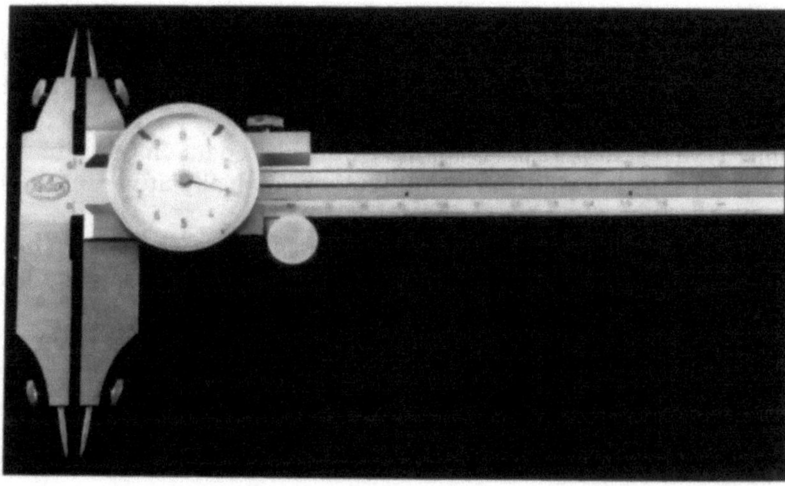

FIGURE 2.24. Helios caliper equipped with needle points for making measurements of bone cortex. Dial can be read to the nearest 0.5 mm.

manner to minimize parallax, and measurements are made of the cortex of the second metacarpal (Figure 2.23), either with a special caliper (Figure 2.24) or with a magnifying glass fitted with a reticule. Garn (1970) and Gryfe et al. (1971) prefer to measure cortex thickness at a point equidistant from the two ends of the bone, whereas Bonnard (1968) takes the reading where the cortex is thickest and includes the 3rd and 4th metacarpals as well as the 2nd. This measurement yields a value for cortex thickness either directly or as periosteal diameter (OD) minus endosteal diameter (ID). Then the cross-sectional area of the mid-cortex is

$$\pi/4\ (OD^2 - ID^2), \tag{43}$$

assuming that the bone cortex is cylindrical at that point. Others prefer to divide Equation 43 by total cortical area [$\pi/4\ (OD^2)$] to yield a value for percent cortical area. Garn (1970) has postulated that metacarpal cortex thickness serves as an index of total skeletal mass.

The problem with such measurements is the fact that cross-sections of the mid-metacarpal cortex vary somewhat in shape and wall thickness. Horsman and Kirby (1972) in their studies of cross-sections of a number of human metacarpal bones, found a variety of configurations—irregular cylinders, rounded rhomboids, variations in wall thickness, etc. In addition, adult metacarpals may exhibit an irregular endosteal surface and small osteoporotic cavities; the latter would of course be unlikely in youthful bone. Hence Equation 43 is an approximation.

Nevertheless, the second metacarpal bone has several advantages over other sites. Its medullary cavity is more nearly centered, it has a more uniform morphology and is closer to a cylindrical form than other bones, and it has a relatively thin covering of soft tissue; a location for measurements is easily fixed. Hence, despite its shortcomings, this bone is the preferred site for radiographic measurements.

Tanner (1962) and Maresh (1966) have published data on bone, muscle, and subcutaneous fat widths measured radiographically at various points on the extremities of children and adolescents.

Poznanski et al (1980) measured the cross-sectional cortical area of the humerus at a location just proximal to the nutrient foramen in prematurely born and term infants for this purpose.

Virtama and Helelä (1969) measured the cross-sectional area of a number of cortical sites in Finnish individuals aged 2–80 years. Their data are presented as $OR^2 - IR^2$ for the mid cortex; when multiplied by $\pi/4$ one obtains cortical area in the same units as presented by Garn. Comparison of the Finnish data with those obtained from Americans by Garn (see Chapter 4) shows that the values for the second metacarpal correspond more closely to those for blacks than for whites in the U.S.A. As will be seen subsequently, blacks tend to have a larger cross-sectional area for this bone than do whites at most ages throughout the childhood and adult years.

Virtama and Helelä (1969) also measured cross-sectional cortical area at the midpoint of the second metatarsal bone. Using selected age groups for comparison of metacarpal and metatarsal cortical areas, namely 10, 20, and 30 years of age, metatarsal values consistently exceeded metacarpal values in females, but only at age 10 in males.

We have developed a variation of this technique to estimate the total volume of the metacarpal cortex. The internal and external diameters of the cortex of the second metacarpal are measured at the point of greatest convexity, together with the diameter at each end of the cortical portion of the bone and the length of the cortex. Then the distal and proximal halves of the cortex are each considered as the frustum of two cones, one defined by the periosteal surface, the other (a hollow cone) by the endosteal surface. The measurement sites are shown in Figure 2.25.

The volume of the outer proximal cone frustum is

$$\frac{\pi}{3}h_1\,(r_2^2 + r_2\,r_3 + r_3^2, \text{ where } r \text{ is radius,} \qquad (44)$$

and the distal outer cone frustum is

$$\frac{\pi}{3}h_2\,(r_2^2 + r_2\,r_3 + r_3^2),$$

The total bone volume is the sum of the proximal and distal cone frusta. The volume of the inner cone frusta are calculated in similar fashion, substituting r_1 for r_2 in the above equations. Then total *cortical volume* is the difference between the volumes of the outer and inner cone frusta.

This method ignores the trabecular bone, and the epiphysis, both of which contribute a portion of the metacarpal weight, and the same is true of the other techniques mentioned earlier. One problem common to all is the indistinct inner surface of the mid-bone cortex in some subjects, which makes for some imprecision in determining the inner diameter.

The measurements are tedious to make, and this technique assumes that the cortex is geometrically equivalent to regular flat-topped cones (i.e., frusta of cones). It has the advantage over the cross-sectional cortical area method (Equation 43) in that cortex length is included in the calculation. While others have made calculations of "bone volume" they did so by merely including cortex length, neglecting the fact that the metacarpal cortex does not have the configuration of a regular cylinder, but one in which the walls taper in thickness towards each end. The same criticism could be made of the index proposed by Exton-Smith et al. (1969), which is metacarpal cortical area divided by the product of periosteal diameter and length.

The technique of *photon absorptiometry* has been used by several investigators (Mazess, 1971; Cohn, 1981b; Dequeker and Johnston, 1982). The bone, usually distal radius and ulna, is scanned transversely by a low-energy photon beam (^{125}I, 27 KeV; ^{241}Am, 60 keV; or ^{153}Gd, 44 and 100 KeV), and the transmission is monitored by a scintillation detector. The arm is encased in a bag of water which simulates soft tissue density. The change in transmission as the beam is moved across the extremity is a function of the bone density in that region, and there is evidence that this is related to total skeletal mass, at least in adults (Manzke et al., 1975;

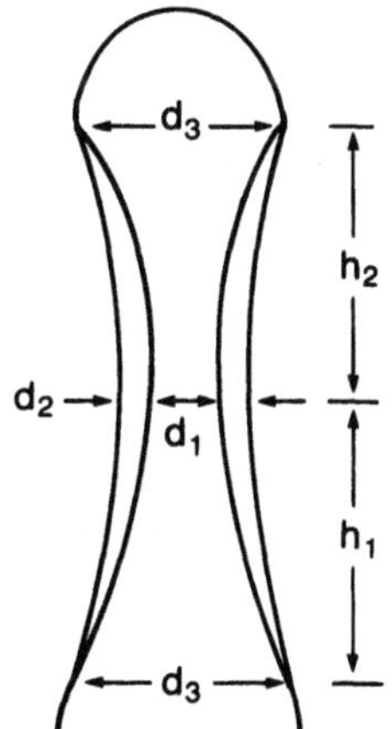

FIGURE 2.25. Diagrammatic representation of metacarpal bone showing location of measurements (d = diameter) necessary for calculating total volume of the cortex from Equation 44. While the shape of the entire cortex resembles a "hyperboloid of one sheet," rather than the frusta of two cones placed end to end, the use of the latter would seem to be a reasonable approximation.

Christiansen and Rödbro, 1975a; Ellis and Cohn, 1975b). Mazess (1971) reports that photon scans of ulna, radius, humerus, and femur correlate well with dry skeletal weight ($r = 0.9$). The radiation dose is only a few millirem.

Christiansen et al. (1975a) found an excellent correlation ($r = 0.96$) between photon scans and bone calcium content in human bone specimens.

Peppler and Mazess (1981) and Mazess et al. (1981a,b) describe a sophisticated method for estimating total bone mineral of the entire body. The body is scanned transversely in 2.5-cm steps over its entire length by radiation from a gadolinium source (^{153}Gd). This radioactive source emits two gamma rays of different energies, 44 kev and 100 kev. The procedure takes about 70 minutes, but delivers only about 1 mrem of radiation. The analysis involves some intricate mathematics requiring a computer. These investigators found a very good correlation between the results of this technique and total body calcium by neutron activation ($r = 0.99$).

They also determined bone mineral content (BMC) of the distal radius by ^{125}I scans, and found a good correlation between BMC and total body Ca ($r = 0.98$), and between BMC and total bone mineral ($r = 0.97$). The correspondence between ^{153}Gd scans and Ca content by neutron activation was also good when only the trunk of the body was measured.

The problems encountered in using the technique of dual photon absorptiometry are those associated with the complicated calibration technique required and the complicated mathematical treatment which is necessary. As is the case for neutron activation, the technique is best suited, at least at the moment, to a research laboratory. However, the dual photon technique has been used by several investigators for the measurement of single bones. Its advantage stems from the fact that the difference in photon absorption by soft tissue and bone is dependent on photon energy as well as tissue density.

The books edited by S.H. Cohn (1981) and Dequeker and Johnston (1982) are excellent sources of information on the various radiogrammetric techniques for estimating bone mass. Included are CAT scans and neutron activation of single bones, Compton scattering techniques, and assessment of trabecular bone and vertebrae, in addition to the procedures discussed above. The application of these techniques to the study of growth and ageing, and their use in the study of disease are described in some detail.

Muscle

Garn (1961) and Tanner (1962) discuss the technique involved in making satisfactory measurements of muscle widths from roentgenograms of the extremities. Since the entire width is measured, this procedure overlooks the variable amounts of fat between the muscle layers, which is so nicely demonstrated by nuclear magnetic resonance.

Subcutaneous Fat Layer

With appropriate x-ray exposures, the subcutaneous fat layer can be identified and measured. The favored sites are the lateral chest wall and the region over the greater trochanter of the femur (Stuart et al., 1940; Garn, 1961; Comstock and Livesay, 1963).

Mention should be made here of the use of ultrasound for estimation of subcutaneous fat thickness (Stouffer, 1963; Booth et al., 1966; Fanelli and Kuczmarski, 1984). These last authors found correlations of 0.80–0.86 between ultrasound and skinfold measurements at seven different sites; however the correlation between the results of ultrasound measurements and body density was only fair ($r = 0.78$).

Muscle, Bone, and Fat

Computerized tomography offers the opportunity of defining these three components in an extremity, in a manner that permits one to calculate the cross-sectional area of each (Heymsfield et al., 1979; Maughan 1984). Figure 2.15 (p. 63) compares the arms of a thin male and an obese female who happened to have the same LBM. The difference in subcutaneous fat thickness and cross-sectional area of fat is striking indeed.

Gotfredsen et al. (1986) have used dual photon absorptiometry (^{153}Gd) to estimate LBM and fat in human subjects. The body is scanned in a rectilinear fashion, the total number of pixels being about 4000. They report good reproducibility (c.v. 2.2%) and that the radiation dose is less than 5 mrad. As mentioned earlier, this technique requires sophisticated instrumentation and calculations.

One awaits with interest the application of nuclear magnetic resonance techniques to the study of body composition in man. This instrument is capable of defining both internal and external fat depots, without exposing the subject to ionizing radiation. The recent literature contains a report of its use to estimate body fat, and total body water in animals (Lewis et al., 1986).

Precision of Various Techniques

Two factors must be considered: technical error *per se,* and biological variability. The former include such phenomena as the random nature of radioactive isotope emissions, the measurement of quantity of injectate, flame photometer errors, etc. For isotope dilution procedures, for instance, four measurements are necessary, each subject to error: composition of the injectate and its volume, assay of the equilibrium sample of body fluid for tracer and/or stable isotope; and in some instances there is a fifth measurement, urinary excretion of tracer. For ^{40}K counting, the subject must be properly situated in the counting chamber, and the instrument must be properly calibrated. Underwater weighing entails measurement

TABLE 2.15. Reproducibility data.

Technique	Coefficient of variation	References
Total exchangeable Na	3.1, 4.9, 3.8%	Price et al. (1969); James et al (1954); Haxhe (1963)
Total exchangeable K	3.3, 2.9, 6.1, 2.5%	James et al. (1954); Haxhe (1963); Price et al. (1969); Davies and Robertson (1973)
^{40}K counting	2–3, 2–3, 1.2–4.8%	Shukla et al. (1973); Forbes et al. (1968); Pierson et al. (1974)
Density	0.0023, 0.0063 g/cm^3	Durnin and Taylor (1960); Lohman (1981)
Cyclopropane, ^{85}Kr uptake	7–8% "uncertainty"	Lesser et al. (1971)
Total body Ca	5.2, 5.5%	Nelp et al. (1972), McNeill and Harrison (1981)
Total body N	3.5%	Ellis et al. (1982)
Creatinine excretion	2–19%	Forbes and Bruining (1976)
Skinfold thickness	6–24%	National Center for Health Statistics (1974)

of residual pulmonary air volume as well as body weight both in air and submerged. Neutron activation requires that the neutron dose delivered to the body must be known as well as the amount of induced radioactivity. In the case of urinary creatinine excretion, urine creatinine concentration and 24-h urine volume are both required; an error of 15 min in collection time represents 1% of a 24-h period. The latter is in reality a component of technical error.

Biological variability must also be taken into account. The body is not a static system: although many individuals are able to maintain their weight within narrow limits over long intervals of time, careful studies have shown that it oscillates around a mean value, with daily variations of a per cent or so. Earlier mention was made (p. 66) of the variation in urinary nitrogen excretion, and Calloway and Kurzer (1982) have documented variations in nitrogen balance with the menstrual cycle. Once considered a "constant," 24-hr urine creatinine excretion is now known to have a coefficient of variation of several per cent under controlled conditions. It is the *total* error—technical plus biologic—that must be taken into account, and this can be evaluated only by replicate assays on individuals, a procedure not always feasible with assays involving radiation exposure.

Table 2.15 gives the variations found by investigators who have assayed their subjects on two or more occasions.

Table 2.16 lists the various techniques which have been used by a sufficient number of investigators to be considered useful and well standardized. The advantages and disadvantages of each are briefly mentioned. None is foolproof, and each one demands practice and attention to detail to achieve worthwhile results. In selecting a technique the purpose of the study to be performed must be considered, along with the availability of

TABLE 2.16. Body composition techniques: advantages and disadvantages.

	Advantages	Disadvantages
Density	Apparatus inexpensive Estimates LBM and fat simultaneously Nonhazardous Can be repeated frequently	Subject cooperation necessary for under-water weighing Unsuitable for young children, elderly Error from intestinal gas
^{40}K counting	No hazard Minimal subject cooperation Can be repeated frequently	Instrument expensive Proper calibration necessary Problem in interpretation in subjects with K deficiency
Metabolic balance	No hazard Suitable for many elements Can detect small changes in body content (<1%)	Measures only *change* in body composition Meticulous subject cooperation Metabolic ward expensive Error from unmeasured skin losses Many laboratory analyses needed
Neutron activation	Minimal subject cooperation Body content Ca, P, N, Na, Cl	Apparatus very expensive Calibration very difficult Radiation exposure
Creatinine excretion	No hazard Estimate of muscle mass	Meticulous subject cooperation Influenced by diet, collection time critical Day-to-day variation (c.v. 5–10%)
Fat-soluble gases	Direct estimate of body fat	Cyclopropane, xenon, ^{85}Kr Apparatus expensive Long equilibration time
Dilution methods	Estimate body fluid volumes Inexpensive Great variety: Na, K, Cl (Br), H_2O	Radiation exposure (some materials) Blood samples needed (some materials) (some require several samples) Incomplete equilibration Na, K; overestimation by D_2O, THO; value for ECF depends on method used; ^{18}O assay requires elaborate equipment
Anthropometry	Cheap Direct estimate of body fat, muscle mass	Poor precision in obese subjects, and in those with firm s.c. tissue Regional variation in subcutaneous fat layer; uncertainty ratio s.c.fat/total fat[a]
Radiography, photon densitometry	Bone density, volume; muscle widths	Limited regions Radiation exposure
CAT scan[b]	Organ size, configuration; s.c. fat; intraperitoneal, pericardial fat; bone	Instrument expensive Radiation exposure
Ultrasound	Organ size No hazard	Poor definition subcutaneous fat layer
3-Methylhistidine excretion	Estimate of muscle mass	Meat-free diet 2 days prior to collection Subject cooperation ? variable contribution from gastrointestinal tissue, ? magnitude of day-to-day variations

There is a recent report of the use of nuclear magnetic resonance to estimate total body water and body fat in infant baboons (Lewis et al., 1986).

[a]Recent observations by CAT scan suggest that the ratio of intra-abdominal fat to subcutaneous fat varies considerably in adults (Borkan et al., 1982).

[b]See, for example, the analysis of the constituents of the human forearm by Maugham et al. (1984) and Heymsfield et al. (1979).

TABLE 2.17. Calculation of lean body mass, fat in adults.

Method	Formula	Remarks
Density	Fraction fat $= \frac{4.570}{D} - 4.142$	Brozek et al. (1963)
	Fraction fat $= \frac{4.95}{D} - 4.50$	Siri (1961)
Total body water	LBM (kg) = H_2O (liters)/0.73	Multiply by 0.95 to correct for ^2H and ^3H exchange with nonaqueous hydrogen
^{40}K counting	LBM (kg) = total K (meq)/68.1 (M), 64.2 (F) BCM (kg) = total K (meq) × 0.00833	Multiply by 1.1 when ^{42}K dilution is used
Creatinine excretion	LBM (kg) = 0.029 Cr (mg/day) + 7.38 LBM (kg) = 0.024 Cr (mg/day) + 20.7	Ordinary diet. Forbes and Bruining (1976) Meat-free diet. G.B. Forbes, unpublished data
Neutron activation	LBM (kg) = total N (g)/33	Change in LBM can be estimated from N balance
Anthropometry (skinfold thickness, body circumferences)	Numerous formulae	Those of Steinkamp et al. (1965) and Durnin and Womersley (1974) are frequently used Many have less than adequate precision
Density—total water	Fraction fat $= \frac{2.1366}{D} - 0.78\frac{H_2O}{W} - 1.374$	Siri (1961)
Neutron activation—body water—^{40}K	LBM (kg) $= 6.25\ N + H_2O + \frac{Ca}{0.34}$	Cohn et al. (1984)
	LBM (kg) $= BCM + ECF + \frac{Ca}{0.117}$	Cohn et al. (1980)

equipment and facilities. Some demand a considerable degree of cooperation by the subject, some require expensive instrumentation or exposure to radiation, while others measure only portions of the body.

Table 2.17 lists the calculations used in arriving at an estimate of LBM and fat in the adult. The coefficients are derived for the most part from carcass analyses, the K/LBM ratio being altered for females on the basis of the reported sex difference in the K_e/H_2O ratio (Moore et al., 1963; Forbes et al., 1968). Some investigators prefer to derive values for LBM composition from a comparison of *in vivo* assays. Cohn et al. (1980) have chosen values of 64.5 and 58 meq K/kg LBM for men and women, respectively, whereas Womersley et al. (1972) suggest that these should be 66.4 and 59.7. Lukaski et al. (1981a,b) derived a value of 62.6 meq K/kg LBM for men. They also offer one of 746 ml H_2O/kg LBM; however, their values of 32.1 mg N and 32.7 g N/kg LBM are close to the one derived from cadaver analysis.

A number of values for total body K/N ratios (meq/g) have been reported: 1.92, 1.88, 1.95 for men; 1.67, 1.73, 1.78 for women (Cohn et al., 1980; Lukaski et al., 1981a; Morgan and Burkinshaw, 1983), although these latter authors derive a value of 1.81 meq K/g N for both sexes by regression analysis. Total body K/N ratios tend to be lower in older adults. The average value from cadaver analysis of adult subjects is 2 meq K/g N (Forbes 1962).

The variations noted above reflect the criteria used to assess LBM, possible differences in analytical technique, or even biological variability. There is a distinct tendency among investigators to use the densitometric technique as the "standard" to which the results of other techniques should be compared. Densitometry, total body water, and body K techniques all are based on the same assumption, namely that the LBM has a constant composition, and hence cannot be considered truly independent of each other. A direct estimate of body fat by recording the uptake of a fat soluble gas (p. 29, 56, 57), whence LBM is gotten by subtraction, is in effect an independent method; so, too, is the CAT scan technique described on pages 62 and 63. Nuclear magnetic resonance can also distinguish water and fat. However, anthropometric techniques lack the precision needed for estimation of the K or H_2O contents of the LBM, or its density.

The values listed in Table 2.17 pertain to young and middle-aged adults. Composition is quite different for the fetus and newborn, and changes occur with aging; these will be discussed in subsequent chapters.

Reference Basis

Having made a determination of body composition by one of the techniques described here, how is one to judge the result? Merely recording a value for total body water or total body potassium for instance, without reference to some criterion of normality is of little help—indeed, the same could be said of height and weight, but these have the advantage of long acquaintance and so "normal" values, roughly speaking, are common knowledge. Body weight is not a suitable reference, for the simple reason that body fat is highly variable, and the frequency distributions of fat content in cohorts are usually skewed. As will be shown in subsequent chapters, normal values for LBM and fat vary with age and sex, and mention has already been made of the influence of stature on LBM, total body calcium, and urinary creatinine excretion. Ideally, one should have gathered data on large numbers of "normal" subjects in one's own laboratory, but this may not always be possible. Lacking such an array of data, the best solution is to compare the results of the assay on the particular subject with values obtained from a group of "normal" subjects of the same sex, age, and stature who have been assayed in one's own laboratory so as to avoid interlaboratory variations. This procedure worked well for us in coming to the conclusion that LBM was indeed normal in the female relatives of boys with sex-linked muscular dystrophy (Borgstedt et al., 1970). In reviewing our own data and those from the literature on body composition in obesity a comparison of such individuals with nonobese individuals of the same age, stature and sex showed unequivocally that they had a supranormal LBM, and also allowed us to estimate the composition of their excess weight (Forbes and Welle, 1983).

3

Body Composition of the Fetus

> Except during the nine months before he draws his first breath, no man manages his affairs as well as a tree does.
>
> *George Bernard Shaw*

During the past 100 years several dozen human fetuses and newborns have been analyzed for water, electrolytes, minerals, and fat by chemical techniques. German scientists were the first to make such studies, with several reports in the late 1800s and early 1900s; Americans came on the scene in the 1930s; then in the 1950s British workers, led by Dr. Elsie Widdowson of Cambridge University, began their extensive observations. Dr. Widdowson and Professor R. A. McCance and their associates were responsible for generating a renewed interest in body composition, and together they have produced a vast amount of data on this topic. Much of the data to be presented in this chapter is the direct result of their work.

For more than a century and a half it has been known that young animals differ from adults in regard to certain aspects of body composition. Water contents are higher and ash contents (per unit weight) lower in the young (*ash* is the residue remaining after removal of water and organic material—the word derives from the mineral ash left after the burning of wood or coal). It is reasonable to ask whether there are changes in body composition during fetal life: does the young fetus differ from the newborn, just as the newborn differs from the adult? Are the changes in the same direction, and how rapidly do they occur?

The many chemical analyses that have been done on fetuses and newborns provide the answers to these questions. References to the original works can be found in publications by Widdowson and Dickerson (1964) and Forbes (1981a). The analyses published by Trotter and Peterson (1969) show that the dry fat-free weight of the skeleton (osseous portion) makes up an increasing fraction of total body weight during fetal life. Between a body weight of 227 g (the smallest fetus analyzed) and term, relative skeletal weight almost doubles, from 1.7% to 3.2% of body weight. What is the situation for other elements? As will be seen subsequently, the change in element content varies: some elements such as Ca increase in concentration during fetal life, whereas others decrease.

Information on body composition during fetal life also has its practical side. It permits one to construct the time course of accretion of nitrogen and minerals from the earliest weeks of gestation to term, and these accretion rates provide a means for judging that portion of the maternal nutritional needs during pregnancy that must be assigned to the fetus. They also serve as a point of departure for estimating the nutrient needs of infants who are born prematurely. Today such infants are living longer than ever before, so adequate nutrition is of paramount importance if they are to thrive.

Fetal Growth

We will begin with the overall growth rate of the fetus. Protected as it is from external influences, and spared the work of breathing and the chores of maintaining its body temperature and of working against gravity, the fetus grows at a rapid rate relative to its body weight. In percentage terms the rate is greater than at any subsequent time, including the growth spurt during adolescence. In addition to the advantages just mentioned, which act to reduce its energy needs, the fetus also enjoys the luxury of having an excellent source of nutrients (the placenta), the most efficient brand of "total parenteral nutrition" yet devised. Under such favorable circumstances, the fetus might grow even faster were it not restrained by the size of the uterus.

There are two ways of looking at the growth rate. Most commonly, this is considered in absolute terms, dW/dt, or change in weight per unit time. During the embryonic and early fetal stages, this rate is very slow; then it rises to reach a maximum at about the 36th week of gestation, whereupon it declines as term approaches.

The second way is to consider relative growth rate,[1] dW/Wdt, or change in weight per unit weight with time. When multiplied by 100, this is "percentage growth rate," which is a more familiar term. In this manner growth velocity is related to body size. The need for such a construction is obvious when one considers that it is common practice to prescribe nutrients for prematurely born infants on the basis of body weight; i.e., so much of a given nutrient per kilogram body weight per day. About a third of the total energy requirement of small infants is needed for growth.

The contrast between these two ways of viewing the growth rate of the fetus and young infant is made clear in the two figures that follow. Figure 3.1 shows the absolute growth velocity from the 16th week of gestation through the 25th week of postnatal life (not shown is the brief period of neonatal weight loss, when velocity is negative). There are two velocity peaks, one at about the 36th week of gestation, the other a week or two after birth. The drop in velocity toward the end of pregnancy may be due

---

[1]Sometimes called a "specific growth rate."

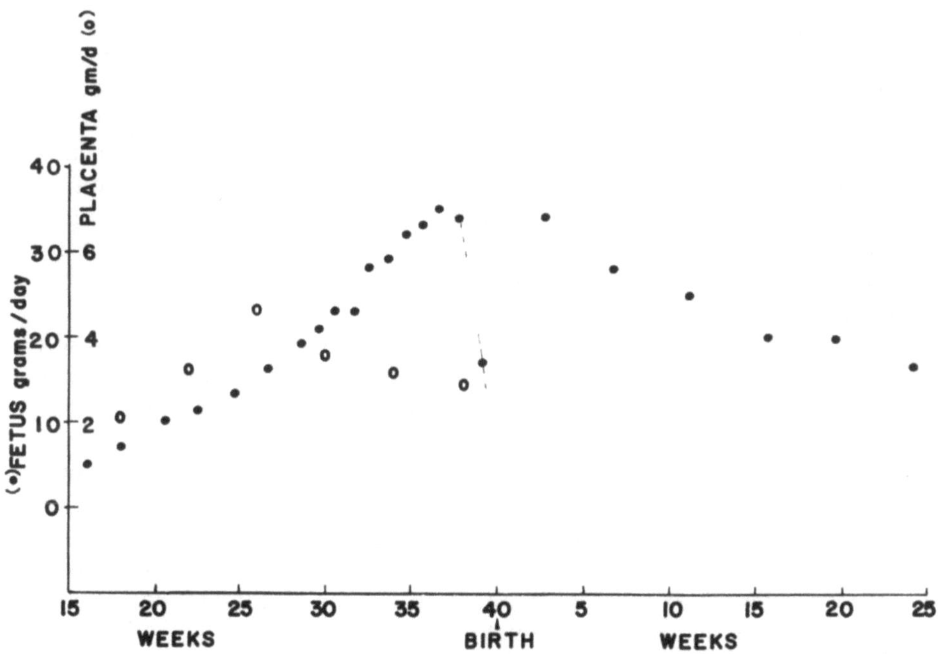

FIGURE 3.1. Velocity of growth from 15th week of gestation to 25th week of postnatal life, based on data of Falkner (1966), Naeye and Dixon (1978) and Fomon et al. (1982) (solid circles). Velocity of placental growth (g/day, open circles), from Hytten and Leitch (1971).

to constraints imposed by the uterus. There may be a limit to uterine size, and it is known that relative uterine blood flow diminishes as gestation proceeds (Hytten and Leitch, 1971). Placental size is also a consideration. The placenta/fetus weight ratio in the rhesus monkey falls progressively during gestation, the equation being placenta, g $= 31.5 + 0.139$ fetus, g ($r = 0.89$) for the period encompassing fetal weights between 50 and 450 g. Hence the ratio falls from 0.77 at a fetal weight of 50 g to one of 0.21 at a fetal weight of 450 g (Hill, 1975).

In the human, the placenta actually weighs more than the fetus early in gestation; by 20 weeks the placenta/fetus ratio is 0.57, then drops to 0.19 at term (Hytten and Leitch, 1971). Late in pregnancy the calcium content of the placenta increases, suggesting some arterial degeneration. It is likely, therefore, that fetal nutrition has been compromised toward the end of gestation, which could account for a slower fetal growth rate. Figure 3.1 includes an estimate of placental growth rate, which peaks shortly after mid-gestation and then declines.

The problem with this hypothesis, however, is the fact that the combined weight of twin fetuses is greater than that of a singleton fetus with a placenta of similar size. For example, the data graphed by McKeown and

Record (1953) show that the ratio of placental to fetal weight is only half as large for twins as for singletons. In light of these data, placental insufficiency *per se* would not appear to be responsible for the slowdown in fetal growth velocity during the last weeks of pregnancy. McKeown and Record (1952) also point out that twins tend to be born about 20 days earlier than singletons, triplets about 15 days earlier than twins, and quadruplets about 10 days earlier yet; the respective average birth weights were 3.45, 4.8, 5.5, and 5.6 kg for these human "litters." They are of the opinion that intrauterine crowding is a factor in the slowdown in fetal growth late in pregnancy.

Once out of the uterus, and after taking a few days to adjust himself to the extrauterine world and to its new source of food, the infant resumes his former rapid rate of growth. Were it not for this perinatal pause, one could conceive of the entire course of growth velocity in Figure 3.1 as one smooth curve, rising rapidly during fetal life to reach a maximum, extending for a brief period into the postnatal phase, and then falling. One could entertain the hypothesis that there is indeed a perinatal growth spurt (analogous to the adolescent growth spurt), which would be unbroken were it not for the exigencies attendant on waning placental function and/or uterine size, and neonatal adjustments.

Incidentally, there is a progressive fall in growth velocity during infancy and childhood, to reach a nadir at about age 4 years, then a gradual increase finally culminating in the adolescent growth spurt. Whereas higher primates exhibit roughly similar patterns of growth velocity, other mammals have a distinctly different pattern. Velocity is slow all during fetal life, with no evidence of a spurt in late gestation; then there is a sharp velocity spurt after birth that then wanes as time goes on.

The course of relative growth velocity is rather different (Figure 3.2). In percentage terms velocity is extremely high in early fetal life—in excess of 10% per day—so that it was not convenient to plot it in the diagram. As shown in the figure, relative velocity continues to decline until about the 28th week of gestation, when there is an intervening plateau lasting about 6 weeks before a further fall just before birth. The rate of about 1.5% per day at the plateau in the curve is much higher than it will be at any subsequent time during infancy and childhood. For example, at the peak of the adolescent growth spurt, which is ordinarily perceived as representing a very rapid phase of growth, the relative growth velocity is only about 0.04% per day.

Changes in Body Composition

Of interest here are the changes that occur in body composition during fetal life. Various aspects are shown in Figure 3.3. Fat-free weight increases progressively save for a slight deceleration in the final weeks.

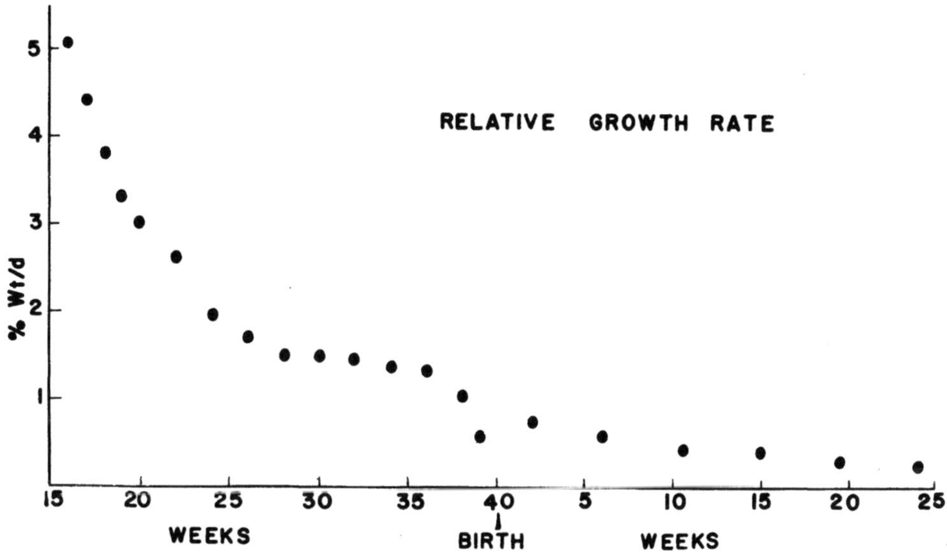

FIGURE 3.2. Relative growth rate (% body weight per day) from 15th week of gestation to 25th week of postnatal life.

Protein and fat contents are roughly parallel.[2] Of interest in this regard is the oft-quoted opinion that there is an upsurge in fat content during the last 10 weeks of fetal life, implying that this is a unique event; but in reality the "upsurge" in protein content is slightly greater, and except for the last 2–3 weeks, when protein content tends to lag a bit, these two body components tend to proceed *en echelon.*

The increase in body fat content in late fetal life has also been documented in live infants. Farr (1966) did skinfold measurements in a series of infants whose birth weight varied from 1.2 to 5.2 kg. The average value for five sites (subscapular, triceps, rib margin, abdomen, and anterior thigh) increased by 1.2 mm for each kilogram increase in birth weight ($r = 0.6$). McGowan et al. (1975) did a similar study using an improved technique for infants born between 36 and 42 weeks gestation. They connected the Harpenden caliper to a transducer and thence to a writing instrument that produced a graph of the jaw width against time. As mentioned in Chapter 2, when calipers are first applied to a fold of subcutaneous tissue, there is a quick decrease in jaw width for a few seconds, followed by a very

[2]The value for body fat content in the human fetus at term, namely 14%, is higher than for the pig (1.1%), rat (1.1%), cat (1.8%), rabbit (2%), and mouse (2.1%); however, the guinea pig (10.1%) approaches the human in this respect, which is in keeping with its advanced physiologic maturity (Widdowson and Dickerson, 1964).

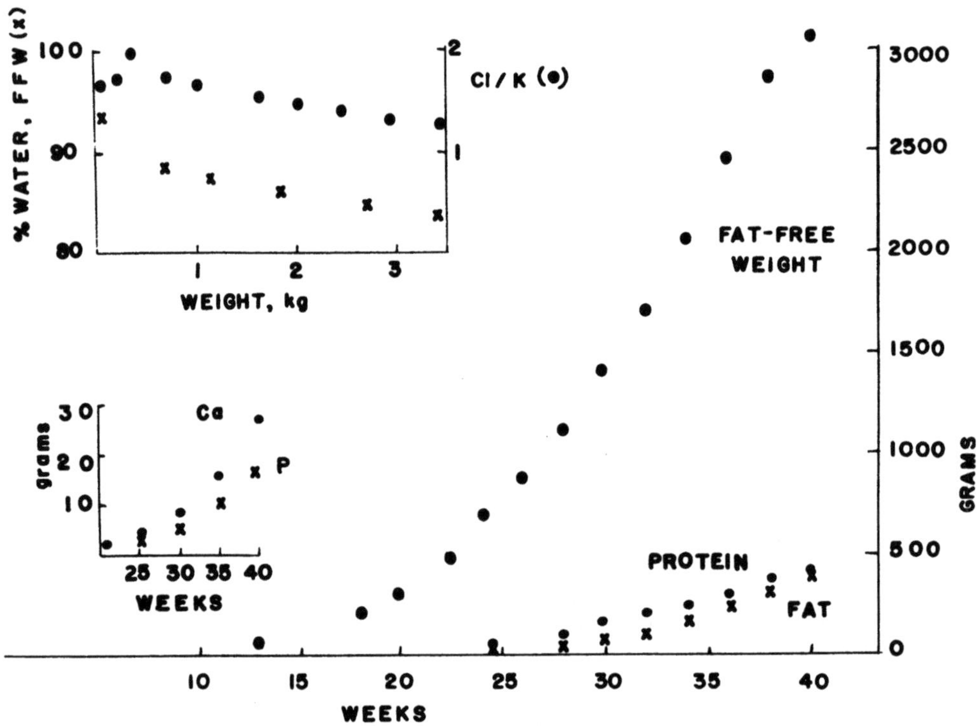

FIGURE 3.3. Time course of fat-free weight, protein, and body fat of the human fetus during gestation. Superior inset shows water content of the fat-free body mass and the body Cl/K ratio as related to body weight. Inferior inset shows body Ca and P contents during gestation. [Based on data of Widdowson and Dickerson (1964) and Zeigler et al. (1976).]

slow decrease. McGowan and co-workers recorded the skinfold thickness value at the onset of this slow phase. They found correlations of birth weight with individual skinfold thickness at the biceps, triceps, quadriceps, subscapular, and flank sites (r = 0.47–0.70). Whitelaw (1979) did measurements of skinfold thickness at eight sites (triceps, subscapular, suprailiac, biceps, all bilateral) on a large number of infants born at 26–42 weeks. From the 26th to the 40th week the sum of skinfolds increased at the rate of 1.4 mm/week, after which there was no further increase. Hence the increase in body fat content noted in Figure 3.3, which is based on cadaver analysis, is also evident in measurements of subcutaneous fat in liveborn infants.

Although such measurements are indicative of an increase in body fat during this period of fetal life, they may not tell the whole story. The reason is the difference in the character of the plots of chemically determined body fat against gestational age (Figure 3.3) and of skinfold thickness against gestational age. The former is curvilinear whereas the latter appears to be a straight line. This difference suggests that the ratio of internal to external fat depots increases toward the end of gestation.

This trend toward increasing skinfold thickness as the fetus grows was to be expected, but what is somewhat surprising is that both Farr (1966) and McGowan et al. (1975) found a sex difference in subcutaneous fat thickness: at all sites and at all birth weights girl infants had thicker skinfolds than boy infants. Although one can easily perceive such a sex difference in adolescents and adults—a difference thought to be associated with hormonal factors—what could account for the difference at such an early age? Fetuses of both sexes are equally exposed to maternal estrogens. However, the fetal testis does produce androgens, which are responsible for masculine differentiation of the external genitalia, and also a Müllerian duct-inhibiting hormone which causes these internal (and feminine) structures to atrophy. Castration of the male fetus in early life results in feminization of the external genitalia. Human male fetuses have higher serum testosterone levels than female fetuses, and this sex difference is present at term and during early infancy. Androgen administration is associated with loss of body fat and an increase in muscle mass in adult males (Forbes, 1985b), and it may well be that the sex difference in testosterone levels is responsible for the smaller subcutaneous fat thickness in male neonates. Since male neonates weigh about 200 g (6%) more than female neonates, and are slightly longer, they undoubtedly have a larger lean body mass. The indications are, therefore, that even at this very early age androgens have an effect on body compositions.

Oakley et al. (1977) did triceps and subscapular skinfold measurements on 1300 infants born at 37–42 weeks gestation, and present graphs showing percentiles arranged by age and sex. Birth weights ranged from 2 to 4.5 kg. Skinfold thicknesses showed a progressive increase with birth weight. When values are plotted according to gestational age, maximum values are achieved at about 38 weeks for the triceps region and at about 39 weeks for the subscapular region. After the 40th week there is a slight decline in thickness at both sites.

Of interest are the variations in skinfold thickness encountered at each gestational age and in each body weight group. For example, at 38 weeks the 3rd percentile for triceps skinfold is about 2.3 mm, and the 97th percentile is about 6 mm. When weight categories are considered, the variability increases progressively with body weight, and in relative terms exceeds the variability in body weight.

These data suggest that the phenomenon of variability in body fat content, which is so evident in children and adults, is already established by late fetal life; hence the factors (whatever these may be) that operate to produce this type of biological variability are in existence very early in human life. That nutrition of the mother is one such factor is strongly suggested by the observation of Frisancho et al. (1977), who found a good correlation between maternal and neonate triceps skinfolds. In a subsequent chapter it will be shown that the variations in body weight among age and sex peers are principally due to variations in body fat content, and that lean weight contributes only a minor fraction of the variability.

Mettau (1978) estimated body fat by the xenon absorption-desorption technique in prematurely born infants who were of normal weight for gestational age, and in infants born at term who were undergrown (more than two standard deviations below normal). There were nine infants in the former group and five in the latter group who were assayed within a week of birth. Mean body fat content was about 7% of body weight in both groups; however, the former were born at 29–36 weeks gestation and the latter at 39–41 weeks. This striking diminution in body fat content is in keeping with the scrawny wasted appearance of these undergrown infants. The causes of intrauterine growth retardation include intrinsic fetal abnormalities, maternal disease, and placental insufficiency; although as was true of Mettau's subjects, in many instances the cause is unknown.

From the graph provided by Southgate and Hey (1976) it would appear that the 67–82% of total body fat is located in subcutaneous sites in fetuses of 180–290 days gestation. These values are far higher than Forbes (1962) found for a single term newborn, namely 42%, and higher than Pitts and Bullard (1968) found in the large number of mammalian species that they analyzed. Unfortunately, Southgate and Hey provide no information on the methods used to determine deep fat and subcutaneous fat in the 38 fetuses that they studied, and only a few of their specimens were delivered at term.

The great bulk of the body fat is contained in white adipose tissue, which serves as an insulating blanket and as an energy storage: when suitably stimulated fatty acids are released into the circulation. There is in addition a special type of adipose tissue known as *brown fat*. This has been detected in fetuses as small as 380 g body weight, and gradually increases in size until it amounts to about 40 g (or about 10% of total body lipid) at term. It is present in several discrete locations, and these are shown in Figure 3.4. Merklin (1974) refers to the distribution as being "vest"-like. Brown fat deposits persist into postnatal life. In a study of children and adults who died suddenly, Heaton (1972) was able to find brown fat in all of the children, and in some adults as old as 70 years.

Brown fat is distinguished from white fat by its multilocular appearance and by the abundance of mitochondria; the cell nucleus is round and centrally located. As indicated in Figure 3.4, some deposits appear to be mixtures of brown and white fat.

Brown fat is the source of nonshivering thermogenesis. It has a rich supply of sympathetic nerve endings. When stimulated by cold exposure, or by norepinephrine administration, its triglycerides are oxidized *in situ*. It thus plays a direct role in heat production. Indeed, under suitable conditions thermographic techniques can detect a temperature difference between the interscapular area, where brown fat deposits are close to the skin, and other areas of the body. Brown fat is important for the survival

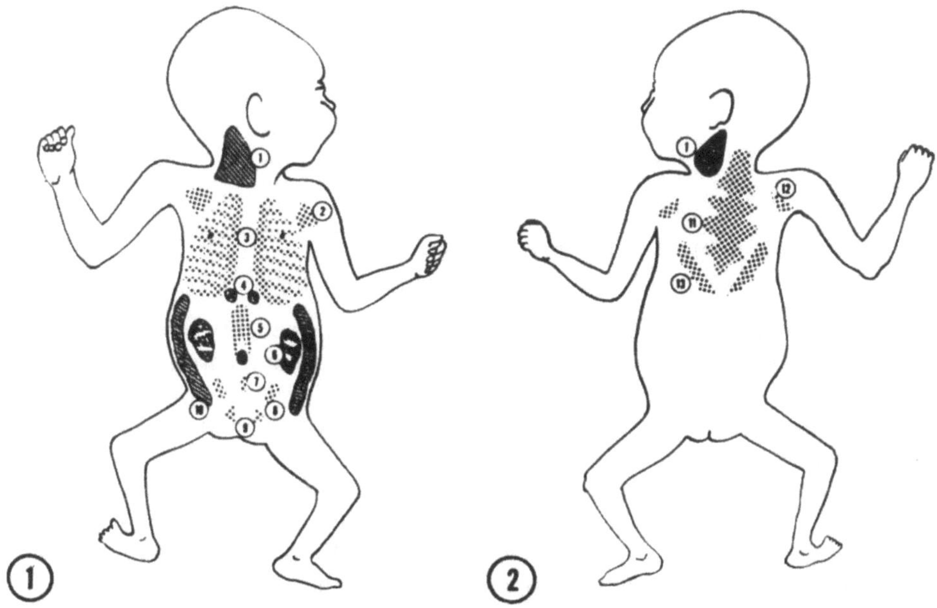

FIGURE 3.4. Distribution of brown fat deposits in the human fetus. (1) Anterior view of the fetus illustrating the position of brown fat bodies and fat cell composition. (2) Posterior view of the fetus illustrating the position of brown fat bodies and fat cell composition. 1, posterior cervical; 2, axillary; 3, intercostal; 4, anterior mediastinal; 5, anterior abdominal; 6, perirenal; 7, urachal; 8, inferior epigastric; 9, retropubic; 10, suprailiac; 11, interscapular; 12, deltoid; 13, lateral trapezial. Fine shading, predominantly multilocular fat cells; Coarse shading, mixed multilocular and unilocular fat cells. [From Merklin RJ (1974). Anat Rec 178, 637–646, with permission.]

of newborn arctic mammals; their large store of this type of fat provides them with a ready source of heat until sufficient mother's milk is available.

The two insets in Figure 3.3 describe other features of compositional changes during fetal life. At the top it is to be seen that the water content of the fat-free body progressively falls as fetal weight increases, though the value of about 83% at term is still considerably above the adult value of 73%. The ratio of total body chloride to total body potassium also declines, which suggests that the ratio of ECF to ICF does also. The value for the Cl/K ratio at term (1.25) far exceeds the adult value of 0.73. Hence in both respects further changes in relative body composition are to occur during postnatal life.

The lower inset shows the changes in body calcium and phosphorus contents as gestation proceeds. The increase in the former is seen to be somewhat greater than that of the latter, which is indicative of the progressive mineralization of the skeleton.

Table 3.1 shows the body content of some other minerals for selected

TABLE 3.1. Element content, human fetus.

Fetal age, weight (weeks, g)	Mg (mg)	Fe (mg)	Zn (mg)	Cu (mg)	Na (meq)	K (meq)	Cl (meq)
24, 690 g	123	33	11	2.3	68	28	48
28, 1160 g	195	84	17	3.5	105	47	77
32, 1830 g	305	145	29	6.5	156	74	113
36, 2690 g	467	203	49	9.6	250	110	173
40, 3450 g	646	290	61	13.8	266	141	176

From Zeigler et al. (1976) and Widdowson and Dickerson (1964).

gestational ages. Except for iron they all show increments of 4- to 6-fold during the last 16 weeks of intrauterine life; the value for iron is about 9-fold.

Prior to 20 weeks the fetus is so small (weight less than 300 g) that it constitutes a negligible drain on mother's nutrition; indeed the drain is small until fetal body weight reaches 1000 g at about the 27th week. After that the growth rate of fetoplacental unit and the uterus calls for an increasing supply of maternal nutrients. Indeed, the mother has to supply nutrients to the entire fetouteroplacental unit, the total weight of which (minus amniotic fluid) is about 5000 g at term, or 1½ times the weight of the fetus alone.

Some Mathematical Constructs

A number of attempts have been made at characterizing changes in fetal body composition in mathematical terms, with varying degrees of success. Shaw (1973) has used an exponential equation to describe the accretion of water, fat, nitrogen, and minerals from the 24th through the 36th week of fetal life. This equation has the form

$$y(t) = y(0)e^{kt} \qquad (1)$$

where $y(0)$ is the value for element content when t is zero, k is a constant, t is time in days, and e is the base of the natural system of logarithms.

For example, the equation for calcium content is:

$$\text{Ca(mg)} = 95 \exp\,(0.0203t)$$[3]

The exponents and coefficients of Equation 1 for the various elements are listed in Table 3.2. An exponential function states in effect that the rate of change of a given parameter is proportional to its size. A unique

[3]The abbreviation "exp" stands for e, the base of the natural system of logarithms, and the parentheses contain its exponent. Many printers prefer this nomenclature since it avoids the use of a superscript.

TABLE 3.2. Coefficients and exponents of Equation 1 for calculating element content at specified gestational age (days).

	k days^{-1}	Y_o
Nitrogen	0.0186	346.7 (mg)
Sodium	0.0151	78.5 (mg)
Potassium	0.0173	47.2 (mg)
Calcium	0.0203	95.3 (mg)
Phosphorus	0.0197	66.6 (mg)
Magnesium	0.0171	6.42 (mg)
Iron	0.0202	13.43 (mg)
Copper	0.0193	79.9 (μg)
Zinc	0.0149	918 (μg)

From Shaw (1973), with permission; suitable for the period 24–36 weeks of fetal life.

property of e is that the derivative of a function such as $y = e^{kx}$ with respect to x is simply ke^{kx}, or ky. Thus using the values for $y(0)$ and k listed in Table 3.2 for calcium, y at 24 weeks is 95 exp $(0.0203 \times 168) =$ 2876 mg, and the accumulation rate is $2876 \times 0.0203 = 58$ mg/day. The fact that all of the element contents listed in Table 3.2 have positive values at zero time (i.e., at conception) means that these exponential equations do not accurately describe element contents in the early fetus. They also fail to take into account the decline in growth rate after about the 36th week. A graph shown in another publication by Shaw (1976) indicates considerable scatter about the calculated regression line for body calcium content.

A convenient way of looking at this problem is to relate element content to body size rather than to age. Plots of element content against body weight on arithmetic coordinates tend to be curvilinear. However, when plots are made on double-logarithmic coordinates these curves straighten out and become linear. Examples are shown in Figures 3.5–3.7, where the linearity for Ca, N, P, Na, and Cl contents versus body weight is immediately apparent (disregard the dotted lines marked "*dy*/*dx*" for the moment; their meaning will be made clear later). In the instance of calcium the regression appears to be valid for the entire fetal period, whereas for the others the fit is not quite as good for body weights less than 300 g. Incidentally, the regression equation for calcium predicts a value of 0.12% for the 1-g fetus, which is close to the value for blood serum (0.10%).

Iob and Swanson (1938) developed log–log plots to relate mineral content to fetal weight in man, and Needham (1950) used this approach to characterize changes in total body ash during growth in a number of vertebrate and invertebrate species. In fact Needham was able to show that the slopes of the plots of log body ash against log body weight did not differ a great deal among the various species, and used such data to develop a concept that he named "the fundamental ground plan of animal growth." Later, with the use of data generated by the work of others, the present author constructed log–log plots of a number of elements versus body weight for

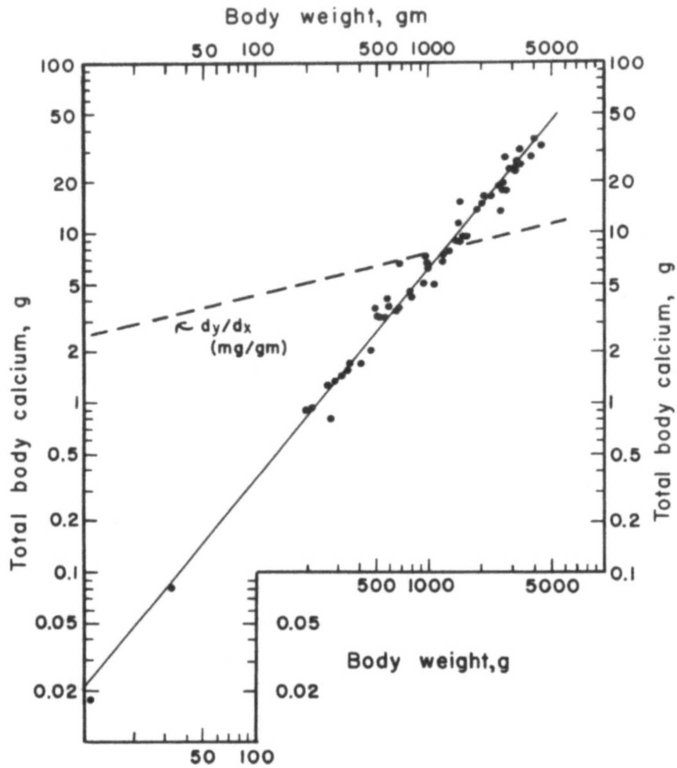

FIGURE 3.5. Plot of log body calcium (g) against log body weight (g) for the human fetus. Regression line calculated by method of least squares ($r = 0.99$). [From Forbes GB. Pediatrics 57, 976–977 (1976), with permission.]

the human fetus, and for postnatal life in several mammalian species (Forbes, 1955).

The existence of these relationships leads us into a discussion of the concept of *allometry* [Greek *allos* (other) + *metron* (measure)], or as some prefer, *heterogony* [(Greek *heteros* (different) + *gonia* (angle)]. Although used earlier by others, this principle was popularized by Julian Huxley some 50 years ago. Huxley (1932) used allometry as a means of relating the size of a portion of the body to the size of the body as a whole. Brody (1945) and Kleiber (1975) proceeded to apply this principle to the size of various organs and also to a number of physiologic functions. The best known relationship is that of basal metabolic rate (BMR) to body weight, which seems to hold over a very wide range (some five orders of magnitude!) of mammalian and avian sizes. As mentioned in Chapter 2, BMR scales to the 3/4 power of body weight, and there is much discussion in the literature as to why this exponent turns out to be 0.75 rather than 0.67, which is the surface/volume ratio of a sphere, and of regular poly-

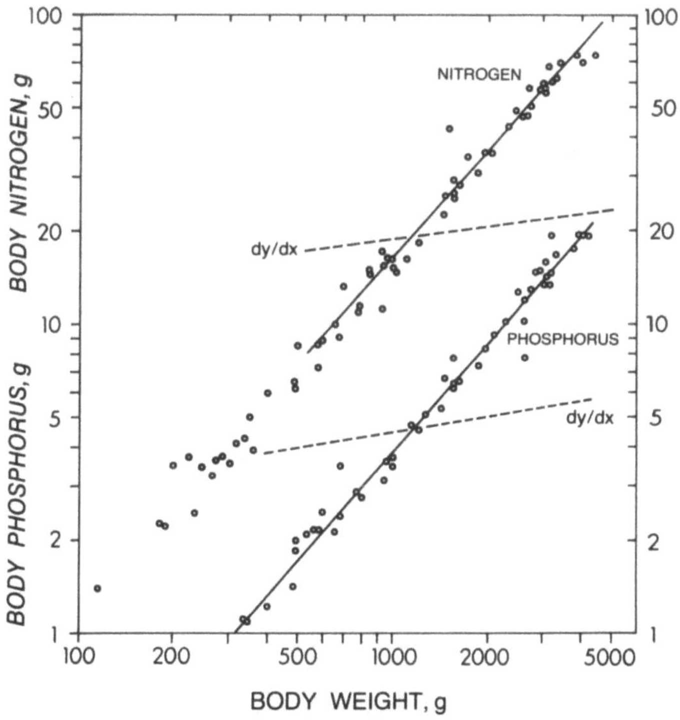

FIGURE 3.6. Plots of log total body nitrogen (g) and log total body phosphorus (g) against log body weight (g) for the human fetus; respective r values are 0.98 and 0.99. [From Forbes GB (1981). In: E. Lebenthal (ed), Textbook of Gastroenterology and Nutrition in Infancy, p 323. Raven Press, New York, with permission.]

hedra. Apparently the "surface area law" for metabolic rate doesn't quite hold true.

Under the premise that nutrient composition of cells should be a function of their volume, and that the uptake of nutrients should be a function of their surface area, the 2/3 power "law" should apply. The problem is that most body cells are not spheres. Muscle cells are cylinders, or long solid hexagons; leukocytes are ameboid; and erythrocytes are biconcave discs, which means that they have a very large surface area/volume ratio. The spherocytes seen in the blood of patients with hemolytic anemia come closest to being true spheres, but these are not normal cells.

Many physiological functions, as well as organ sizes, have been described adequately in allometric terms. The slopes of the lines vary considerably among these various parameters. A recent book by Schmidt-Nielsen (1984) is entirely devoted to the topic of allometry; he makes the point that body size is an important determinant of both form and function. It should be noted here that most of the allometric functions described in the literature are concerned with interspecies data, an example being the

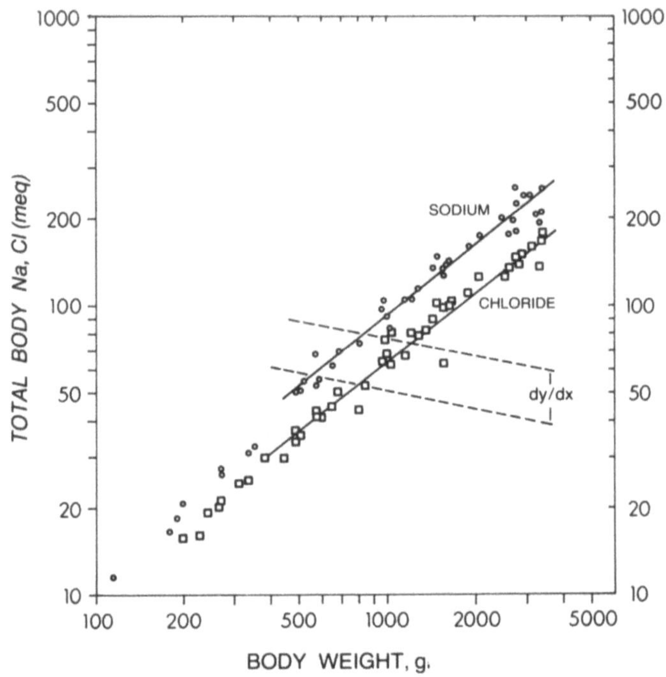

FIGURE 3.7. Plots of log total body sodium (meq) and log total body chloride (meq) against log total body weight (g) for the human fetus; respective r values are 0.97 and 0.98. [From Forbes GB (1981). In: E. Lebenthal (ed), Textbook of Gastroenterology and Nutrition in Infancy. Raven Press, New York, with permission.]

BMR–weight relationship noted above. Moreover, most involve relationships among adult animals of various species. Indeed, when an individual species is studied, as in the relationship of a given function or organ size to body size during growth, it often turns out that the log–log function is either not linear, or if it is, the slope differs from that obtained by plotting the adult members of many species.

For the purposes of this chapter, we will concentrate on a single species, the human fetus, and deal with the relationship of element content to body size as the fetus grows. Figures 3.5–3.7 show that log body contents of Ca, P, N, Na, and Cl are each linearly related to log body weight during most of the fetal period, and thus conform to the allometric principle. Hence one is justified in describing these relationships by the equation to follow:

$$\log y = \log b + k \log x, \tag{2}$$

where y is the mass of the element (Ca, P, etc.) and x is the mass of the body as a whole. Log b is the value of log y when log x is zero, that is when $x = 1$ (since log zero is minus infinity, there is no zero value for

y), and k is the slope of the regression line. Equation 2 has the form of a linear regression equation.

Taking the antilogs of both sides of Equation 2, we have a power function,

$$y = bx^k \tag{3}$$

b now takes on the attributes of a proportionality coefficient, since it defines the ratio y/x^k for all values of x encompassed by the data used to construct Equation 2. When $x = 1$, b becomes body content in whatever units are chosen. Since we will in effect use mass units in the equations to follow, the total body content (not concentration) of a given element can be viewed in the same manner as an organ, as an entity *per se*.[4] If milliequivalents are used in Equation 3 instead of milligrams, the coefficient b would change but not the exponent k.

Some interesting results are achieved by differentiating Eq. 3, and rearranging as a differential equation:

$$dy = bkx^{k-1}\,dx \tag{4}$$

This says that the change in y is a function both of x and change in x.

Dividing Equation 4 by Equation 3 we have

$$\frac{dy}{y} = \frac{bkx^{k-1}\,dx}{bx^k},$$

which simplifies to

$$\frac{dy}{y} = k\frac{dx}{x} \tag{5}$$

Hence the specific growth rate of the part (y) is a constant fraction, or multiple (k), of the specific growth rate of the body as a whole (x); when multiplied by 100 this is percentage growth rate.

The constant k can be looked upon as a *constant differential growth* ratio, i.e., as defining the ratio of two specific growth rates. When $k > 1$, the specific growth rate of the part (y) is greater than that of the body as a whole (x), so that the body becomes enriched with the particular element in question. When $k < 1$, the concentration of y, i.e., y/x, falls

[4]There are two problems inherent in the use of allometric equations. While the relation of the size of a part of the body, or even several parts, to the size of the body as a whole can thus be described, it is obvious that not every part can be. This is because the summed values for y ($y_1 + y_2 \ldots y_n$) as used in Equation 3 cannot be made to equal x^1. The second problem is that logarithmic plots automatically reduce the *apparent* variability of any set of data (for example, 2 seems closer to 3 than 100 is to 1000). Hence the apparent linearity of log–log plots can be deceptive. This is also true for semilog plots, such as those used by Shaw (1973).

as growth proceeds. When $k = 1$, the two specific growth rates are equal. Plots on double-logarithmic coordinates will be steeper than 45° in the first instance, less than 45° in the second, and at 45° in the third.

The reader may have noticed that *time* does not appear in any of the equations above. Since one is used to looking at growth as a velocity, change in size per unit time, such an omission seems strange. One can, of course, differentiate both sides of Equation 3 with respect to time, but all that happens is that *dt* appears in both denominators; and if this is carried through to Equation 5, the result is

$$\frac{dy}{ydt} = \frac{kdx}{xdt}$$

This says that the change in y per unit of y with respect to time is equal to the change in x per unit x with respect to time multiplied by the constant k.

The absence of *time* in Equations 2–5 shows that it is body size that is important in determining fetal body composition. One practical consequence is that estimates of gestational age, which is sometimes difficult to determine, are not required. Element content is a function of body weight; *time* is only incidental, as one of the correlates of body weight.

An additional feature concerning the change in body composition as growth proceeds is inherent in Equation 5. This equation states that the relative change in body composition (dy/y) will decline in concert with the decline in the relative growth rate of the body as a whole; as the latter slows down, as it does in late fetal life and even more so during childhood, so too will the change in body composition occur more slowly. Hence mammals mature, in terms of body composition, most rapidly when they are young.

Table 3.3 lists the equations for the various elements together with their derived differential equations (Equations 3 and 4, respectively). These equations were calculated by least-squares regression analysis from data published by a number of investigators (referenced in Forbes, 1981c). In each instance the correlation coefficient exceeded 0.90, indicating a satisfactory fit of the data. The symbol n is the number of fetuses analyzed for each element; fetal weights (W) ranged from 300 to 4000 g. Log–log plots of Ca, P, N, Na, and Cl against body weight are shown in Figures 3.5–3.7.

The equations fall into three groups: those with exponents (k) > 1 (Ca, P, Mg, N, Fe, Cu, S), those with exponents < 1 (Na, Cl, Zn), and one whose exponent is close to 1 (K). The body becomes enriched with those elements in the first group as growth proceeds, whereas the opposite trend occurs in the case of elements in the second group. From these equations one can calculate body element content at any given body weight, and also accretion rate at any given body content and growth rate.

Trotter and Peterson (1969) plotted their data on fetal skeletal weight

TABLE 3.3. Fetal body composition and accretion rates.

n[a]	Nutrient	Body Content[b]	Relative accretion rate
58	Ca, g	$0.00117W^{1.241}$	$dCa = 0.00145W^{0.241}dW$
61	P, g	$0.00125W^{1.164}$	$dP = 0.00146W^{0.164}dW$
72	N, g	$0.00646W^{1.136}$	$dN = 0.00733W^{0.136}dW$
35	Na, meq	$0.362W^{0.806}$	$dNa = 0.292W^{-0.194}dW$
39	Cl, meq	$0.284W^{0.784}$	$dCl = 0.223W^{-0.216}dW$
62	K, meq	$0.0449W^{0.989}$	$dK = 0.0444W^{-0.011}dW$
28	Mg, mg	$0.0761W^{1.145}$	$dMg = 0.0871W^{0.145}dW$
16	Cu, mg	$0.00185W^{1.086}$	$dCu = 0.00201W^{0.086}dW$
16	Zn, mg	$0.0431W^{0.868}$	$dZn = 0.0375W^{-0.132}dW$
21	Fe, mg	$0.0168W^{1.203}$	$dFe = 0.0202W^{0.203}dW$
7	S, mg	$0.0713W^{1.117}$	$dS = 0.0796W^{0.117}dW$

Ferrell et al. (1982) have published data on the body composition of 81 f fetal calves, varying in gestational age from 120 to 255 days, and in weight from 0.64 to 41.6 kg. Plots of element content (Ca, P, K, Mg, Zn, Na, and Fe) against body weight are linear on double-logarithmic coordinates, with correlation coefficients of 0.97–0.998. The exponents for Ca (1.21), P (1.15), K (0.98), and Mg (1.13) are all very similar to those for the human fetus; however, those for Zn (1.13) and Na (0.96) are higher than in the human fetus, and that for Fe (1.06) is lower. Individual data kindly provided by Dr. C. L. Ferrell.
[a]Only those fetuses larger than 300 g were used in calculating the regression equations.
[b]*W* in grams.

against body weight on double-logarithmic coordinates, the range of body weights being 227–3369 g. The dry fat-free weights of the osseous portion of the skeleton were determined, and the resultant graph proved to be linear. The calculated regression line was: skeletal weight = $0.00724\ W^{1.17}$. The exponent k is reasonably close to that for total body calcium (Table 3.3), which is to be expected. There were no significant differences that could be ascribed to sex or race.

Returning now to Figures 3.5–3.7, the dotted lines labelled "dy/dx" are based on Equation 4, which says that the change in y with respect to x is a function of a power of x.[5]

Element Accretion

Using data on fetal growth rates (Naeye and Dixon, 1978) one can use the equations listed in Table 3.3 to calculate element accretion rates at any specified gestational age > 20 weeks. Taking calcium as an example, body weight at 25 weeks (Figure 3.3) is 800 g, and weight gain (Figure

[5]This is also linear on double logarithmic coordinates. Rearranging Equation 4 we have

$$\frac{dy}{dx} = bkx^{k-1};$$

taking logarithms of both sides of this equation the result is

$$\log (dy/dx) = \log (bk) + (k - 1) \log x,$$

which is linear.

TABLE 3.4. Element accretion at selected times in fetal life.[a]

	Gestational age					
	25 Weeks		30 Weeks		35 Weeks	
Body weight, g	800		1480		2450	
LBM, g	790		1400		2250	
Fat, g	10		80		200	
Weight gain, g/day	15	(19)[b]	22	(15)	35	(14)
Accretion rates						
Ca, mg/day	109	(136)	185	(125)	333	(136)
P, mg/day	66	(82)	106	(72)	184	(75)
N, mg/day	273	(341)	435	(294)	741	(302)
Na, meq/day	1.20	(1.5)	1.56	(1.05)	2.25	(0.9)
Fe, mg/day	1.18	(1.47)	1.96	(1.32)	3.45	(1.41)
Mg, mg/day	3.44	(4.30)	5.52	(3.73)	9.45	(3.86)
Zn, mg/day	0.23	(0.29)	0.32	(0.21)	0.47	(0.19)

[a]Based on equations shown in Table 3.3.
[b]Values in parentheses are in terms of per kg body weight per day.

3.1) is 15 g/day. Substituting these values in the equation for calcium accretion rate (Table 3.3), this turns out to be 109 mg/day; when referred to body weight the rate is 136 mg/kg/day.[6] Table 3.4 lists such values for a number of elements at three selected gestational ages. Accretion rates reach their maxima at about the 35th week of gestation, at a time when the growth rate of the body as a whole is greatest.

The daily accretion of calcium at the 35th week, namely 330 mg, constitutes a sizeable fraction (one-third to one-half) of the usual maternal intake of this element. Values for iron are also close to one-third of the usual dietary intake, and perhaps five times the usual amount that is absorbed by the nonpregnant woman. On the other hand, zinc, magnesium, and nitrogen accretions amount to $< 10\%$ of the usual maternal diet. Fortunately, calcium absorption from the maternal gastrointestinal tract is increased during pregnancy (Heaney and Skillman, 1971), and so too is iron absorption (Apte and Iyengar, 1970).

The values shown in Table 3.4 represent *net* transfer rates from mother to fetus. Isotope studies show that substances such as sodium and deuterium cross the placenta very rapidly by diffusion. Some substances such as calcium and certain amino acids are also actively transported. Studies in the rhesus monkey show that radiocalcium is also transferred from fetus to mother as well as in the expected, opposite direction; the amounts transferred from mother to fetus are 6–10 times greater than those needed for growth of the fetal skeleton (MacDonald et al., 1972). This distinction

[6]
$$\frac{d\text{Ca}}{dt} = 0.00145\,W^{0.241}\,\frac{dW}{dt}$$

$$d\text{Ca, g/day} = 0.00145 \times 800^{0.241} \times 15 \text{ g/day}$$
$$= 0.00145 \times 5.01 \times 15 = 0.109$$
$$109 \text{ mg Ca/day} \div 0.80 \text{ kg} = 136 \text{ mg/kg/day}$$

between flux and *net* transfer is much the same as for other tissues of the body. In Chapter 2 mention was made of the extreme rapidity with which radiopotassium enters muscle cells following intravenous injection; yet the *net* accumulation of K by the entire muscle mass of a growing infant is no more than 20–30 mg per day, or about 1% of total muscle K content. Although the use of radioactive tracers has established the existence of rapid transport and/or diffusion of ions across cell membranes, ion fluxes cannot be construed as representing *net* transfer rates.

Consequences for the Prematurely Born Infant

The fetus *in utero* enjoys the privilege of being fed parenterally: its gastrointestinal tract need not function in the nutritional sense (it is unlikely that swallowed amniotic fluid adds to its net nutrition); there are no endogenous fecal losses; it does not need to maintain its body temperature. Once born the infant is abruptly cut off from this generous supply of nutrients; and if it is to emulate the intrauterine growth rate and body composition it must have an adequate supply of energy and nutrients by vein or by mouth.

In practice it turns out to be rather difficult to supply all of these elements by vein or mouth in the quantities that would have been supplied by the placenta had the prematurely born infant stayed *in utero*. This is particularly true for calcium and phosphorus. Human milk has a relatively low content of Ca, P, N, and Na compared to cow's milk, and so is not suitable for feeding small prematurely born infants. Modern infant formulas offer an advantage in this respect, although they, too, are not able to supply all of the Ca and P needed; nor is the infant's gastrointestinal tract able to absorb all of the various elements presented to it.

The amounts of nitrogen and minerals that must be retained by the infant each day (i.e., absorbed minus excreted) to provide for normal growth and body composition are given in Table 3.4; or one can calculate the amounts from the equations listed in Table 3.3, which take into account the growth rate. At any given body weight the amounts of nitrogen and minerals to be retained are directly proportional to the growth rate.

Now that prematurely born infants are surviving in greater numbers and for longer periods of time, the provision of proper feeding in an attempt to ensure normal body composition has become a real challenge.

The Fundamental Chemical Ground Plan of Animal Growth

Joseph Needham (1950) noted that plots of total body ash could be related to body weight on double-logarithmic plots for a number of vertebrate and invertebrate species, and that the slopes of the regression equations were much the same for many species. Hence he coined the heading above.

There does seem to be a tendency for the young of several mammalian species to have a high content of water, Na, and Cl and a reduced content of Ca in comparison to the adult; and it is known that the turnover rates of water and of several elements are higher. Metabolic rates per unit body weight are also greater. Fast growing malignant tissues have higher water and Na contents than the body of the host (Waterhouse et al., 1955; Toal et al., 1961). Young human erythrocytes have a different composition than older cells, from which they can be separated by centrifugation ("old" cells tend to collect at the bottom). Young cells have higher content of water and potassium and a lower content of sodium, and a higher rate of glycolysis (Bernstein, 1959).

Taking a clue from Needham's (1942) work, we examined the possibility that allometric relationships existed in mammals other than man (Forbes, 1955). It turned out that such relationships did indeed exist for the rat, cat, rabbit, and pig at least for water, Na, Cl, and K,[7] and furthermore that the exponents and coefficients of Equations 3 did not differ a great deal among these four species or for man. Moreover, these allometric functions also hold during postnatal life.

The point is not the mere change in body composition during growth—a fact established in 1857 by the young Albert von Bezold—rather it is the regularity of the change among various species, large and small. Within limits, the process of compositional change as animals mature proceeds in regular fashion in accordance with the change in body weight; in Needham's (1950) words there does indeed seem to be a "chemical ground plan of animal growth."

One could speculate that the larger concentration of water and the larger relative volume of extracellular fluid (ECF) in the young permits a more rapid transportation of nutrients to the cells of the body; this means of course that intracellular fluid (ICF) volume is relatively smaller. Young cells have a larger surface area-to-volume ratio, so that transfer of materials from the ECF to the cells and from cells to the ECF is facilitated. Such a circumstance favors a higher level of metabolism, and rapid growth. A further speculation is that the decline in the ratio of ECF to total body water with increasing body size places a limit on eventual size, in that the ratio of ECF to ICF becomes just sufficient to support normal metabolic functions.

Cell multiplication, as judged from changes in total organ DNA content,[8] is the predominant feature of growth in the fetal period, and in some organs such as muscle it continues during childhood. Body cells begin to increase in size, as judged by protein/DNA ratios, during late fetal life, and this process of cell hypertrophy will continue throughout the growth period.

---

[7]Dr. Elsie Widdowson kindly provided the data used in these calculations.

[8]Based on the assumption that cells have but one diploid nucleus, which contains 6.2 pg of DNA.

Hence the growth process at the cellular level, generally speaking, has three phases: hyperplasia early, hypertrophy later, and an intermediate transition phase in which both phenomena coexist. The timing of these three phases differs from organ to organ (Rozovski and Winick, 1979). One could conceive of this change in mode of tissue growth as a factor in determining the change in ECF/ICF ratio during growth. As cells multiply they retain a relatively high surface/volume ratio, and hence a high ECF/ICF ratio; and as they hypertrophy the ratio falls.

Effect of Maternal Malnutrition

Apte and Iyengar (1972) analyzed 41 fetuses born to malnourished Indian women for Ca, P, N, Mg, Fe, water, and fat. They ranged from 230 to 3340 g in weight and in estimated gestational age from 20 to 40 weeks. The mothers had a mean body weight of 42 kg, and had consumed diets estimated to provide about 1400 kcal and 40 g protein per day. Many had anemia and clinical signs of vitamin B-complex and vitamin A deficiency, and it was estimated that the intake of other nutrients was also inadequate. Eighteen of the infants were born at term; the weight range for 17 of these was 1980–3340 g (average 2440 g); the 18th infant weighed only 1090 g.

When the entire group of 41 fetuses and newborns are considered, it turns out that body element content can be described as a log–log function of body weight, in accordance with Equation 2. The equations for Ca, P, N, Mg, and Fe are listed in Table 3.5; correlation coefficients are 0.90–0.96.

The coefficients and exponents differ somewhat from those given in Table 3.3, and it is of interest to compare the predicted body contents with those that can be derived from the equations in that table. When this is done for two selected fetal weights, 2500 and 3000 g, the infants born to the malnourished women have only 61% as much Fe and about 82–86% as much N, P, and Mg as the others; on the other hand Ca content is only slightly lower. It is evident that severe degrees of malnutrition can influence the rate of acquisition of several important elements by the fetus.

TABLE 3.5. Equations for element content for fetuses born to malnourished women.

Element	Equation	Correlation coefficient	Element content at 2500 g	Element content at 3000 g	Percent of value from Table 3.3
Nitrogen, g	$= 0.003273W^{1.194}$	0.96	37.3	46.5 g	82
Calcium, g	$= 0.001091W^{1.244}$	0.90	18.4	23.1 g	95
Phosphorus, g	$= 0.001968W^{1.1088}$	0.90	9.8	11.8 g	86
Magnesium, mg	$= 0.07691W^{1.126}$	0.95	510	630 mg	86
Iron, mg	$= 0.03467W^{1.057}$	0.95	126	157 mg	61

W in grams.

However, the six heaviest term infants had an average body fat content of 11.7%, a value not too far below normal.

Since nitrogen content of these undernourished fetuses was lower than that reported by others for fetuses carried by women who were presumably not undernourished, they probably had a lower lean weight. Hence, at a given body weight one could anticipate that they had a larger content of body fat. The table provided by Apte and Iyengar (1972) shows that this is indeed the case. When calculations of body fat content were made for fetuses of 1000, 1500, 2000, and 2500 g body weight, the fetuses born to these undernourished women had 1.3–2.6 times as much body fat as those analyzed by Widdowson and Dickerson (1964). Hence, it would appear that maternal undernutrition in humans exerts its effect primarily on the lean tissues of the fetus.

It is known that both maternal pre-pregnant weight and maternal weight gain during pregnancy have a direct effect on birth weight and also on placental weight. Infants born during the Dutch famine of 1944–45 were about 300 g (10%) lighter than normal, and the placentas were about 15% lighter than normal (Stein et al., 1975). Understandably, body composition studies were not made.

Zlatnik and Burmeister (1983) measured the urine urea/creatinine ratio in 870 pregnant women, and used this to judge the adequacy of their protein intake, the assumption being that a low ratio is indicative of an inadequate intake. The range of values was rather large, the 90th percentile being 2½ times that of the 10th percentile value. However, the authors failed to find a correlation between the maternal urea/creatinine ratio and either birth weight, length, or ponderal index, or with placental weight. Rush et al. (1980) failed to find an effect of protein supplements during pregnancy on birth weight of neonates born to Harlem blacks. On the other hand Lechtig et al. (1975) did find a beneficial effect on birth weight when either caloric or protein supplements were given to malnourished pregnant women in Guatemala.

Cheek (1975) has summarized data on the effects of low-protein diets during pregnancy. In the rat, fetal cell number is reduced in viscera and brain; in the monkey, changes in brain composition were in the main not significant, whereas in muscle there was some reduction in protein/DNA ratio. The reduction in birth weight was the most striking finding.

A number of investigators have studied the effect of procedures designed to interfere with the blood supply to the fetus. Hohenauer and Oh (1969) ligated one uterine artery on the 17th day of pregnancy in rats, and then permitted the mother to carry her fetal litter to term (21 days). The pups that had resided in the uterine horn with the compromised circulation weighed only 80% as much as their normal littermates. The growth-retarded pups contained less nitrogen and fat than the normal pups; however, there was no difference in the percentage of water, calcium, or phosphorus.

In the monkey, placental insufficiency produced by compromising the placental circulation resulted in a severe deficit in fetal body weight but only a slight reduction in brain weight. Total RNA and DNA were reduced in cerebellum but not in cerebrum; muscle and liver were both reduced in size, and there was some reduction in total RNA and DNA in these organs. However, the relative amounts of DNA and RNA were normal in all three organs, showing that the circulatory insufficiency affected total organ size, and not its composition (Cheek, 1975).

This procedure of interrupting placental circulation presents a serious insult to the fetus, one that is much more severe than is likely to ensue from dietary deprivation of the mother.

Some *In Vivo* Observations

There are two recent studies aimed at assessing skeletal size in prematurely born infants. Greer et al. (1983) used photon absorptiometry of the radius at a point one-third the distance from the distal end to measure bone mineral content, expressed as mg/cm. The radiation source was ^{125}I; the estimated radiation dose was 13 mrad. Measurements were made on 114 infants ranging in gestational age from 22 to 42 weeks. The results conform to an exponential increase in bone mineral content (BMC):

$$\text{BMC} = 3.47 \exp(0.082A),\ r = 0.92\ (A \text{ is age in weeks}).$$

Poznanski et al. (1980) measured the cortical thickness of the humerus on chest radiographs of prematurely born and full-term infants ranging in weight from 500 to 3600 g. Measurements of the inner and outer diameters of the cortex were made at a point just proximal to the nutrient foramen; cortical area was calculated in a manner analogous to that used for the second metacarpal, namely cortical area = $\pi/4(\text{OD}^2 - \text{ID}^2)$. These authors found a straight line relationship between cross-sectional cortical area and birth weight ($r = 0.88$), with a slope as estimated from their graph of about 0.005 mm^2/g body weight.

Although these methods can be of help in assessing changes in skeletal size and mineralization resulting from growth and nutritional intervention in these tiny infants, one has reservations as to how accurately they reflect the intrauterine progression in body calcium content. The data of Poznanski et al. (1980) show that humerus cortical area per unit body weight actually decreases as body weight increases, and the data of Greer et al. (1983) show that the same is true for radius bone mineral content. On the contrary, the chemical analysis data show that total body calcium per unit body weight actually increases as the fetus grows (Figure 3.5), and the same is true for total skeletal weight (Trotter and Peterson, 1969). This discordance suggests that the regions of the skeleton chosen for the measurements of bone density and cortical thickness do not accurately reflect total body calcium content or total skeletal weight. The histological appearance of immature bone indicates the presence of a lot of trabecular bone, whereas the method used by Poznanski et al. (1980) examines only cortical bone.

Cassady and Milstead (1971) have compared normally grown mature and prematurely born infants with undergrown infants born at term, with regard to total body water and ECF volumes. The mature infants had a lower percent water than the other two groups (688, 809, and 790 ml/kg, respectively), and thus presumably a higher content of body fat. Values for the undergrown infants were about the same as those for normally grown premature infants of comparable weight, and the ratio of ECF to total water did not differ.

Mention should be made of the increasing use of ultrasound to monitor fetal growth *in utero*. This technique can estimate crown–rump length, cranial diameter, femur length, total intrauterine volume, and placental thickness. A number of anatomical abnormalities can also be detected. Nuclear magnetic resonance is also being used for this purpose. It permits sequential assays of uterine size, fetal fat and bone, and individual organ size. The application of these two techniques is destined to provide a wealth of information on fetal growth.

4

Body Composition in Infancy, Childhood, and Adolescence

Growth is the only evidence of life.
John Henry, Cardinal Newman

The preceding chapter dealt with changes that occur during fetal life, and mention was made of the fact that fetal tissues at term still differ in composition from those in the adult. The achievement of "chemical maturity" will require further changes in relative organ size, fluid compartments, and body fat content during the period of postnatal growth. The culmination of these processes results in what is known as the "standard man," or "reference man," a convenient construction for those investigators interested in such matters as energy and nutrient needs, radiation doses, trace element content, and so forth for population groups. However, it will become clear subsequently that there is considerable variation in body composition among ostensibly healthy individuals, just as there is variation in stature, weight, and hair color. Moreover, sex differences in body composition, which are present to a slight degree in early life, become significant during adolescence, and once established remain throughout most of the adult years. So in truth one should also consider the existence of a "standard woman."

We shall begin by considering the changes in relative organ size during postnatal life.

Relative Organ Size

The contrast between the full-term neonate and the adult is evident from the data shown in Table 4.1. The neonate has a relatively larger brain, skin, and viscera than the adult, but a smaller muscle mass. Although total skeletal size, relatively speaking, is the same, a large proportion of the newborn skeleton consists of cartilage. This can readily be appreciated from roentgenograms of the wrist since none of the carpal bones are visible at this age. As would be expected, the ratio of surface area to body weight is 3–4 times greater in the newborn than that of the adult; this means that heat loss and insensible losses of water and electrolytes are proportionately greater. Since brain and viscera contribute a large share to met-

TABLE 4.1. Relative organ size (%)[a].

	Newborn	Adult
Brain	13	2
Liver	5	2.6
Kidneys	1	0.5
Heart	0.7	0.5
Skin	15	7
Muscle	25	40
Skeleton	13	14[b]
Bone	4	7
Cartilage	9	1.6
Body fat	10–15	15–25
SA/*W*	0.10	0.027

[a]Compilations of Widdowson and Dickerson (1964), Forbes (1960), International Committee on Radiation Protection (1975).
[b]Includes marrow, periarticular tissue, see Table 5.2.

abolic rate, the relatively large size of these organs in the neonate is responsible in large part for the fact that BMR per unit body weight is greater in the newborn than in the adult.

According to Widdowson (1981), skeletal muscle accounts for 27% of total body N, 33% of K, and 28% of water in the newborn; values for the adult are 47%, 62%, and 51% respectively. However, my own compilation yielded a value of 49% for K in the adult (Forbes and Bruining, 1976); assuming this to be correct, about the same proportion of total body content of all three moieties—N, K, and H_2O—reside in skeletal muscle at this age.

Skin contains 22% of total body N, 16% of total body K, and 18% of total body water in the newborn; these values are lower in the adult because skin makes up a smaller proportion of body weight, namely 10%, 5%, and 7%, respectively. The large proportion of total body water in newborn skin accounts for the observation that skin turgor suffers more in infants when they become dehydrated.

Organ Composition

There have been a number of studies on the changes in nitrogen and electrolyte content of individual organs during growth. Many organs have been studied in this manner (Widdowson and Dickerson, 1964), but we will consider here just a few of the major organs—muscle, brain, and bone. For obvious reasons, data from animals are available in greater detail than those from humans.

Growth changes in muscle composition are nicely illustrated by the data published by Vernadakis and Woodbury (1964) for the rat. Figures 4.1 and 4.2 show the progression of changes in water, Na, K, and Cl contents from day 1 to 45 days of age in the rat. During this interval muscle K

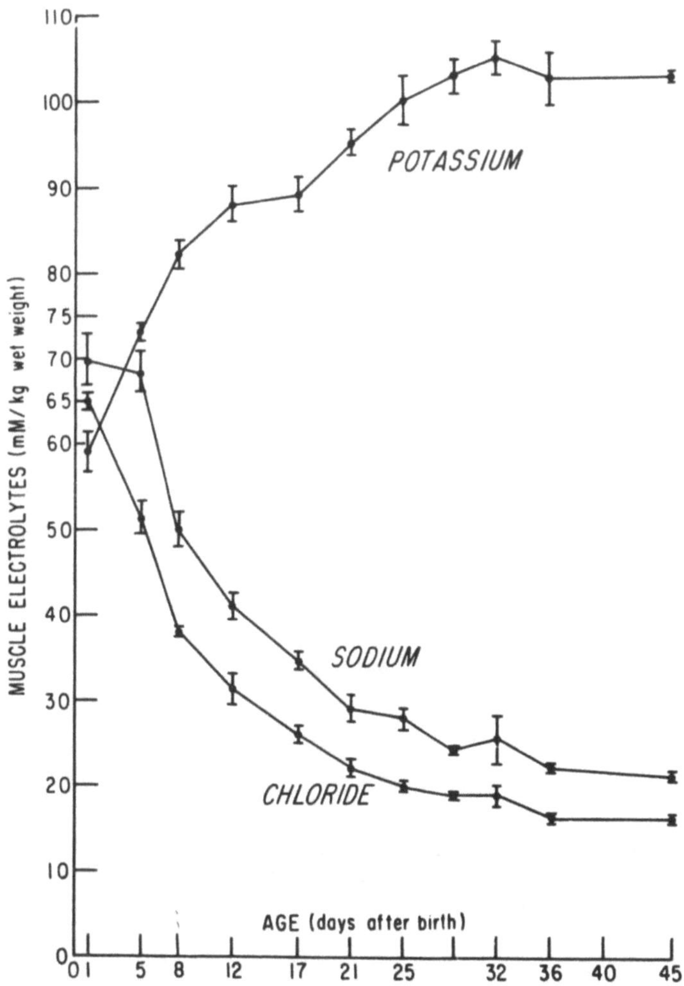

FIGURE 4.1. Muscle electrolyte concentrations in rat skeletal muscle as a function of age. [From Vernadakis and Woodbury (1964), with permission.]

concentration rises—as does N concentration (not shown)—while Na and Cl concentration and water content decline. Incidentally, the rat is relatively mature by age 45 days. Calculated extracellular fluid (ECF) volume as estimated from Cl space declines and intracellular fluid (ICF) volume increases during this period; the ECF/ICF ratio is 1.7 at birth and 0.18 at 45 days. However, calculated ICF potassium concentration does not change with age.

Two additional findings should be mentioned. The ratio of the two principal intracellular elements, potassium and nitrogen, falls somewhat as growth proceeds, from 4.5 meq K/g N on day 1 to 3.5 meq/g on day 45.

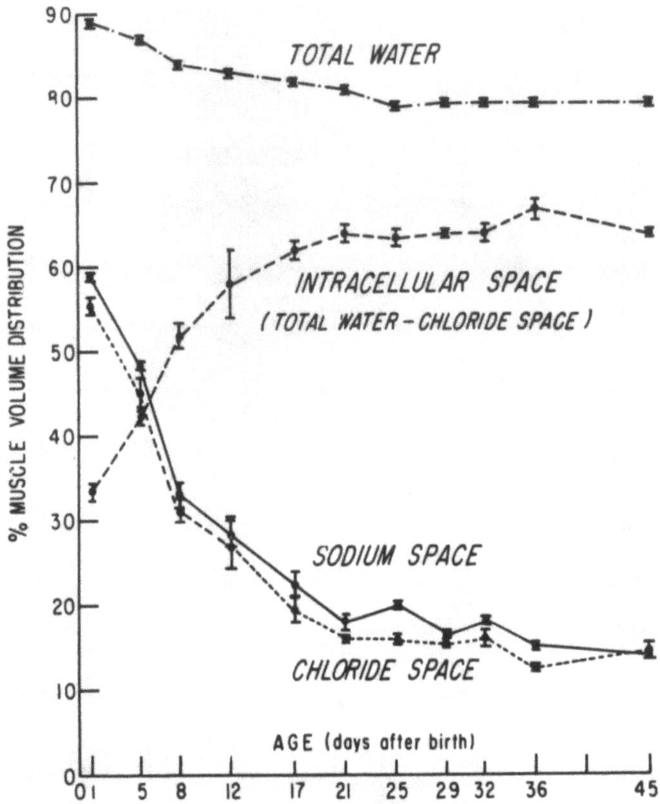

FIGURE 4.2. Total water and calculated intracellular space, and sodium and chloride spaces in rat skeletal muscle as functions of age. [From Vernadakis and Woodbury (1964), with permission.]

This may represent a change in the nuclear/cytoplasmic ratio. In addition, calculated ICF Na concentration declines, from about 18 meq/kg wet weight on day 1 to zero on day 45, while serum Na does not change appreciably.

These changes in muscle electrolyte content and in relative fluid spaces during growth have been documented for other species—pig, man, chicken, cat. There is a uniformity about the direction of change, although the magnitude of the several changes that take place during the period between birth and "maturity" differs somewhat among species. However, the big difference is the speed of attainment of "mature" body composition. Changes occur quickly in fast-growing species: the rat achieves in 45 days a maturation process that requires several years in man (Nichols et al., 1972).

The reason(s) for the higher calculated intracellular Na in young muscle as compared to adult muscle is not known. The possibility that it is due to the higher nucleus/cytoplasm ratio in the young is suggested by the work of Itoh and Schwartz (1956), who examined calf thymus, liver, and kidney, and found that nucleus Na concentration was 2–3 times higher than that in cytoplasm. It is doubtful that the age difference could be accounted for by a difference in serum water concentrations or in the magnitude of the Donnan factor, on which the calculation of interstitial fluid Cl "space"—and hence ICF Na concentration—depends. The possibility of increased binding of Na by intracellular organic compounds cannot be excluded.

Figure 4.3 shows that the changes in muscle composition during growth are mirrored in its histological structure. In the fetus the fibers are small and widely separated by extracellular material, and the nuclei are numerous and occupy a considerable proportion of the cell volume. As growth proceeds, cell nuclei are less prominent, and apparent ECF shrinks in size, the muscle bundles occupying more of the total volume. These changes are evident in the increasing protein/DNA ratio during growth (Cheek, 1968). The average of several species listed by Widdowson and Dickerson (1964) shows that the concentration of P and Mg—both intracellular elements—is about 40% higher in adult muscle than in the newborn.

The compositional changes in *cardiac muscle* and in *liver* are in the same direction as are those for skeletal muscle. However, the magnitude of the changes in cardiac muscle is considerably less. Liver presents problems by virtue of its variable content of glycogen: although this does not affect K content (glycogen accumulation is accompanied by K), the content of nitrogen and other minerals would be expected to vary inversely with glycogen content. *Skin* presents a somewhat different picture: whereas water and Na contents decline during growth, and N and Mg increase, Cl and P contents do not change appreciably, and K content falls. The increase in N content is due to the increase in collagen, for noncollagen N does not change with growth.

Brain

Brain is unique in certain respects. Whereas the growth changes in water, Na, Cl, N, P, and Mg are in the same direction as those for muscle, the Cl "space" is not a true indicator of ECF volume in this tissue. The data of Vernadakis and Woodbury (1962) show that after age 12 days in the rat the Cl "space" is far greater than the histologically identifiable interstitial fluid space, with much of the former being occupied by glial cells. The concentration of Cl in these cells is about one-half the plasma Cl value. This difference between brain and other tissues in the location of Cl is supported by the work of Amberson et al. (1938). These workers perfused cats with Ringer's solution in which the chloride had been replaced by sulfate. Beef erythrocytes and gum acacia were added in order

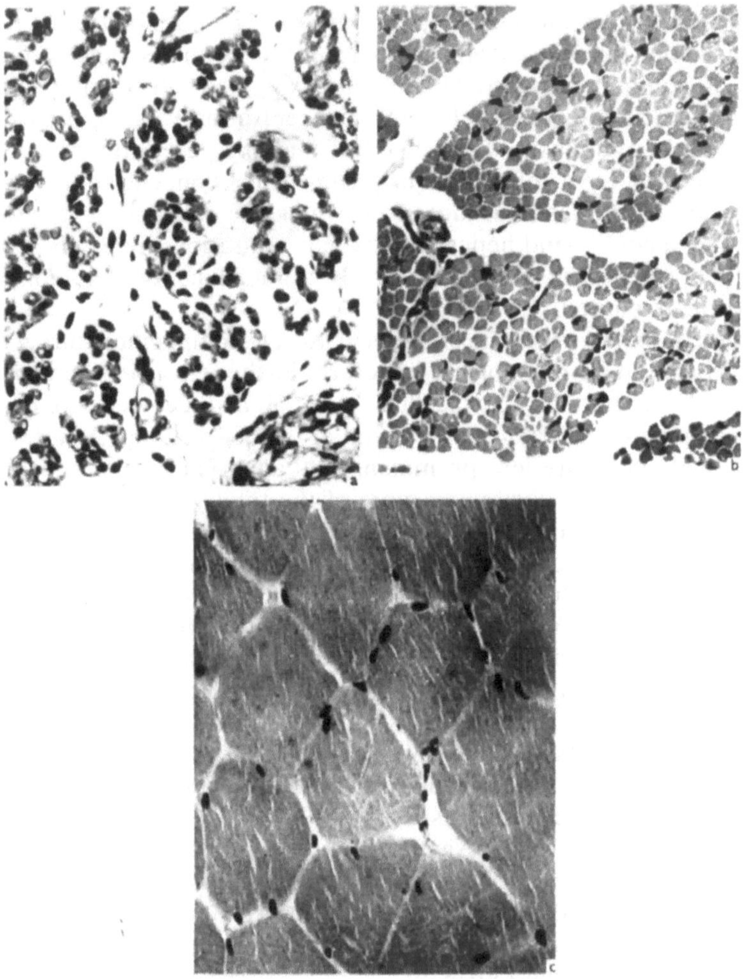

FIGURE 4.3. Transverse sections of human quadriceps muscle. × 420. (a) fetus, 20 weeks' gestation; (b) full term newborn; (c) adult. [From Dickerson and Widdowson (1960), with permission.]

to avoid anemia and changes in intravascular oncotic pressure. This procedure reduced serum Cl to very low levels, and for many tissues Cl concentration was directly related to serum Cl, indicating that almost all of the tissue Cl was diffusible. The one important exception was brain, which gave up only about one-quarter of its Cl content in the face of a serum value that had been reduced to about 7% of normal. Sciatic nerve, on the other hand, lost about 5/6 of its original Cl content. The large amount of Cl in glial cells acts to lower their resting electrical potential, according to Vernadakis and Woodbury (1962).

Skeleton

Skeleton presents several features that make for difficulty in assessing its composition. This organ consists of cartilage, marrow, trabecular and cortical bone, and its surface is covered with periosteum. As noted earlier, cartilage makes up a sizeable proportion of the total in the young infant, and in the growing individual there is in addition a provisional zone of calcification at the growth plate, together with variable numbers of osteoprogenitor cells. In the adult, the cortex may contain small osteoporotic cavities filled with fatty tissue. The character of the marrow tissue changes with age, being full of white and red cell precursors in the young, and full of fat in the adult (so-called "red" and "white" marrow). Hence the composition of the "skeleton" will depend on how many of these components are included in the sample taken for analysis. For those elements such as Ca, Sr, and F which are almost exclusively present in bone crystal, sample preparation may not be of great moment, but for those present in RBC and ECF such as Na, K, and Cl, it is of the utmost importance to exclude soft tissue and blood from the sample.

Bone crystal is considered to be hydroxyapatite ([$Ca_{10}(PO_4)_6(OH)_2$)]. The crystals are bounded by a thin layer of water (the "hydration layer"), and are intimately associated with an organic phase, which is largely collagen. The apatite crystals have the ability to adsorb certain ions to their surface, and this proclivity is stronger in young bone than in adult bone. In addition, certain ions may be incorporated, in minute amounts, into the crystal lattice through the process of recrystallization. The amounts of such ions present in bone are a function, therefore, of the age of the animal and the amounts that are ingested and appear in the body fluids.

Figure 4.4 shows the water, calcium, and fluoride contents of rat cortical bone that has been cleaned of marrow, trabecular bone, and periosteum.

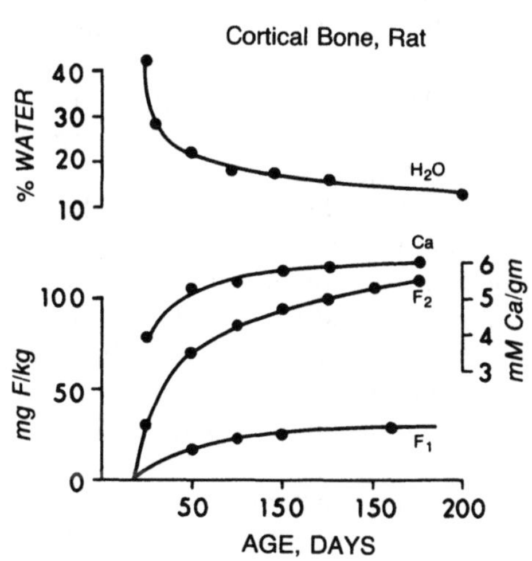

FIGURE 4.4. Composition of rat cortical bone (wet weight basis) as a function of age. **Top:** percent water. **Bottom:** upper line shows mM Ca/g bone (scale at right); lower line (F_1) is mg F/kg bone for animals fed normal diet (46 mg F/kg diet); middle line (F_2) is mg F/kg bone for animals whose drinking water contained 25 mg F/liter. (scale at left) (GB Forbes and FA Smith, unpublished data.)

Young bone has a higher water content and a lower Ca content than does adult bone. The higher water content facilitates the uptake of various ions, for studies with injected radioisotopes have shown that the uptake of Na, Mg, Zn, Sr, Ba, and Pb occurs much more readily in the bones of young animals than in the adult.

The curves for fluoride illustrate these phenomena: the slopes are steep at an early age, and then flatten as the animal ages; and an increase in fluoride intake enhances bone F content.

These same phenomena pertain to human bone. During the period 1950–1970 large numbers of bone samples were obtained for analysis of ^{90}Sr, a fission product of nuclear weapons explosions. This isotope gradually falls to earth—hence the term "fallout product"—enters the food chain, and following ingestion is deposited in bone. The results are usually expressed as picocuries ^{90}Sr per gram Ca. Continued weapons testing during the 1950s resulted in a steady increase in the bone ^{90}Sr/Ca ratio to reach peak levels in 1965; then with signing of the Nuclear Test Ban Treaty (in 1963) the level soon began to fall. Throughout this period the ^{90}Sr/Ca ratios in infant bone were 3–10 times those of adult bone (NCRP Report, 1975). The analogy with fluoride (Figure 4.4) is evident: ^{90}Sr levels in bone reflect the intake of this element, and uptake is greater in infant bone as compared to adult bone.

The ratio of calcium to phosphorus in human cortical bone remains at about 2.2 (g/g) over the age span from mid-fetal life to the adult (Widdowson and Dickerson 1964). On a molar basis this ratio is 1.7, which is close to the Ca/P ratio of 1.67 in hydroxyapatite.

Bone contains other elements in trace amounts—Al, Mn, Ra, Pu, and some of the rare earth elements. The effect of age on bone content has not been well studied for these elements.

Bone has the highest sodium content of any tissue in the body, and in contrast to soft tissues, bone Na content increases with age (optic lens is the only other tissue showing this age effect). The Na content of young cortical bone is about 0.10 m*M*/g wet weight; that of the adult bone about 0.25 m*M*/g. Values for dental enamel are the highest, namely 0.29 m*M*/g (Forbes and McCoord, 1963). Epiphyseal cartilage contains 0.15–0.17 m*M*/g (Manery, 1961).

Howell's (1960) data nicely illustrate the variations in composition of different regions of costal cartilage. The concentrations of Ca, P, Mg, Na, and S are all higher in the zone of hypertrophied cells than in the resting cell region, and Ca and P contents show an abrupt increase in the zone of provisional calcification. The exchange of radiocalcium, and perhaps other radioisotopes, takes place more readily in trabecular than in cortical bone; radiosodium very quickly exchanges with cartilage sodium.

All of these considerations make it necessary to specify the nature of the skeletal sample chosen for analysis, and to use age-matched controls in experiments designed to test the effect of treatments on bone composition.

Neonatal Adjustments

Blood Volume

Most investigators have calculated this from plasma volume and a corrected venous hematocrit. The reason is that plasma volume is easier to measure than total red cell mass,[1] and it is difficult to determine both simultaneously. The problem in calculating total blood volume from either plasma volume or red cell mass lies in the variability in the ratio of venous to total body hematocrit. The other problem in interpreting body fluid volumes in neonates is the fact that until recently most of them were not fed until 8–12 hr after birth; those who are breast-fed are not apt to receive very much milk until the second or third day of life, so most will lose weight for several days, in amounts of 5–10% of initial weight. The result is that plasma volume and red cell mass per unit body weight and hematocrit will undoubtedly change during the first 24–28 hr of life purely from the effects of dehydration. Yet another factor concerns the time of clamping of the umbilical cord; if this is delayed for a few minutes the infant may receive additional blood from the placenta. All of these factors should be taken into account in evaluating reported data on "blood volume" of neonates.

Mollison et al. (1950) determined plasma volume (Evans Blue dye) and total red cell mass (^{32}P-labelled RBC) in 39 normal neonates on the first day of life. Venous hematocrit ranged from 40 to 66%; and the total body/venous hematocrit ratio was 0.87 + 0.03 (SD), so two standard deviations include values between 0.81 and 0.93. Although the variability in this ratio is not as great as that reported in adults (see Chapter 2, p. 19), it is sufficient to make calculations of blood volume from either plasma volume or total red cell mass rather imprecise.

Usher et al. (1963) measured plasma volume and hematocrit at intervals during the first 24 hr of life in a group of infants whose umbilical cords were clamped at birth and in another group whose cords were clamped after an interval of 5 min. The infants were not fed for the first 12 hr of life, and then were put to the breast. For the first group plasma volume rose from 46 ml/kg at ½ hr of age to 54 ml/kg at 24 hr, and hematocrit fell from 48% to 44% (all average values). Plasma volume values for the second group (delayed clamping of the cord) were higher at ½ hr of age (70 ml/kg), fell to 40 ml/kg at 4 hr, and then rebounded to 45 ml/kg at 24 hr; hematocrit values averaged 48%, 64%, and 62% at these same time periods. It seems as though the situation was more stable in those infants whose umbilical cords were clamped immediately, and there must have been some circulatory adjustments to account for the changes in plasma volume recorded for those in whom clamping was delayed. Usher and co-

[1]This term avoids the ambiguity of "red cell volume" since this is commonly used in describing the size of individual erythrocytes.

TABLE 4.2. Plasma volume and red cell mass in neonates.

Author	Plasma volume (ml/kg)	Red cell mass (ml/kg)	Hematocrit (%)	Remarks
Mollison et al. (1950)	41.4±4.5	42.2±7.6	40–66	^{32}P, Evans Blue dye
Bratteby (1968)		42.0±6.5	55±5.7	Values for day 1 only, n = 14 ^{51}Cr
Whipple et al. (1957)	58		49	Cord clamped promptly
	68		56	Cord "stripped" Evans Blue dye
Sisson et al. (1959)	56		52	Evans Blue dye
Usher et al. (1963)	44±1.7[a]		48	Cord clamped promptly; age 4 hr
	40±1.8[a]		64	Cord clamped at 5 min; age 4 hr
	54±2.2[a]		44	Cord clamped promptly; age 24 hr
	45±1.4[a]		62	Cord clamped at 5 min; age 24 hr ^{131}I-albumin
Yao et al. (1969)	39.6±4.8[b]		50.2±2.9[b]	Cord clamped promptly
	42.6±7.1		60.7±3.8	Cord clamped at 3 min
	38.4±8		51.0±4.4	Cord clamped promptly; prematures
	41.9±9.5		57.5±5.5	Cord clamped at 3 min; prematures ^{131}I-albumin

[a]SEM
[b]SD
All assays done on first day of life; variability ±SD except where indicated otherwise.

workers were of the opinion that the larger hematocrit value at 24 hr of age for the latter group offered an advantage, since these infants begin postnatal life with a larger store of hemaglobin.

Whipple et al. (1957) had investigated this problem a few years earlier. Infants whose umbilical cords were not clamped until 3 min after birth had higher hematocrit values than those whose cords were clamped immediately, but plasma volumes per unit weight were slightly less. The highest values for plasma volume were recorded in those infants whose umbilical cords were "stripped," i.e., when the blood in the cord was mechanically propelled toward the infant.

Data on plasma volume, venous hematocrit, and total red cell mass are given in Table 4.2. Only those measurements made within the first 24 hr of life are included. Of note is the considerable degree of variability for all three measurements. Since some body weight is usually lost during the first few days of life, principally due to loss of water, plasma volume would be expected to increase relative to body weight, and this is shown in the data of Usher et al. (1963) as listed in the table. The drop in hematocrit may be related to circulatory adjustments, or possibly to red cell hemolysis which is known to occur in normal newborns and which leads to mild hyperbilirubinemia.

Body Composition Changes During the Early Neonatal Period

Although many infants have a satisfactory suck reflex soon after birth, an adequate intake of fluid and energy is usually not established for 2 or 3 days. Hence it is customary for neonates to lose weight—often 5–10% of their initial weight—during the first few days of life; if all goes well birth weight is usually regained by the end of the first week of life in the full-term infant. For the prematurely born this may take 2–3 weeks, the delay being due in large part to the difficulty in establishing an adequate caloric intake. What is the nature of the tissue lost during those first few days?

Wagen et al. (1985) did serial studies of prematurely born infants (average weight 1600 g) during the first 11 days of life. Total body water was estimated by deuterium dilution and ECF volume by sucrose dilution. During the first 3 days of life body weight fell by 5.2%, total body water by 4.8%, and body solids (i.e., weight minus H_2O) by 6.4%. By 11 days of age all of these losses had been made good, due to provision of adequate energy and fluid. These authors conclude that body water and solids were lost, and then regained, in proportion to their presence in the body, and that the initial weight loss was due to catabolism, not dehydration. A further analysis of their data shows that the water loss was all from the extracellular compartment.

Brans et al. (1974) approached this problem by means of an anthropometric technique. They measured skinfold thickness at 15 sec and again at 60 sec after applying the calipers. The difference between these two readings divided by the first reading (which they called a ΔSFT) was considered to reflect the relative water content of the skin and subcutaneous tissues, whereas the second reading was taken as indicative of the amount of subcutaneous fat. These investigators did find that the ΔSFT was proportional to the degree of postnatal weight loss, and that the 60-sec reading was proportional to birth weight.

Cheek et al. (1984) measured total body water and bromide space on one occasion in each of 107 neonates whose ages ranged from birth to 4 weeks of age. During the first 6 days of life total body water per unit body weight did not change, although Br space declined by about 8%. Assuming that these infants had lost some weight during the first few days of life, water and solids were lost in proportionate amounts, and the authors conclude that there was a shift of water from the extracellular to the intracellular compartment. Unfortunately, no information is given on fluid or energy intake.

The nature of the "solids" lost by neonates during the first few days of life can be estimated from the measurements of urinary nitrogen and electrolytes. McCance and Widdowson (1954), Nicopoulos and Smith (1961), and Auld et al. (1966) have collected urine on term infants who were given minimal intakes of water during the first 48 hr of life. It is important to consider the urinary losses of nitrogen and electrolytes in relation to the amount of weight that was lost. The ratio of nitrogen lost in urine to weight *loss* ranged from 0.7 to 1.7 g/kg in these three studies, and the ratio of K loss to weight *loss* was 9.2 to 10.2 meq/kg. The urinary K/N ratio ranged from 6.2 to 15 meq/g. The value for normal premature infants was 7.3 meq K/g N, but those who were distressed excreted larger amounts of nitrogen in their urine, so that the ratio was 2.5 meq K/g N (McCance and Widdowson, 1955). The ratio of Na loss to weight *loss* was 8.8 meq/kg in the infants studied by Auld et al. (1966) and 9.2 meq/kg in those studied by Nicopoulos and Smith (1961).

Anderson et al. (1979) did 5-day nitrogen balances on prematurely born infants (average birth weight 1670 g) who were given glucose water intravenously. Fluid and energy intakes were 150 ml and 60 kcal/kg/day. Weight loss amounted to 12.2 g/kg/day and they excreted an average of 132 mg N/kg/day; the ratio of N loss to weight loss was 10.8 g N/kg, which is considerably in excess of the ratios noted above for infants given little or no fluid. Possibly some of these infants were distressed. It is unfortunate that urinary potassium was not measured.

The amounts of nitrogen lost in urine per unit weight lost by those infants who were fasted and thirsted for 2 days is much smaller than the concentration of nitrogen in the newborn body itself, namely 20 g/kg (Table 4.3, p. 144). Even if one doubles the urinary N values to account for cutaneous and fecal losses, these data show that body protein stores are not consumed to any significant degree, and that the "solids" that are lost must consist largely of carbohydrate and fat. This ability of the neonate to conserve body protein is not shared by the adult. McCance and Widdowson (1954) also studied six young adult males who were given no food and only 350 ml water daily for 3 days. These subjects lost an average of 6% of their body weight, and on days 2 and 3 of the experiment they lost an average of 20.3 g N and 62 meq K.[2] Assuming that one-half of the total weight loss occurred on days 2 and 3 of the experiment, about 11 g N and 34 meq K were excreted per kilogram of weight *lost;* both values are far in excess of those listed above for the neonate, but the K/N ratio (~ 3 meq/g) is much lower, and approximates the K/N ratio of the body itself. Other studies of nonobese fasting adults show that they lose about 20 g N for each kilogram of weight loss (Forbes and Drenick, 1979).

The losses of Na in the neonate's urine can be accounted for by the documented decline in ECF volume, but the reason for the disproportionate loss of K is not readily apparent. The recorded ratio of urinary K loss to N loss in the three studies of neonates cited above ranges from 6 to 15 meq/g, a value far in excess of the total body ratio of 2.3 meq/g at this age. A possible explanation is that it reflects the loss of glycogen from liver. It is known that each gram of glycogen deposited in liver is associated with 0.35 meq K (Fenn, 1939). (This explains the transient hypokalemia that follows the administration of glucose and insulin.) Loss of liver glycogen during fasting would be expected to liberate potassium for excretion in the urine, but no nitrogen. The mobilization of as little as 5 g of glycogen during 2 days of fasting would provide 1.75 meq K, an amount sufficient to provide for the elevated urinary K/N ratio noted above.

The lack of data on cutaneous and fecal losses of nitrogen and electrolytes makes it impossible to give precise estimates of the effects described above. However, it is safe to conclude that the weight that is lost during the first few days of life is composed principally of water, and non-nitrogenous material such as carbohydrate and fat. Longer periods of fasting would undoubtedly involve a significant loss of protein, but in those first 2 days of life while the neonate is adjusting to his new lifestyle this most important body constituent remains relatively intact.

A possible explanation for the difference in urinary nitrogen excretion per unit weight loss between neonate and adult is that some of the N loss in the latter is from gastrointestinal tissue. These tissues have obviously been fully functional in the adult prior to starvation, but not so in the fetus whose gastrointestinal tract has not been burdened by food. Hence, fasting in the adult might be expected to result in some involution of gastrointestinal tissues, but not in the neonate.

Another possibility is the low basal metabolic rate during the first few days of life. According to Heim (1981) this is only 33 kcal/kg/day for a day or two after birth; by the end of the first week of life it has risen to 48 kcal/kg/day, a value that is characteristic of infants throughout the first year of life. It may be that the low rate of protein catabolism noted during the first 2 or 3 days of life is the result of this subnormal metabolic rate.

[2]No data are given for day 1; initial body weights assumed to be 60 kg.

Kagan et al. (1972) studied a group of infants weighing 1340–1850 g by means of antipyrine and thiosulfate dilution. At 6 days of age water content averaged 73.1% of body weight (SD 4.0) and ECF 39.7% body weight (SD 2.50). The average ECF/total water ratio was 0.54. They were fed several types of milk and then restudied at 28 days of age. Those fed larger amounts of sodium and total ash had greater increases in total body water and ECF volume, and one of the high Na formulas led to a slight increase in ECF/total water ratio. There was no effect on total dry weight. Apparently, body water content and distribution can be altered somewhat by diet at this young age.

Several groups of investigators have studied body composition changes in growing premature infants by means of nitrogen and energy balance. Whyte et al. (1983) determined that 11% of the weight gain consisted of protein and 34% as fat in those fed human milk in quantities sufficient to provide 2.6 g protein/kg/day; the values for those fed formula providing the same amount of protein were 10% and 28%, respectively. Earlier Reichman et al. (1981) had found values of 11% protein and 32% fat. Both estimates of fat accretion exceed the intrauterine rate for the same gestational age range, and Reichman's data on skinfold thickness also show a larger increase in this measurement than is characteristic of newborn premature infants in the same age range. Catzeflis et al. (1985) fed a slightly higher protein diet (3.04 g/kg/day) and found that 77% of the weight gain consisted of lean tissue and 23% was fat. These investigators also estimated protein synthesis and degradation rates, but the interpretation of these results is difficult because the values calculated from ^{15}N enrichment of urinary ammonia differed from those calculated from urea enrichment. The energy cost of the weight gain varied from 2.8 to 4.0 kcal/g in these three studies; hence these tiny infants make very efficient use of their food. (Comparison values for other nutritional situations will be presented in Chapter 7.)

Although the results of these studies, which are most difficult to perform, are of interest, the balance periods were brief (3–6 days), and none of the three reports includes mention of the possibility of cutaneous losses of nitrogen. Although it is true that premature infants are rarely observed to sweat, it is likely that there is some insensible loss of nitrogen.

The excessive fat accretion (in comparison to intrauterine standards) shown in two of the above studies suggests that extrauterine nutrition is less efficient at promoting increments in lean weight than is intrauterine nutrition. Alternatively, the possibility exists that such excessive fat accretion is merely an attribute of early extrauterine life, and that the premature infant once born emulates the behavior of the full-term infant, who rather quickly begins to increase its stores of fat (Fomon et al., 1982). A third possibility is that dietary protein intake was suboptimal.

The rate of accretion of body calcium by the young infant is a matter of some interest, particularly so in view of the rather small amounts of Ca present in human milk. Metabolic balance studies performed some years ago by Genevieve Stearns and her associates (1939) indicated that estimated percent body Ca actually declined somewhat during the first few months of infancy, in infants fed at the breast. From a value of about 0.8% Ca at birth, estimated body Ca content dropped to about 0.6% at 16 weeks of age, and then rose to 0.75% at age 40 weeks. In those fed milk formulas, body calcium as a percent of body weight remained fairly steady at 0.7–0.8% for the first 16 weeks, and then rose to 1.2% at age 40 weeks. These findings reflect the difference between the mineral content of human milk (280 mg Ca, 140 mg P/liter) and cow's milk (1200 mg Ca, 960 mg P/liter). Nevertheless, even with formula feedings it appears that the nonosseous portions of the body grow at a faster rate, at least in early infancy, than does the bony skeleton.

Recent studies of bone density by means of photon absorptiometry of the radius and ulna (Greer et al., 1981) serve to confirm Stearns's earlier work. In a group of infants in whom serial studies were done during the first 3 months of life, these investigators found that "bone mineral content" actually declined in breast fed infants, but remained at birth level in those fed milk formulas.

Both studies show an effect of dietary calcium on body Ca accretion, and serve to emphasize the fact that human milk is in actuality a low calcium food.

Rutledge et al. (1976) did serial ^{40}K assays on infants during the first few months of life. As would be expected, estimated body K rose progressively, and there was no difference between breast-fed and bottle-fed infants in this respect.

Changes in Relative Body Fluid Volumes During Growth

Earlier mention was made of the decline in total body Cl/K ratio as the fetus grows (see Chapter 3), and the fact that there is a further fall in this ratio during postnatal life. The values are about 2.0 meq Cl/meq K at 20 weeks' gestation, 1.24 at birth, and 0.72 in the adult.[3] Since chloride is almost exclusively an extracellular ion and almost all of the body potassium is located inside cells, the fall in the Cl/K ratio suggests that the ECF/ICF ratio falls progressively as the fetus grows.

It so happens that several investigators have made simultaneous assays of bromide space and total body water in a number of human subjects. Taken together they range in age from the newborn to the adult. The fact that all the investigators to be cited used the same indicators for estimating ECF volume and total body water avoids the problem of differences in "space" estimates afforded by different indicators (see Chapter 2, p. 14).

[3]According to Widdowson and Dickerson (1964), the total body Cl/K ratios for other adult mammals is even lower, ranging from 0.44 to 0.66 in the monkey, dog, rabbit and cat; values for the newborns of these species are 1.03–1.10.

The bottom section of Figure 4.5 shows plots of ECF volume against total body water for the various groups of subjects, as determined by regression analysis. Each has a positive intercept on the *y*-axis, and there is a reasonable correspondence among the regression slopes. Hence I felt justified in calculating a regression for all of the subjects, and this is shown as the dashed line in Figure 4.5. These data pertain to normal, nonobese subjects.

One can use these data to calculate the ratio of ECF to total body water and the ratio of ECF to ICF for any given total body water content. When

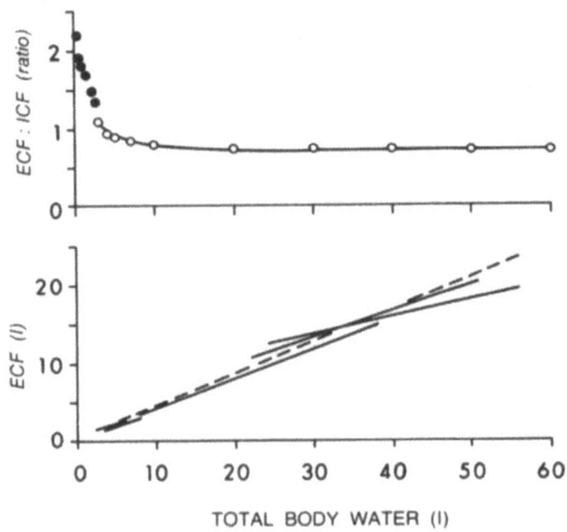

FIGURE 4.5. **Top:** Plot of ECF/ICF ratio against total body water from 20-week fetus to adult. Postnatal data (○) calculated from combined regression line shown below; fetal data (●) calculated from total body Cl/K ratio shown in Figure 3.3. Line drawn freehand. **Bottom:** Plots of ECF volume (liters) based on Br "space" (stable Br or ^{82}Br) against total body water (liters) by deuterium or tritium dilution; combined sexes. Data sources as follows:

Author	*n*	Age range (yr)	H_2O range (liters)	Regression slope	Intercept (liters)	*r*
Cheek (1968)	15	0.18–1.2	3–8	0.395	0.07	0.91
Owen et al. (1962)[a]	70	0.04–0.73	2.4–5.7	0.398	0.362	0.85
Cheek (1968)	39	3.5–16	8–38	0.396	0.134	0.98
Moore et al. (1963)[b]	34	23–84	22–51	0.341	3.47	0.90
Forbes and Brown[c]	32	13–46	24–56	0.218	7.29	0.81
Combined data	190		2.4–56	0.414 (±0.005)	0.306 (±0.115)	0.986

[a]Published values for Br "space" multiplied by 0.795 to correct for Donnan factor (0.95), serum water (0.93), and uptake by erythrocytes (0.90) (see Chapter 2, p. 8, 15); 35 subjects, some assayed repeatedly.
[b]Published values for Br "space" corrected by 0.96, because Moore et al. (1963) used a correction of 6% for erythrocyte Br rather than 10%.
[c]Unpublished data.

such calculations are made it turns out that both ratios decline in curvilinear fashion as total body water—itself an index of body size—increases during the process of normal growth.

The values for the ECF/ICF ratio can be extended back into the fetal period by using the determined total body Cl/K ratios (see Chapter 3, p. 106). Assume ECF chloride concentration to be 115 meq/liter, multiply total body Cl by 0.90 to correct for red cell chloride, and assume ICF potassium concentration to be 140 meq/liter; then ECF volume in liters is total body Cl (meq) $\times$ 0.9/115, and ICF volume is total body K (meq)/ 140.

The top section of Figure 4.5 is a plot of the calculated ECF/ICF ratio over the age range from the 20-week fetus (total body water 0.3 liters) to the adult (body water 50 liters). Although the values for the term infant as calculated from total body Cl and K and those derived from bromide and deuterium dilutions do not quite coincide, the trend is clear: there is a fairly smooth decline in the ratio with increasing body size, which is rapid at first and slower later on.

Cassady and Milstead (1971) measured bromide and antipyrine spaces in full-term infants, prematurely born infants, and those with intrauterine growth retardation. Total body water averaged 688 ml/kg ($\pm$16 SEM) in the first group, 809 ml/kg ($\pm$11) in the second group, and 790 ml/kg ($\pm$13) in the third group. The average ECF/ICF ratio was 1.13 in the normal full-term infants and 1.17 in the other two groups. Hence this method for estimating body fluid compartments produced values similar to those shown in Figure 4.5. Cheek et al. (1982) found no difference in body water compartments between full-term infants delivered vaginally and those delivered by caesarean section.

Thus the picture for the body as a whole is much the same as that for muscle (see Figures 4.1 and 4.2): early in life ECF volume exceeds ICF volume[4]; and by the time growth is complete ICF has come to occupy a majority position. In keeping with our earlier discussion of the prime influence of body size on element content (Chapter 3), the course of this transition can be related to total body water, itself an index of body size. The coefficients of the equation for the combined data suggest that the ECF/total water ratio reaches a lower limit of about 0.41 in large humans; hence the lower limit for the ECF/ICF ratio is about 0.41/0.59 = 0.70. It may be that this is the lowest ratio which is compatible with normal cell function.

Also worthy of note is the fact that blood flow rates are generally different in the young. Smith and Nelson (1976) quote data that show rates for the foot and calf region of 14 ml/100 ml/min in the premature infant, 7 in term infants, 3.4 in children, and 2–5 in the adult. Animal studies

---

[4]Indeed, some prematurely born infants who are otherwise normal appear to be edematous.

demonstrate newborn/adult ratios for organ blood flow (ml/100 g tissue/min) of 1:1 in central nervous system, 0.4:1 in kidney, 1:1 in gut, and 8:3 in muscle. These values attest to the well established metabolic activity of the brain and gut in the young, the fact that renal function is proportionately less, and that blood flow to muscle is greater.

The data compiled by Fomon et al. (1982), who calculated ECF from total body K and total body water for children from birth to age 10 years, show slightly higher ECF/ICF ratios for the younger years than Figure 4.5, but good correspondence later on. At age 10 years, the ratio by their method is 0.70 for boys and 0.83 for girls.

Although the proximate cause of the decline in the ECF/ICF ratio during growth is not known, two contributing factors may be mentioned. Growth in the fetal period is by cell multiplication; in late postnatal period it is by cell hypertrophy. As body cells enlarge, their surface area/volume ratio declines, and this could account for some of the decline in the ECF/ICF ratio. A second contributing factor could be the decline in serum aldosterone levels that occurs during growth. Sippell et al. (1980) measured serum aldosterone in infants and children ranging in age from the newborn to 15 years. From an average value of 2.5 ng/ml in the newborn, the level fell to 0.87 at 7 days of age, to 0.64 at 2 weeks to 3 months, then to 0.28–0.36 for subjects aged 3 months to 15 years; the adult value is 0.11 ng/ml. Thus there is a rough parallelism between the drop in ECF/ICF ratio and the drop in serum aldosterone level during growth. Since this hormone acts to enhance the renal tubular reabsorption of sodium, the high serum levels in early life could play a role in maintaining a larger ECF volume at that time. Weldon et al. (1967) have measured aldosterone secretion rates in human subjects whose ages ranged from the newborn to the 40-year-old adult. The average value for the newborns was 23 μg per 24 h, but after 10 days of age there was no change with age, the averages being 72–91 μg per 24 h over the entire age span. Assuming no change in aldosterone degradation rate, the decline in serum aldosterone concentration with age as reported by Sippell et al. (1980) could have been anticipated.

Two further comments are in order. Ninety-one of the subjects in Figure 4.5 were males, and 109 were females. The regression parameters for the former are intercept 0.282 liters and slope 0.409 liter ECF/liters total H_2O, and for the latter these are 0.323 and 0.418, respectively. Thus there is no indication of a sex difference in the relationship between ECF volume and total body water. Nor is the situation any different in the obese individual. Cheek et al. (1970) measured bromide space and total body water in 19 obese individuals, and Forbes and Brown (unpublished) did so in 13 obese individuals. They ranged in age from 7 to 35 years, in body weight from 48 to 196 kg, and in total body water from 24 to 69 liters. A number of them were massively obese (BMI >45); none had edema. The calculated regression line for these 32 subjects is ECF (liters) $= 0.961 \pm 1.87 + 0.408$ H_2O (liters) ± 0.044 ($r = 0.86$). This does not differ significantly from the regression equation for the 190 nonobese subjects shown in Figure

4.5. Hence it appears that the obese state does not alter the ratio of ECF to total body water. It should be noted here that Shizgal et al. (1979) failed to find a difference between obese and nonobese adults in the ratio of Na_e to K_e; moreover, Egusa et al. (1985) found values for the ratio of plasma volume to fat-free weight in obese males to be similar to that for nonobese males.

Growth of Blood Volume

A knowledge of blood volume is essential for the surgeon and for others who treat patients who have sustained severe trauma or hemorrhage. Many investigators have made measurements on normal individuals of various ages, either with indicators of plasma volume or total red cell mass. As mentioned earlier, although the calculation of blood volume from either assay and venous hematocrit may not be very precise, such values have clinical usefulness.

Hawkins (1964) has collated data on blood volume from a number of sources. His compilation shows an average value of 7% of body weight for the infant, 8% for the child, 8.5% for young adult men, and 7% for young adult women. Beyond age 30, the average male value is 7.5%, and the female 6.5% of body weight. Since adipose tissue has a much poorer blood supply than muscle or viscera, relative blood volume will vary inversely with adiposity. Indeed, Allen et al. (1956) found that blood volume per unit body weight varied directly with body density, and hence inversely with adiposity.

Blood volume is also a function of stature—actually, Allen et al. (1956) found it to be related to the cube of height—and with this in mind the

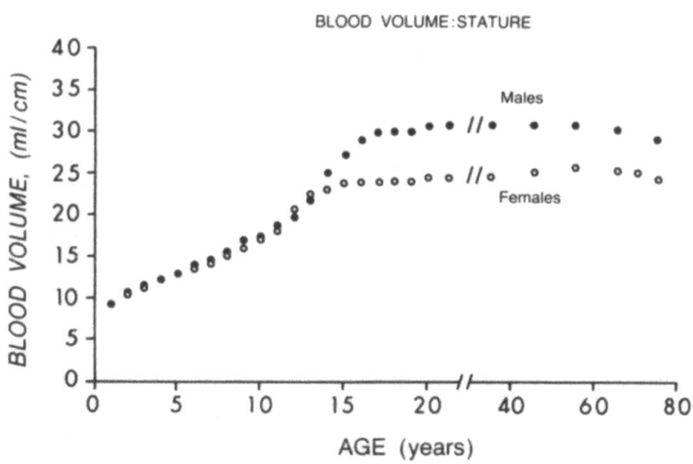

FIGURE 4.6. Ratio of blood volume to stature (ml/cm) for infants, children, and adults. [Based on data of Hawkins (1964).]

graph shown in Figure 4.6 was constructed from the data provided in Hawkins' (1964) review. The average value for young infants is about 10 ml/cm of height; this ratio rises during childhood, and then there is an adolescent spurt that culminates in a value of 30–31 ml/cm in the male and 24–25 ml/cm in the female. Using an average stature of 178 cm for adult males, blood volume is about 5500 ml; for females (163 cm) it is 4100 ml, so the sex ratio is 1.35:1. Based on average venous hematocrit values of 46% for males and 42% for females, total red cell mass is 2200 ml and 1500 ml, respectively, and plasma volume is 3300 ml and 2600 ml, respectively (hematocrit × 0.87).

These should be looked on as average values for normal individuals. Hawkins (1964) did not include an estimate of inter-individual variability, a phenomenon that in all likelihood exists.

Total Body Content

Table 4.3 lists total body content of a number of elements for the newborn and for a 4½-year-old child. While this child died of tuberculous meningitis and hence undoubtedly suffered some major metabolic insults, this subject represents the only one available between the newborn and the adult, and is presented as an indication of changes in body composition during the early years of life.

Taking these data at face value, while body weight increased 4-fold during 4½ years, Fe and Cu contents increased about 3-fold, Mg and Zn about 6-fold, N by 8-fold, P by 9-fold, and Ca by 10-fold. The increase in body content for all elements listed, save Fe and Cu, exceeded the increase in total body weight.

Element content for the adult will be listed in Chapter 5.

Table 4.4 lists data on average lean body mass (LBM) and fat for newborn girls and boys, and gives the components of the LBM. As mentioned earlier girls are slightly fatter at birth than boys, and have slightly less

TABLE 4.3. Total body content.

Element	Newborn (3.4 kg)		4½ year old (14 kg)	
Nitrogen	66 g	2.36 mol	535 g	19 mol
Calcium	28 g	0.70 mol	295 g	7.4 mol
Phosphorus	16 g	0.52 mol	147 g	4.7 mol
Sodium	10.4 g	240 mmol	—	
Potassium	3.8 g	150 mmol	23 g	910 mmol
Chlorine	4.5 g	160 mmol	—	
Magnesium	0.76 g	32 mmol	5 g	208 mmol
Iron	0.32 g	5.7 mmol	0.9 g	16 mmol
Zinc	0.053 g	0.81 mmol	0.31 g	4.7 mmol
Copper	0.014 g	0.22 mmol	0.046 g	0.72 mmol
Iodine	0.057 mg	0.45 mmol		

From Widdowson and Dickerson (1964).

TABLE 4.4. Newborn at term.

	Boys	Girls
Body weight	3545	3325 g
LBM	3059	2830 g
Fat	486 (13.7%)	495 g (14.9%)
Components of LBM		
Water	80.6	80.6%
Protein	15.0	15.0%
Osseous mineral	3.0	3.0%
Density	1.063	1.064 g/ml
Potassium	49.0	49.0 meq/kg
Ratio ECF/total water	61[a]	61%[a]

From Fomon et al. (1982).
[a]The data in Table 4.5 and those of Cassady and Milstead (p. 141) indicate a value of 54%.

lean weight. According to the calculations offered by Fomon et al. (1982), there is no sex difference in the composition of the LBM. The values for density of the LBM were estimated from calculated total body calcium as derived from photon absorptiometric measurements of the radius-ulna.

Table 4.5 presents data on certain aspects of LBM composition over the age range of the 24-week fetus to the adult. Water content progressively falls, while potassium and nitrogen contents both rise, as does calculated body density. Of interest is the fact that the K/N ratio (meq/g) progressively falls, from about 2.9 in the 24-week fetus to 2.0 at birth; then there is a further fall to 1.8 in the young adult and finally to 1.7 in the older adult (the sex difference in the adult years is of questionable significance). The adult ratios, which were obtained by whole body counting for potassium and by neutron activation for nitrogen, are slightly below the value of 2.0 meq K/g N obtained by cadaver analysis.

Growth of Muscle Mass

Skeletal muscle is estimated to be about 25% of body weight in the neonate and about 40% of body weight in adult men. In absolute terms these values are 850 g and 28,000 g, respectively. Hence in a period of 20–25 years, by which time muscle mass is at its maximum, the average increment is 1200 g per year, or 3.3 g per day. However, this growth process does not take place in a uniform manner. As is true for total body weight, growth of muscle mass is relatively slow during childhood; then there is a spurt during adolescence; and as will be shown later, this spurt is more intense and of longer duration in boys than in girls, so the former achieve a larger final value as adults.

Anthropometric and radiographic techniques have been used to estimate the growth of muscle mass. As mentioned in Chapter 2, computerized tomography provides an excellent method for assessing the cross-sectional areas of limb muscles, but this technique has for obvious reasons not been

TABLE 4.5. LBM composition.

Age	Male				Female			
	H_2O (%)	K (meq/kg)	N (%)	Density (g/ml)	H_2O (%)	K (meq/kg)	N (%)	Density (g/ml)
Fetus 24 week	89	40	1.4		89	40	1.4	
Fetus 32 week	86	46	1.8		86	46	1.8	
Birth	81	49	2.4	1.063	81	49	2.4	1.064
5 year	77	64	3.0	1.078	78	62	2.9	1.073
10 year	75	67	3.1	1.085	77	64	3.0	1.075
18.5 year	73.6		3.2	1.093				
Young adult	73	68.1		1.10	73	64.2		1.10
			K/N (meq/g)				K/N (meq/g)	
Young adult			1.88				1.73	
Old adult			1.78				1.65	

Data from Ziegler et al. (1976); Fomon et al. (1982); Haschke (1983); Cohn et al. (1984).

applied to large numbers of infants and children.[5] Maresh (1966) and Tanner (1962) have both published data on width of limb muscles as determined from roentgenograms in infants, children, and adolescents, and these nicely illustrate the changes that occur during growth, and the sex difference that develops during adolescence. Although it may represent only a minor problem, this technique, as well as the anthropometric techniques to be mentioned, does not delineate the amount of fat that lies between the muscle bundles.

Anthropometry has been the most widely used technique for estimating skeletal muscle mass, and although this has been applied to only selected muscle groups, namely the arm, thigh, and calf, it does offer a rough estimate, and it does possess the advantage of being easily applied to large numbers of individuals.

Several anthropometric measurements were made on large numbers of individuals who participated in the 1971–74 national nutrition survey. Of interest here are the measurements made at the mid region of the upper arm. From measurements of the arm circumference and triceps skinfold thickness one can calculate the cross-sectional area of the muscle-bone portion and the area occupied by subcutaneous fat (see p. 75). Since the subjects were largely recruited from lower socioeconomic groups the results may not be completely representative of the general population; nevertheless, the data do indicate the relative proportions of lean and fat in the upper arm region, and the age and sex trends.

Figure 4.7 shows values for arm muscle + bone cross-sectional area (M + B area) for males and females from age 1½ years through age 25 years.[6] Although there is a slight sex difference in the 50th percentile values during infancy and childhood, with the onset of adolescence this difference becomes clearly evident and then progressively increases. The male/female ratio is 1.7:1 at age 18 ½ years.

In studying a group of suburban school children in this manner, we took this process a step further: mid-arm circumference and triceps and biceps skinfold thicknesses were measured with the elbow held in extension, that is, with the arm in the relaxed position, and then again with the elbow flexed and the biceps muscle under maximum tension (the "make a muscle" maneuver). From these measurements the cross-sectional arm M + B area was calculated for both the "flexed" and "relaxed" positions of the elbow. The difference between these two areas is shown in Figure 4.8 as a function of age. Adolescent boys not only have a larger cross-sectional arm M + B area than girls, but they can achieve twice the increase in

---

[5]Nuclear magnetic resonance is also capable of assessing the size of any one of a number of muscle groups, including the amount of interstitial fat.

[6]Many authors use the term "muscle area," including the author who compiled and reported the data shown in Figure 4.7. I prefer "muscle + bone area" since this is what it actually is.

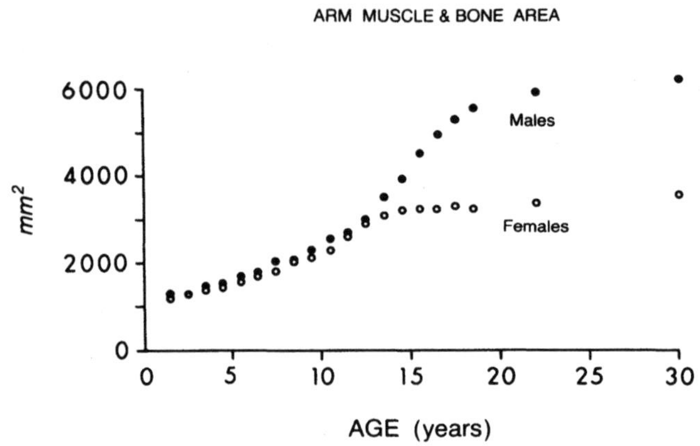

FIGURE 4.7. Cross-sectional muscle + bone (M + B) areas of the arm for white males and females aged 1½ through 25 years. (●), males (♂) females. Calculations based on mid-arm circumference and triceps skinfold thickness measurements made during the 1971–74 national nutrition survey. [Data compiled by Frisancho (1981).]

area when the biceps muscle is put under tension. Data such as these serve to emphasize further the sex difference in muscle bulk that develops during adolescence.

Fuchs et al. (1978) used a formula containing height and *flexed* arm circumference to estimate LBM in a group of adult men, with rather good results (Chapter 2, p. 84). It would be of interest to derive such a formula for children and adolescents.

Urinary creatinine (Cr) excretion is generally considered to be an index of skeletal muscle mass. As mentioned earlier (see Chapter 2, p. 59),

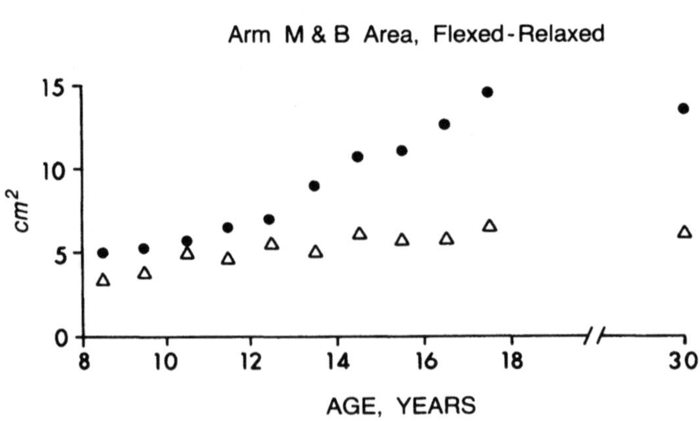

FIGURE 4.8. Increase in mid-arm muscle + bone area achieved by full flexion of the arm. Normal children, adolescents, and young adults: (●), males (n = 181); (△), females (n = 189). (Author's data.)

the ratio of muscle mass to Cr excretion varies somewhat with the size of the former, larger individuals having a somewhat lower muscle/Cr ratio than smaller individuals. For example, this ratio has been calculated as 18 kg muscle/g urine Cr for an excretion rate of 1 g Cr per 24 hr, and 16.2 kg/g at an excretion rate of 2 g Cr per 24 hr. Be that as it may, urinary Cr excretion (assuming reasonable dietary control and adequate urine collection) can provide a satisfactory clue to the amount of skeletal muscle.

Figure 4.9 depicts values for urinary Cr excretion per unit height for various age groups. On average, infants excrete 1–2 mg Cr/24 hr per centimeter of stature; in mid-childhood this value stands at 3–4 mg Cr/cm; and by age 16–18 years it has risen to 10.7 mg Cr/cm in boys and 7.5 mg Cr/cm in girls. The development of a clear-cut sex difference during adolescence is clearly evident. Not only do boys excrete more Cr per day than girls [the ratio is 1.57:1 in the data of Clark et al. (1951)], but the excretion rate corrected for stature is also greater.

Bistrian and Blackburn's (1983) tabulation shows that Cr excretion per unit height is somewhat less in adults than Clark et al. (1951) found in 17-year-old subjects. The latter authors made repeated assays on free-living normal subjects; the former do not specify diet, or the status of their subjects. The lower values could be explained if the urine samples were collected from hospitalized subjects who were consuming meat-free diets.

Our subjects were normal individuals who were housed on the Clinical

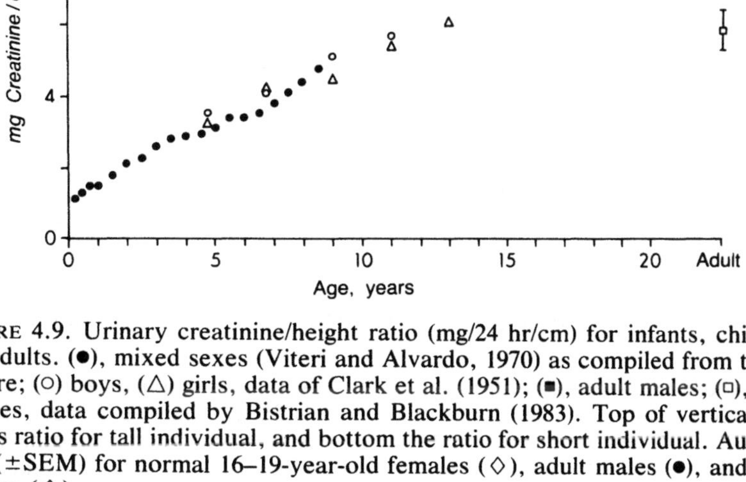

FIGURE 4.9. Urinary creatinine/height ratio (mg/24 hr/cm) for infants, children, and adults. (●), mixed sexes (Viteri and Alvardo, 1970) as compiled from the literature; (○) boys, (△) girls, data of Clark et al. (1951); (■), adult males; (□), adult females, data compiled by Bistrian and Blackburn (1983). Top of vertical bars shows ratio for tall individual, and bottom the ratio for short individual. Author's data (±SEM) for normal 16–19-year-old females (◇), adult males (●), and adult females (◇).

Research Center and fed a meat-free diet; these values are more in keeping with those of Clark et al. (1951).

Cheek (1968) calculated total muscle cell number from the DNA content of a muscle biopsy and the total mass of skeletal muscle as estimated from urinary Cr excretion. The assumptions are that muscle cells have but one nucleus, which contains 6.2 pg DNA. Muscle cell size is taken as the protein/DNA ratio. Cheek's data show that muscle cell number is about 0.2×10^{12} in early infancy, about 3×10^{12} in 15–17-year-old boys, and about 2×10^{12} in similarly aged girls. Muscle cell size, as judged from the protein/DNA ratio, approximately doubles during this interval. According to these data, skeletal muscle growth involves both cell multiplication and cell hypertrophy.[7] The biopsy data of Brooke and Engel (1969) show that muscle fiber cross-sectional area increases from 15 μm in early infancy to 60 μm at age 15 years. The ratio of type I to type II fibers was approximately equal throughout this period.

Growth of Skeletal Mass

Trotter and Peterson (1970) determined total skeletal weight on a large number of human cadavers, and Figure 4.10 shows the results of their analysis. As might be expected there is a fair amount of variation in the dry fat-free weight of the skeleton at any given age; generally speaking, skeletons of blacks tend to be heavier than those of whites. The sex difference during adolescence and in the adult years is clearly evident, as is the decline in skeletal weight with aging. In this series of 150 skeletons maximum size appears to have been attained in the third decade of life. It is of interest that estimates of skeletal size made by Garn and Wagner (1969) from metacarpal cortex thickness provide a reasonable fit to the data, at least up to age 18 years.

McNeill and Harrison (1981) have estimated total body calcium in a small number of children and adolescents by the technique of total body neutron activation. They found an excellent correlation ($r = 0.97$) between total body calcium and height raised to the 2.89 power. In adults less than 55 years of age this exponent was determined to be 3.10, so the relationship for both age groups is close to the cube of height. As will be shown subsequently, LBM is also a function of height cubed, and mention has already been made of the fact that this is true of blood volume.

There are a lot of data available on metacarpal cortex thickness and on photon absorptiometric measurements of the radius-ulna both for children and adults. The former measurement is relatively easy to make, and requires only a single roentgenogram of the wrist taken at a tube distance

[7] The report by Hansen-Smith et al. (1978) casts some doubt on the validity of such calculations. They did biopsies of the quadriceps muscle in malnourished and recovered children, and found that myogenic nuclei represented only two-thirds of the total cell nuclei seen, the remainder being vascular and interstitial cell nuclei.

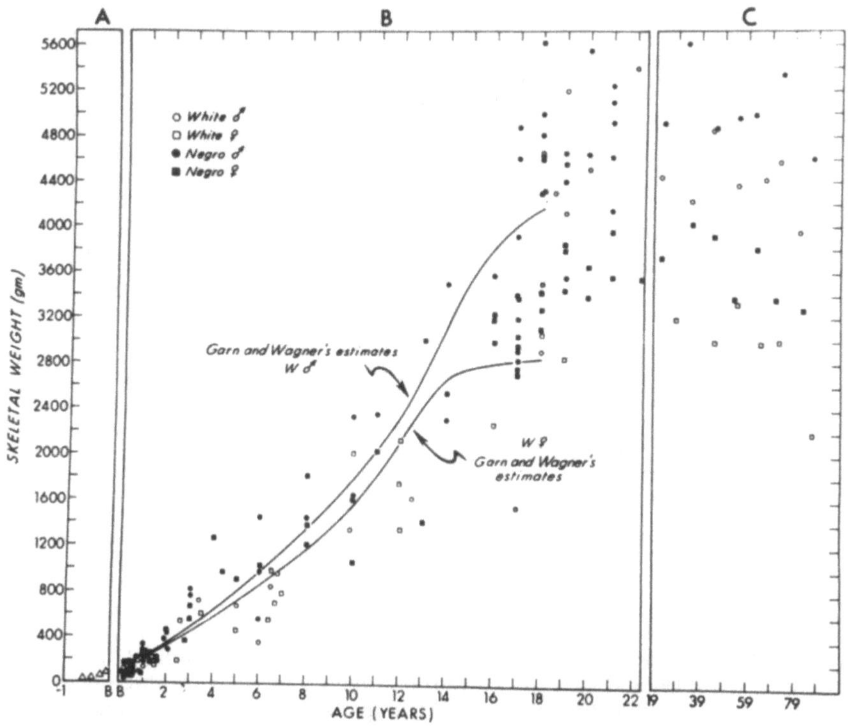

FIGURE 4.10. Weight of dry-fat free skeleton as a function of age. (A) Fetal period. (B) Birth to 22 years (n = 150). (C) Adult. Individual points with superimposed curves of skeletal weights estimated by Garn and Wagner (1966) from metacarpal roentgenograms. [From Trotter and Peterson (1970), with permission.]

of 60 cm so as to minimize parallax. The latter requires somewhat more elaborate instrumentation and careful calibration of the apparatus. Although there is still some question as to how accurately either technique serves as an estimate of total skeletal size (ossified portion), the results provide a reasonable picture of bone growth.[8]

Figure 4.11 illustrates the course of bone growth during childhood and

[8]There is an important distinction between these estimates of skeletal mass and what is commonly known as *bone age*. The latter is an index of skeletal maturation based on the roentgenographic appearance of ossification centers, configuration of the epiphyses, and degree of closure of the epiphyseal growth plates in various regions of the skeleton. The *bone age* value is listed as the degree of maturation that characterizes the average child at that age. For example, a 10-year-old whose skeletal maturation corresponds to that of the average 8-year-old is said to have a bone age of 8 years. Although it is true that this index of skeletal maturation is based in large part on the degree of skeletal calcification, it does not necessarily agree with other estimates of skeletal mass. For instance, although adolescent boys usually have a greater skeletal mass than adolescent girls, the girls usually have a more advanced skeletal maturation for a comparable chronologic age.

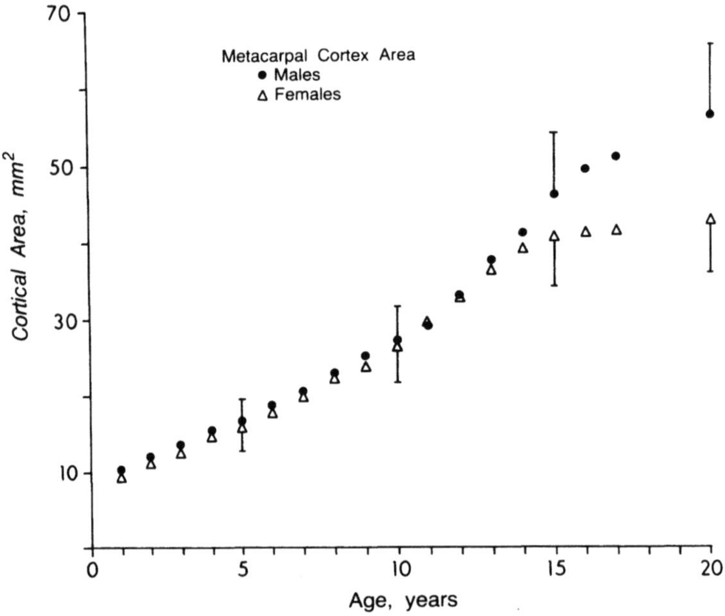

FIGURE 4.11. Cross-sectional area of the second metacarpal bone for U.S.A. white boys (n = 2983) (●) and girls (n = 3136) (△) aged 1 to 20 years. Means for age, with standard deviations for selected age groups. Data from the 10-state nutrition survey 1971–74 (Garn et al. 1976).

adolescence as depicted by the cross-sectional area of the second metacarpal cortex. The measurements were made on several thousand individuals who participated in the 1971–74 national nutrition survey. The standard deviations indicate the degree of variability encountered.

The data shown in Figure 4.11 pertain to whites. At almost all ages shown the values for blacks exceed those for whites, and those for Spanish-Americans are less than those for whites. For example, at age 17 years values for black boys are about 3% greater than whites, and those for Spanish-American boys are about 4% less. Roughly similar differences obtain for 17-year-old girls.

Virtama and Helelä (1969) did measurements on roentgenograms taken at various sites in Finnish children and adolescents. Their age curves for the second metacarpal cortex are very similar to those shown in Figure 4.11, the actual values being close to those for American blacks.

Gryfe et al. (1971) presented their data on British children in somewhat different form, the formula being $OD^2 - ID^2/ODL$, where L is the length of the second metacarpal bone. According to the authors the product in the denominator offers a correction for total metacarpal size. The problem with this construction is that the cortex of this bone is far from being a regular cylinder. Fortunately, the authors also present values for the numerator alone, which are in constant proportion to those obtained by the usual formula [$\pi/4$ (OD^2 − ID^2)]. The male/female ratio is about 1.25:1 at age 18 years.

Christiansen et al. (1975a,b) have published their photon absorptiometric measurements of the distal radius-ulna in Danish children, and Ringe et al. (1977) have made similar measurements in German children. All show the same general trends illustrated in Figure 4.11; namely an adolescent spurt together with a definite sex difference developing during adolescence.

All of these data constitute somewhat of a puzzle for the nutritionist. Why, for example, do blacks have thicker metacarpals than whites despite the widespread occurrence of intestinal lactase deficiency in this race? The Finns are thought to be big milk drinkers and yet their metacarpals are no thicker than American blacks'.

Growth of the LBM

A sufficient number of children and adolescents have had body composition assays to permit the construction of a growth curve for the LBM, defined here as body weight minus ether-extractable fat. ^{40}K assays have been made on several thousand normal children and adolescents (and some infants too) both in the U.S.A. and abroad; large numbers of adolescents and some older children have had body density measurements, and a few score have had measurements of total body water and ECF volume.

Fomon et al. (1982) have compiled normative data on body composition from several sources for infants and children up to age 10 years. Figure 4.12 shows average values for LBM and body fat in fetus and infant. Note that the abscissa scale is in post-conception years. The brief pause in growth during the first few days of life has been neglected. Also, the scale of the graph is such as to obscure the slight deceleration in LBM growth during the last few weeks of fetal life. The rapid increase in LBM during the late fetal period continues for a time after birth; then there is a deceleration to be followed by a fairly steady but slower rate of growth during the childhood years. The available data show a persistent sex difference in LBM from birth throughout childhood, boys on average having a slightly higher LBM. Sex differences in body fat content will be discussed in the next section of this chapter; for the moment it should be mentioned that although the graph shows that boys have a slightly greater total amount of body fat, values for percent body fat are lower.

The continued growth in LBM during the first 10 years of life is shown in Figure 4.13. Included are values for total body weight, and for completeness an insert showing percent body fat. Note the rapid increase in percent body fat during the first year of life. Boys weigh slightly more than girls until age 8 years, whereupon the girls take the lead, a situation soon to be reversed once adolescence is well under way. However, they still lag behind in LBM, their weight advantage being due to an increase in body fat content.

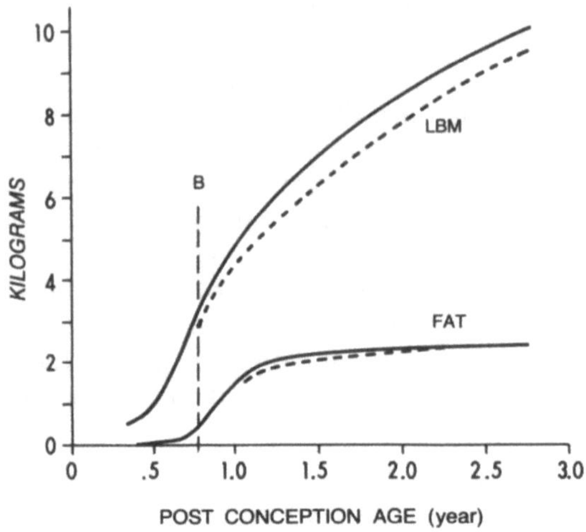

FIGURE 4.12. Average values for LBM and fat in fetus and infant. Age in post-conception years (for example age 3.0 years corresponds to 2.25 postnatal years). (———), boys, (— — —), girls. Graph drawn from data compiled by Ziegler et al. (1976) and Fomon et al. (1982).

The trend during these years is toward a "maturation" of LBM composition, a process that is brought to completion during adolescence. Looked at in this manner, the ratio of the percent of LBM occupied by water in the newborn to that in the adult is 1.10; the ratio for protein is 0.73, for K 0.74, and for density it is 0.96. By age 10 years these ratios

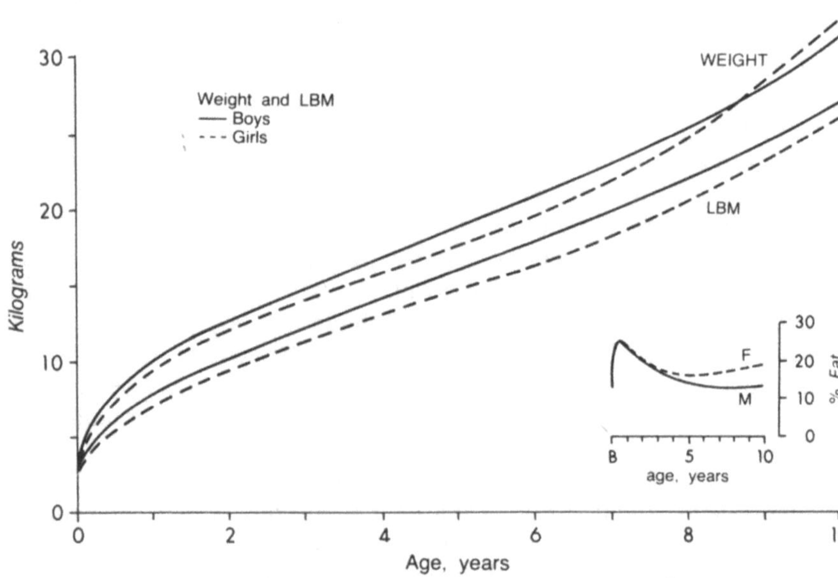

FIGURE 4.13. Graph of average body weight and LBM for boys and girls from birth to age 10 years. Inset shows percent body fat. Based on data published by Fomon et al. (1982).

have changed: that for water is now 1.04, for protein 0.93, for K 0.99, and for density 0.98. These changes reflect the decrease in the ECF/total water ratio, and the increasing contribution of muscle and bone mass to the LBM during the course of infancy and childhood.

Although there is a sex difference in LBM throughout infancy and childhood, it is not of great magnitude. It is the adolescent period that sees the development of a significant sex difference, which once established persists throughout the adult years. Between his 10th and 20th birthdays the average boy's LBM is augmented to the extent of 33 kg; the increment in girls is only half as much, namely 16 kg.

Figure 4.14 shows the author's data, which were obtained by ^{40}K counting of white suburban school children and University personnel. The hatched areas enclose the 95% confidence limits for our subjects. The rapid increase in the male LBM is clearly evident. By age 15 years the male/female ratio is about 1.23:1, and by age 20 years it has increased to 1.45:1. Not only is the adolescent spurt in LBM more intense in the male, but it also lasts longer. Maximum LBM is achieved by the girl at about 18 years, but in the boy not until age 20 or so.

It should be emphasized that the male/female ratio for LBM, namely

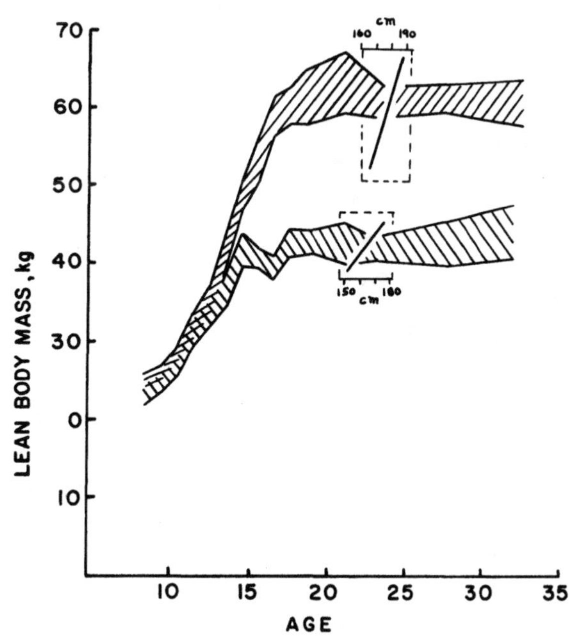

FIGURE 4.14. Ninety-five percent confidence limits for LBM. Inserts show calculated regression lines for LBM against height in subjects 22–25 years old: slope for males is much steeper (0.69 kg/cm) than for females (0.29 kg/cm). [From Forbes (1978), with permission.]

1.45:1 at age 20 years, is above that for body weight (1.25:1) and considerably above that for stature (1.08:1). Hence sexual dimorphism is more pronounced for LBM than for the other two commonly used measures of body size. This pronounced sex difference accounts in large part for the sex difference in nutritional requirements and in athletic performance.

The insets in Figure 4.14 show the calculated regression lines for LBM against stature in subjects 22–25 years old. The slope for males (0.69 kg/cm) is much steeper than that for females (0.29 kg/cm). This points up the fact that LBM is indeed related to stature, and this is true at all ages thus far examined (Forbes, 1974). Hence taller children and adults have a distinct advantage in LBM over their shorter age peers, and undoubtedly in strength since muscle constitutes about one-half of the LBM.

Indeed, Parizkova (1977) has shown in longitudinal studies that peak LBM velocity coincides with peak height velocity in adolescent boys.

The relationship of LBM to stature is shown in Figure 4.15. The progression in this ratio and the development of a distinct sex differential during adolescence run parallel to the urine Cr/height ratio shown earlier in Figure 4.9. Per unit stature late adolescent and adult men have about 30% more LBM than women.

This figure also includes values for percent body fat as a function of age. The fat "spurt" during early male adolescence is clearly shown, and will be discussed in a subsequent section of this chapter.

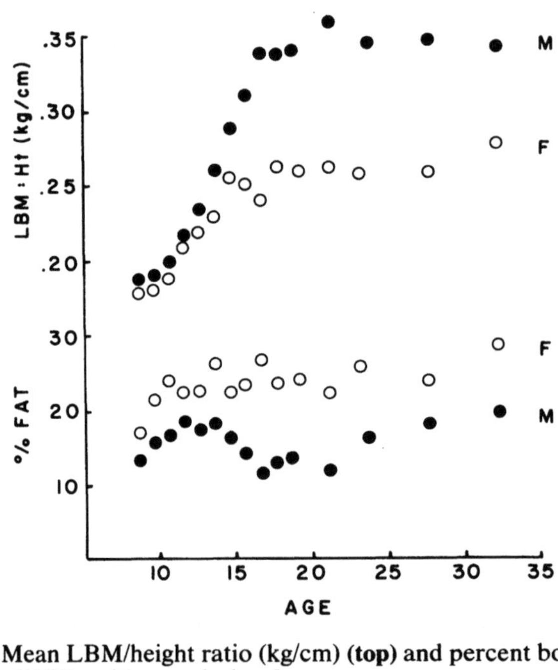

FIGURE 4.15. Mean LBM/height ratio (kg/cm) **(top)** and percent body fat **(bottom)**. [From Forbes (1978), with permission.]

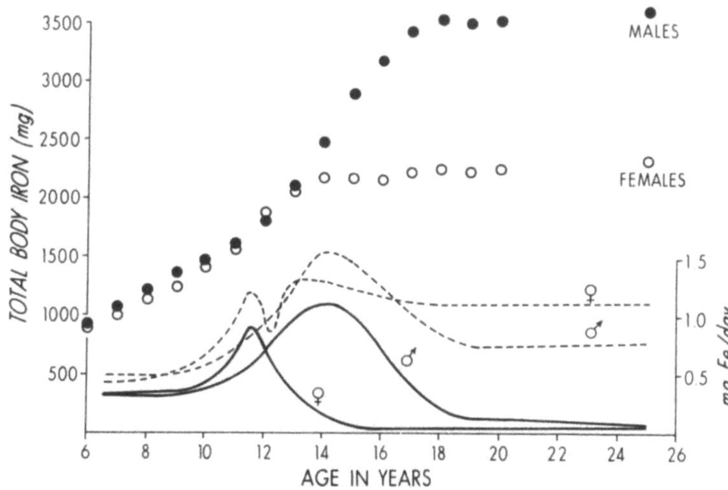

FIGURE 4.16. Total body iron (mg) during childhood and adolescence. (●), boys; (○), girls, scale at left. Lower section (scale at right), mg Fe/day: solid lines show calculated iron accretion rate, dotted lines show amount of Fe absorbed in order to satisfy this requirement. Graph prepared from data compiled by Hawkins (1964). [From Forbes (1981), with permission.]

Total Body Iron Content

Changes in total body iron during growth can be estimated from data on blood volume and hemoglobin content (hemoglobin contains 0.34% Fe) together with an estimate of storage and nonstorage tissue iron (most of the latter is in myoglobin). Hawkins (1964) has compiled such data from several sources, and the graph in Figure 4.16 is based on his published data.

The curves for estimated total body iron are similar in shape to those for LBM (Figure 4.14), and clearly show the development of a sex difference during adolescence. The male/female ratio is 1.32:1 at age 15 years, and 1.57:1 at age 20. This sex difference is largely due to the greater muscle mass of males, and their greater blood hemoglobin concentration.

The lower portion of the figure shows the calculated velocity, that is the increment in total body iron per unit time. As is the case for LBM, the adolescent spurt in body iron is more intense and of longer duration in boys. At the peak of the adolescent spurt the boy is accumulating about 1.2 mg Fe per day, and the girl about 0.8 mg per day.

The dotted lines in the lower portion of the figure represent an attempt to calculate the amount of iron that must be absorbed from the gastrointestinal tract each day in order to satisfy these growth increments. The difference between the dotted and solid lines includes estimates of Fe losses in urine and through the skin, and an average loss in menses. It is known that menstrual losses vary, but for purposes of illustration a value of 0.6 mg Fe per day (loss per menstrual period averaged over the entire month) was chosen as a reasonable approximation.

Assuming such estimates to be correct, it is of nutritional significance that the daily requirement for absorbed dietary iron in the boy actually exceeds that in the menstruating girl during the mid-teen years. The reason is of course the more rapid increase in LBM in the boy, an increase of such magnitude that his need for iron for growth is greater than that for the menstruating, slower growing girl.

Growth of Body Fat

The full-term infant is born with a store of body fat (10–15% of body weight) which is greater, relatively speaking, than that of other mammals.[9]

Mettau (1977, 1978) estimated body fat content by xenon absorption in a number of newborn infants. Those who were appropriate size for gestational age and born at 30–35 weeks had 80–180 g of fat, and those who were small at term had about the same amount. In 22 infants the amount of fat was correlated with body weight ($r = 0.82$), the calculated regression equation being: fat (g) $= 0.15W$ (g) $- 133$.

Taylor et al. (1985) estimated body volume in infants in a new instrument that operates on the principle of Boyle's law. This instrument consists of two chambers. The subject is placed in one and an inanimate object of roughly the same volume is placed in the other; both are subjected to the same change in air volume, whence the induced pressure change is read. There are a number of technical problems. These authors have estimated body density in this manner, the results being a fairly wide range of density values (1.041–1.089 g/cm^3) in a series of low birth weight infants. Derived values for percent body fat will depend on values chosen for the density of the LBM at this age.

Fat accumulation begins promptly after birth, and calculations by Fomon et al. (1982) show a remarkable increase during the first year of life; by the end of this year it is estimated that the infant body is 20–25% fat (Figure 4.13). Thereafter there is a decline in percent fat, to a low level in the mid-childhood years, and then an increase in the adolescent girl.

One way to depict this progress is shown in Figure 4.17, which illustrates the time course of triceps skinfold thickness during the first 20 years of life. These measurements reflect the changes in body fat content that occur during infancy, childhood, and adolescence, and nicely illustrate the sex

[9]An exception is certain species of arctic mammals, where newborns have a large store of brown fat; and it is this generous amount that supplies the energy needed for homeothermy, and hence survival, in subzero weather until an adequate milk intake is established.

In human newborn infants who died shortly after birth Heim (1982) found that brown fat stores were much more abundant in those who were kept in a neutral thermal environment than in those kept at lower temperatures.

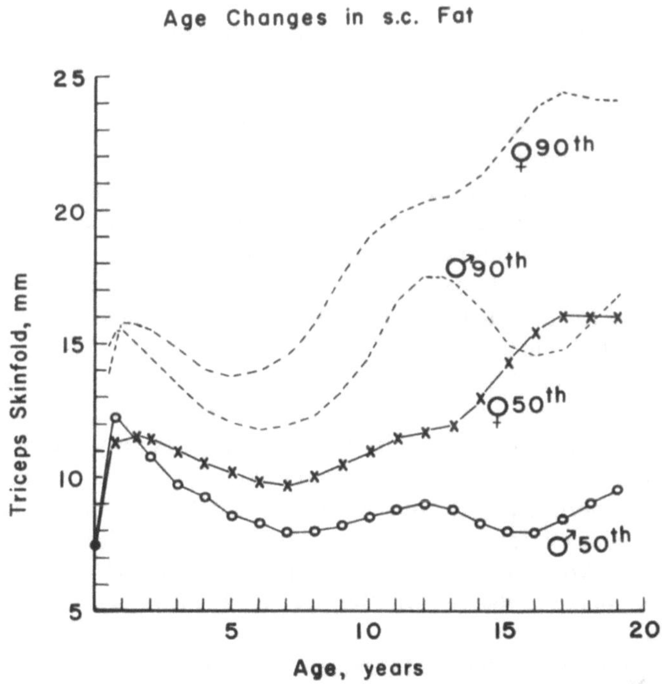

FIGURE 4.17. Skinfold thickness at the mid-triceps region for British children: 50th and 90th percentiles for boys and girls. [Redrawn from Tanner and Whitehouse (1975).]

difference in this aspect of body composition. Also evident is the considerable variability in skinfold thickness during most of these years, and the tendency for skewness in the data, the 90th percentile being far above the 50th percentile.

The data derived from the national nutrition surveys conducted in the U.S.A. (Frisancho, 1981; Cronk and Roche, 1982) are roughly the same as those depicted in Figure 4.17. The latter authors provide data for blacks as well as whites, and it should be noted that the 50th percentile values for black males are less than those for white males at all ages between 6 and 50 years. This is true for both the triceps and subscapular regions. Black females have smaller triceps skinfolds than whites for ages 6–25 years, but thereafter the values for blacks are higher.

Cronk and Roche (1982) also list values for subscapular skinfold thickness. In both races the 50th percentile subscapular measurement in males is less than the triceps until about age 18 years; thereafter the subscapular skinfold is thicker, at least until age 50 years. Data for subscapular skinfold are not given for females.

Another way of looking at the size of the subcutaneous fat depot is to calculate the cross-sectional areas of subcutaneous fat for the mid-portion of the upper arm from measurements of arm circumference and triceps skinfold thickness.[10] Frisancho (1981), who compiled these data from the national nutrition survey, makes the point that such areas are more indicative of the subcutaneous fat mass than is triceps skinfold thickness itself, the reason being that arm fat areas are proportional to arm circumference as well as to skinfold thickness. Hence for a given skinfold thickness the person with the larger arm will have a larger arm fat area. The result is that the sex difference in arm fat areas is not as great as the difference in skinfold thickness shown in Figure 4.17. Indeed, the 50th percentile values for the male fat areas during early adolescence—at the height of their "fat spurt"—are not too far below those of their female age peers, whereas the difference in skinfold thickness is rather large.

There is a modest increase in triceps skinfold thickness and in arm fat area during the early "teen" years in the boy; this lasts only a short time, for by late adolescence these measurements have decreased a great deal, indeed to levels approaching those seen in late childhood. This short-lived "fat spurt" is also evident in measurements of total body fat in Czech adolescents (Parizkova, 1977) and in British adolescents (Figure 4.17) as well as in those reared in the U.S.A. (Novak, 1973, Figure 4.15, and Figure 4.18 to follow).

A publication from the National Center for Health Statistics (1974) contains an interesting graph relating the half-yearly increments in triceps skinfold (SF) thickness to height velocity for children and adolescents (Figure 4.18). For boys velocity for triceps skinfold is positive in the earlier years, with a peak at about age 10; then velocity declines and actually becomes negative, with a nadir that nicely coincides with peak height velocity at about age 13½ years. The velocity curve for girls is quite similar in shape, the entire curve being shifted upwards, so that the nadir of the velocity dip just touches zero; but here, too, the nadir coincides in time with peak height velocity at about age 12 years. The later years of adolescence witness a positive triceps skinfold velocity once again. Tanner's (1974) published curves for radiographically determined subcutaneous fat thicknesses at various sites are similarly related to peak height velocity.

This interesting coincidence, in which it appears that triceps skinfold and height velocity curves resemble mirror images of each other in *both sexes*, raises the possibility that a single hormonal influence is responsible for both phenomena.

It is known that menarche usually occurs shortly after the peak in height

---

[10]Cross-sectional arm fat area = total arm area minus arm muscle + bone area, where total area is (arm circumference)$^2/4\pi$ and arm M + B area is $1/4\pi$ (circumference $- \pi$ SF)^2. Hence arm fat area is SF/4 (2 circumference $- \pi$ SF).

It is unfortunate that most observers have used only the triceps skinfold thickness for this purpose. As mentioned in Chapter 2, an average of biceps and triceps skinfolds would be more appropriate since these two measurements usually differ.

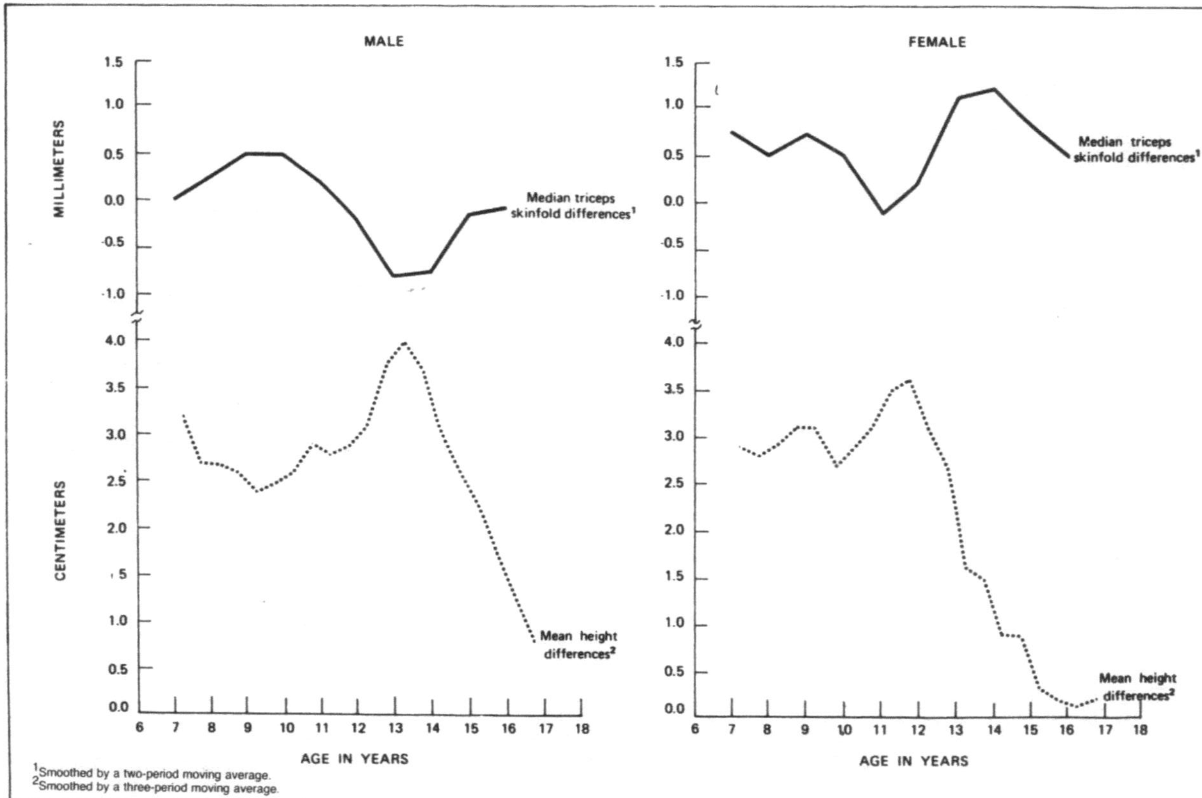

FIGURE 4.18. Velocity curves for triceps skinfold thickness and for stature, based on data from U.S.A. children and adolescents. [Reproduced from National Center for Health Statistics (1974).]

velocity, and it would appear that as height velocity decelerates there is an acceleration in the rate of change of triceps skinfold thickness, perhaps due to increased estrogen production.

Knittle et al. (1979) made estimates of adipocyte size and number in infants and children, the adipose tissue biopsy being done at the gluteal area. Figures 4.19 and 4.20 show the results for obese subjects and for those considered to be normal. Adipocyte numbers increase about 10-fold during the period 6 months to 18 years in normal infants and children, while the obese subjects have about twice as many adipocytes at all ages. The pattern for adipocyte size roughly resembles the curves shown in Figure 4.17: there is a rapid rise during infancy, then a fall in mid-childhood, to be followed by a rise during adolescence. Of considerable interest is the observation that the infantile upsurge in body fat content appears to be due to an increase in cell size rather than cell number. As is the case for cell number, the obese children had on average larger adipocytes in the gluteal area at all ages examined by these authors. The length of the vertical bars, which represent standard errors of the mean, attest to the variability in both fat cell size and number in the population. While these data represent the combined sexes they serve to illustrate the age trend in adipose tissue.

The Frisch Hypothesis

The human mammal has a much longer period of infancy and childhood than other species, and students of growth and endocrine function have long wondered about the nature of the timing mechanism that finally brings

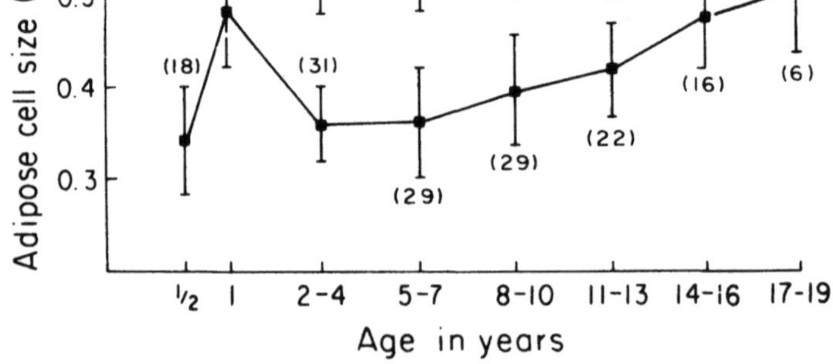

FIGURE 4.19. Adipose cell size as a function of age, combined sexes. (○), obese; (■), nonobese. Means ±SEM. [From Knittle et al. (1979), with permission.]

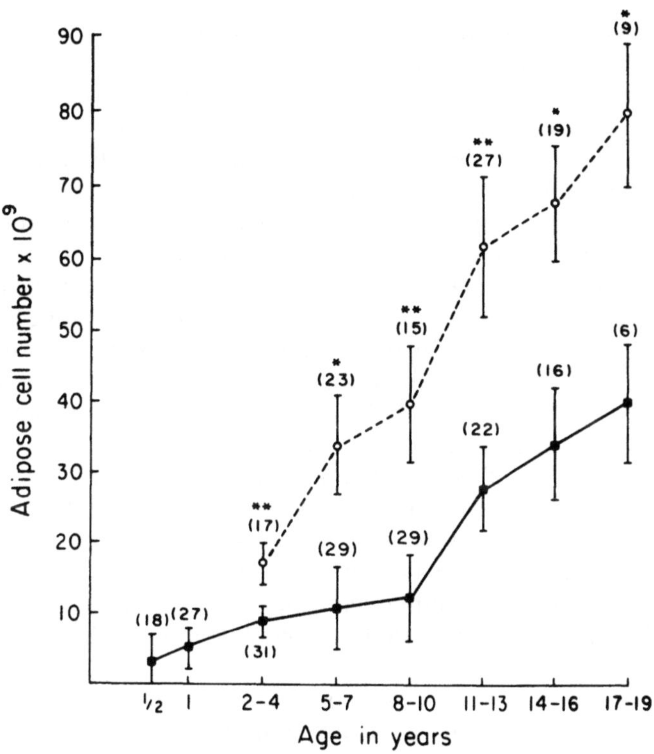

FIGURE 4.20. Adipose cell number as a function of age, combined sexes. (○), obese; (■), nonobese. Means ±SEM [From Knittle et al. (1979), with permission.]

about sexual maturity a full decade after birth. In a series of publications Rose Frisch and her associates (Frisch and Revelle, 1970; Frisch, 1974) put forth the hypothesis that menarche is somehow "triggered" by the attainment of a certain body weight (48 kg) and by the acquisition of a certain amount of body fat (17% of body weight), and that continued menstruation requires a body fat content of 22%. This hypothesis generated widespread interest in the possible role of body fat as a determinant of mature sexual function in the female. It was pointed out that some conversion of androgens to estrogens takes place in adipocytes, and indeed excessively obese males tend to have a reduced serum testosterone and an increased serum estradiol compared to nonobese males and an increased conversion of androstenedione to estrone (Kley et al., 1980). Ballet dancers, who are inclined to be thin, were said to have a high incidence of oligomenorrhea, and of course girls with anorexia nervosa are usually amenorrheic. Girls who menstruate early tend to weigh more than their premenarchal age peers, and those who menstruate late tend to weigh less than their postmenarchal age peers. Indeed, the extensive data compiled by Zacharias et al. (1976) show that girls who mature early are shorter on average than late maturers, while body weight at menarche tends to be independent of menarchal age. Obese girls tend to have an earlier menarche than their nonobese peers.

All of these considerations arose at about the same time that investigators were embarked on studies of adipocyte size and number in obese individuals (cell size is determined on adipose tissue biopsies, and cell number from estimates of total body fat). The obese state was characterized either as "hyperplastic" or "hypertrophic," and the former type was considered to be more resistant to treatment, as well as a factor in recidivism of the obese state. The adipocyte was thus imbued with properties, such as a "trigger" for menarche and a prognosticator of treatment success, both of which are far removed from its known functions. Never mind that girls who develop anorexia nervosa usually stop menstruating *before* they lose weight, that ballet students are apt to resume regular menses on vacation from ballet school (Abraham et al., 1982), that normal infants acquire a lot of body fat (up to 25%) long before puberty (see Figure 4.13); adipose tissue possesses functions—so the reasoning must have been—that transcend those of energy storage and insulation. Could it be that white adipose cells—which incidentally do not contain nerve endings—find a way to communicate in some mysterious manner with the hypothalamus so as to modulate gonadotropin production and even appetite control?

What was generally overlooked is that Frisch did not perform body composition assays or even measure skinfold thickness but rather *calculated* body fat from weight and height by means of the Mellits-Cheek equation. The origin of this equation and its shortcomings have already been mentioned (see Chapter 2, p. 87–90), but further discussion is appropriate now. From this equation one calculates total body water from height and weight; division by 0.73 yields estimated LBM, whence fat = W − LBM.[11]

Frisch emphasized her finding that *calculated* lean weight, and *calculated* LBM/weight ratio showed less variability at menarche than body weight *per se,* and so contended that it was the acquisition of a certain body composition that acted as the "trigger" for menarche. What she failed to appreciate was the fact that the measured LBM/W ratio shows less variability at almost all ages among girls and women than does body weight (G.B. Forbes, unpublished data). Hence this lesser variability is not unique to the perimenarchal period.

---

[11] For example, in the case of girls over 111 cm in height, the equation is:

$$\text{Total } H_2O(\text{liters}) = 0.252W + 0.154H - 10.3$$

hence

$$\text{LBM} = H_2O/0.73 = 0.35W + 0.21H - 14.1$$

and subtracting LBM from weight,

$$\text{Fat} = 0.65\,W - 0.21H + 14.1$$

The Frisch hypothesis has not been immune to criticism. Scott and Johnston (1982) pointed out that recorded body weight at menarche varies from 28 to 97 kg; indeed a glance at one of Frisch's published graphs (Frisch et al., 1973) shows a wide range of calculated total H_2O/W ratios (i.e., LBM/*W* ratios) at menarche. The imprecision of the Mellits-Cheek equation for estimating body composition was documented in Chapter 3 (Table 3.3), and recently Johnston (1985) has shown that it predicted values for body fat of 25–28% in girls and young women whose measured body fat (by densitometry) ranged from 12 to 27%, so that the equation overestimated body fat in thin women. His data actually show a good correlation ($r = 0.92$) between prediction error and percent body fat. Hence girls and women of a given height and weight do not necessarily have the same body composition, an observation made (for males) by Albert Behnke several decades ago.

In their analysis of data from many thousands of women Garn and La Velle (1983) point out that pregnancy actually can occur in women whose body weight is below the "critical value" for menstruation of 48 kg. Carlberg et al. (1983) studied two groups of female athletes; one group menstruated regularly, while the second had fewer than five menstrual periods per year. Exercise patterns were the same in both groups. Average body fat content (by densitometry) was 16% in the first group (range 8–27%) and 13% in the second group (range 7–18%). Hence neither group met the criterion of a "critical" level of body fat. C. M. Young et al. (1968) measured body density and total body water in a group of 104 girls aged 9–16 years. Body density was about the same in the premenarchal girls as in those who had menstruated. Percent body fat, as estimated by tritium dilution, did not show a definite rise until after menarche.

Relationship of Hormone Status to Body Composition

Serum levels of various hormones have been measured in children and adolescents. The years immediately preceding adolescence include an increase in the levels of certain adrenal hormones, namely dehydroepiandrosterone (DHEA), its sulfate, androstenedione, and androsterone sulfate, and a rise in somatomedin levels. At this stage the levels are about the same in boys and girls. It is conceivable that these hormones, which are known to have anabolic properties, could be responsible for the modest rise in LBM that occurs in late childhood a year or two prior to the adolescent growth spurt.

The fact that serum testosterone levels (both the free and bound hormone) rise during adolescence in temporal concert with the spurt in LBM suggests that the two phenomena are related. Testosterone has strong anabolic properties: it produces positive nitrogen balance when administered to men and women (Kochakian, 1976), and one can easily discern the increase in muscle mass that occurs when hypogonadal boys are given this hormone. Large doses produce striking increases in LBM, and a decline in body fat content, in adult males (Forbes, 1985).

Serum testosterone levels rise more rapidly, and reach higher values, in boys, and it is reasonable to conclude that their rapid increase in LBM, together with a fall in percent body fat later on, are largely the result of the action of this hormone.

Serum estradiol levels rise during female adolescence (there is also a very modest increase in boys), and this rise is temporally related to breast enlargement, widening of the hips, and the increase in body fat content. The effect on body fat is easily discernible in agonadal girls (44–XO karyotype) who are treated with estrogens. Although estrogens undoubtedly have an effect on skeletal mass, as evidenced by the occurrence of postmenopausal osteoporosis, it is not at all certain that they have a significant effect on the LBM as a whole. They do of course affect uterine size. In acute experiments on castrated female rats, Cole (1952) found that estradiol injections caused an increase in ECF volume in both muscle and liver, but a decrease in ICF volume. While there are no extant data on body composition, castrated girls tend to grow tall, and girls who have a late menarche tend to be taller than their postmenarchal age peers.

Both Winter (1978) and Sizonenko (1978) have published data on changes in hormone levels during childhood and adolescence.

Krabbe et al. (1984a,b) have done sequential measurements of serum testosterone, DHEA, androstenedione, and alkaline phosphatase, and bone mineral content (BMC) (by photon absorption) of the forearm bones in adolescent boys. They found a correlation (r = 0.73) between age of maximum serum testosterone and age of maximum BMC and maximum alkaline phosphatase. The mean difference in age between maximum testosterone and maximum BMC was 4.7 months. The adrenal androgens did not appear to have a major influence on bone mineralization during the male pubescent spurt.

This same group of investigators (Krabbe et al., 1978) also reported a good correlation (r = 0.73) between blood hemoglobin and serum testosterone in adolescent boys. The rise in testosterone coincided in time with the rise in hemoglobin, and as would have been expected the changes in adolescent girls were minor in degree. It is entirely likely, therefore, that the sex difference in blood volume that develops during adolescence is related to testosterone production. In this connection, it should be noted that testosterone, and its congeners, have been used in the treatment of patients with hypoplastic and aplastic anemia.

Further studies by this same group (Krabbe et al., 1984b) showed that in the 18-month period preceding the age of peak height velocity in boys serum testosterone levels rose 5-fold, while the increase in DHEA was only 30%, and the increase in bioassayable somatomedin was about 40%.

Nocturnal surges in plasma growth hormone levels become more pronounced during adolescence in both sexes, and the 24-h integrated levels are higher (Finkelstein et al., 1972).

The adolescent growth spurt thus represents the combined action of several hormones, both pituitary and gonadal, the latter being dependent on the secretion of follicle stimulating and luteinizing hormones by the former. The sex difference in the magnitude of the LBM spurt is in all likelihood the result of the sex difference in testosterone production. Studies of adults given large doses of testosterone support such a conclusion (Forbes, 1985). The total dose of testosterone needed to produce an increment in LBM equal to the normal male–female LBM difference at age 20 years is about the same as the sex difference in testosterone production during the adolescent years (see Chapter 9).

Nutritional Requirements

Data on LBM growth provide a basis for estimating the requirements of nitrogen and minerals for sustaining normal growth. The sex difference in LBM that develops during adolescence is an indication that the growth requirements during those years will also differ. Table 4.6 provides some estimates of the increments of body Ca, Fe, N, Zn, and Mg for the second decade of life, together with those at the peak of the growth spurt. Those for calcium are based on measurements of metacarpal cortex thickness and photon absorption of the radius and ulna, and those for iron on the data shown in Figure 4.16. For the others it was assumed that they paralleled the growth of the LBM. During this decade the average boy will accumulate 766 g Ca, 2.08 g Fe, 1168 g N, 0.98 g Zn, and 16.1 g Mg. For the average girl these are much less, namely 402 g Ca, 0.84 g Fe, 584 g N, 0.66 g Zn, and 8.4 g Mg. As indicated earlier, some additional accumulation of calcium will occur during the third decade of life, but this is minor compared to the amounts gained during the second decade. The

TABLE 4.6. Daily increments in body content due to growth.

		Average for age period 10–20 years (mg)	At peak of growth spurt (mg)	Maintenance need[a] (mg)
Calcium	M	210	400	800
	F	110	240	800
Iron	M	0.57	1.1	10
	F	0.23	0.9	18
Nitrogen[b]	M	320	610	8960
	F	160	360	7360
Zinc	M	0.27	0.50	15
	F	0.18	0.31	15
Magnesium	M	4.4	8.4	350
	F	2.3	5.0	300

[a]Adult allowances.
[b]Multiplying by 0.00625 equals grams of protein.

table is constructed so as to indicate daily increments in nitrogen and minerals.

The sex difference in growth needs for these elements is striking indeed, and this is due to the more rapid accumulation of lean tissue by the adolescent boy. To these must be added the maintenance needs, and the problem here is that there are no data available for adolescents. If we assume for the moment that these are similar to those for the adult as defined by the National Research Council (1980), it turns out the growth increments for Fe, N, Zn, and Mg constitute only a small fraction of the total requirement. Growing adolescents don't require very much more of these elements than the adult, even at the peak of their growth spurt. However, this conclusion must be tempered somewhat because the adult allowances as listed in Table 4.6 probably exceed true minimum requirements.

The reason for the rather small growth requirement is that the relative growth rate of the human adolescent is a very modest one. At the peak of the growth spurt the average gain in weight is only 0.035% per day, and the gain in height is only 0.015% per day. Even a fast-growing adolescent would not be expected to exceed the values shown in the table by more than a factor of 1.5.

The situation for calcium is different, for here the growth increment for the adolescent boy at the peak of his growth spurt is fully one-half of the adult daily requirement, while that for the girl is about one-third. It is difficult to translate these into dietary calcium needs because there are no data on gastrointestinal absorption of calcium for adolescents. If by any chance adolescents were more efficient at absorbing calcium than adults, in whom the average is about 40% of the amount ingested, the additional dietary requirement would be less. The timing of the growth spurt varies somewhat among adolescents, and it is for this reason that the average daily increments in element content for the entire second decade are listed in the table.

5

The Adult

> The first forty years of life give us the text:
> the next thirty supply the commentary.
>
> *A. Schopenhauer*

Reference Man

In an attempt to characterize the "standard" or "reference" man a task force has drawn up a list of the quantities of various elements to be found in the adult human body. These data are derived from average organ size and the results of chemical analysis of organs by various investigators. They represent the best estimates of total body content based on an extensive review of the literature.

Table 5.1 lists the multitude of elements to be found in the body of a male adult, as determined by the International Committee on Radiation Protection (1975). Since on average the lean body mass (LBM) of the adult woman is about two-thirds that of the average man, the body content of the elements listed will be, with the exception of O_2, C, and H_2 which are present in fat, two-thirds to three-fourths of the amounts listed. Hence I have added some values for certain elements for the average woman as determined by body composition assays. Many elements are present in very small amounts, and for some the quantities are dependent on diet or occupational exposure. The values listed are averages. The report of the International Committee also gives the 80% range for many of the elements, and for some this range for individual organs turns out to be two- to threefold. This reflects sampling errors together with technical error, and to some extent the state of health of the donor at the time of death. For example, one would anticipate variations in fluoride content based on source of drinking water, and to a considerable extent on age, for fluoride continues to be deposited in bone throughout life. There are wide variations in lead exposure. Total body iron and zinc can be altered by disease. Hence one must view the values for trace elements in the table as averages only. However, the variations in body content for the major elements, i.e., those present in quantities of 100 g or more, is relatively small for individuals of a given body size, keeping in mind the fact that the amount of, say, potassium will depend on that size. Since adipose tissue is largely composed of neutral fat, major element content is better correlated with lean weight than with total weight.

TABLE 5.1. Reference man: total body content of some elements.

Element	Amount (g)		Adult woman Amount (g)	
Oxygen	43,000	(1340 mol)		
Carbon	16,000	(1333 mol)		
Hydrogen	7000	(3500 mol)		
Nitrogen	1800	(64 mol)		
Calcium	1100[a]	(27 mol)	830[a]	(21 mol)
Phosphorus	500[a]	(16 mol)	400[a]	(13 mol)
Sulfur	140	(4370 mmol)		
Potassium	140	(3600 mmol)	100	(2560 mmol)
Sodium	100	(4170 mmol)	77	(3200 mmol)
Chlorine	95	(2680 mmol)	70	(2000 mmol)
Magnesium	19	(780 mmol)		
Silicon	18	(641 mmol)		
Iron	4.2	(75 mmol)		
Fluorine	2.6	(137 mmol)		
Zinc	2.3	(35 mmol)		
Copper	0.07	(1.1 mmol)		
Manganese	0.01	(180 μmol)		
Iodine	0.01	(79 μmol)		
Rubidium	0.32	(3.7 mmol)		
Strontium[b]	0.32	(3.6 mmol)		
Bromine	0.2	(2.5 mmol)		
Lead	0.12	(0.58 mmol)		
Aluminum	0.06	(2.2 mmol)		
Cadmium	0.05	(0.45 mmol)		
Boron	< 0.05	(< 5 mmol)		
Barium	0.02	(0.14 mmol)		
Tin	< 0.02	(<0.17 mmol)		
Nickel	0.01	(0.17 mmol)		
Gold	< 0.01	(< 50 μmol)		
Molybdenum	< 0.01	(< 100 μmol)		
Chromium	< 0.002	(< 38 μmol)		
Cesium[b]	0.0015	(11 μmol)		
Cobalt	0.0015	(25 μmol)		
Uranium	9×10^{-5}	(0.38 μmol)		
Beryllium	4×10^{-5}	(4.4 μmol)		
Radium	3×10^{-11}	(0.13 pmol)		

[a]These are newer data derived from neutron activation analysis of young adults (Cohn et al., 1976). The Ca/P ratio (g/g) is about 2:1. The values in the original table were 1000 g Ca and 780 g P for the adult male.
[b]Radioactive ^{90}Sr and ^{137}Cs are fallout products from nuclear explosions. Both are found in food and water, and both have a long physical half-life (28 years and 27 years, respectively). The current body burden of ^{137}Cs is about 2 nCi; although this level is considerably less than it was during the 1960s, it is expected to decline very slowly in future years. This isotope emits γ-rays, so the body content can be assayed in the whole body counter. Such measurements have been used during the past two decades to monitor fallout from the atomic bombs.

The adult man weighing 20 times the newborn has 36 times as much Ca, 30 times as much P, 27 times as much N, 25 times as much Mg, 13 times as much Fe, 43 times as much Zn, and 5 times as much Cu as the newborn infant.

A number of trace elements have also been found in newborn infants; these include Sr, F, Pb, Br, ^{137}Cs, ^{90}Sr, and Mn. The placenta is permeable to ions, so any element present in maternal blood and that is not completely bound to plasma proteins will find its way into the fetus.

Data for newborn infants and for one 4 ½-year-old boy are given in Chapter 4 (Table 4.3).

Mercury and arsenic both are present in human urine in milligram quantities, the amounts depending on the exposure. The trace elements Sn, Cr, F, Mo, and Ni are now considered to be essential for health, at least in animals, as are others omitted from the table, namely Se, Va, Si, and As.

Average organ weights for an adult man are listed in Table 5.2 compared to those for the newborn infant. The adult weighing 20 times the newborn has 33 times as much muscle and 23 times as much skeleton as the newborn. The ratios for skin, liver, and kidney are about 10-fold, for heart about 19-fold, whereas the newborn brain is proportionally larger, namely 30% of the adult weight.

The relative proportions of body compartments for the average adult man and woman are shown in Figure 5.1. This was constructed from a number of sources. It clearly shows the effects of sex and of age on the magnitude of LBM, fat, and skeletal size. Men have a larger LBM and less body fat than women, and older adults of both sexes have a smaller LBM and more body fat than young adults. Indeed in young adults the sex difference in LBM (male/female ratio 1.4) is relatively greater than the difference in stature (ratio 1.08) or in body weight (ratio 1.25). The ratio for urinary creatinine excretion (an index of muscle mass) is 1.5. This means that the sex difference in body weight is not a proper indicator of the sex differences in such parameters and functions as BMR, urinary creatinine excretion, blood volume, and energy requirements, to mention

TABLE 5.2. Reference man: gross organ size (g).

	Adult[a]	Newborn[b]
Weight	70,000	3400
Skeletal muscle	28,000	850
Adipose tissue	15,000	
(fat 12,000)		500
Skeleton	10,000	440
(cortical bone 4,000, trabecular bone 1,000, marrow 3,000, cartilage 1,100, periarticular tissue 900)		(cartilage 140)[c]
Skin	4900	510
Liver	1800	170
Brain	1400	440
Heart	330	17
Kidneys	310	34

[a]International Committee for Radiation Protection (1975).
[b]Widdowson and Dickerson (1964).
[c]Swanson and Iob (1940).

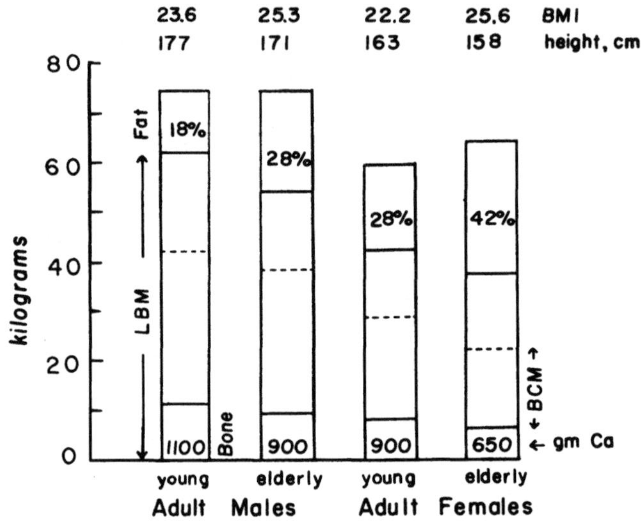

FIGURE 5.1. Body composition of average men and women: fat, LBM, skeletal weights, body cell mass (BCM), and total body calcium for individuals aged 20–25 years and 60–70 years, as compiled from various sources. Skeletal weights, including marrow, from Borisov and Marei (1974); total body Ca from Cohn et al. (1980); mean heights and weights from National Center for Health Statistics (1979). Body mass index (BMI) is weight/height^2.

but a few. Likewise, the decline in LBM with advancing age means that these functions will be similarly affected.

Figure 5.1 also includes estimates of body cell mass (BCM), i.e., the amount of body tissue calculated to contain potassium at a concentration of 120 meq/kg (see Chapter 2, p. 49). This moiety is preferred over LBM by some investigators, since it defines "that component of body composition containing the oxygen-containing, potassium-rich, glucose-oxidizing, work-performing tissue . . . and is the only mass of tissue in the body that contains all of the cellular elements concerned with respiration, physical and chemical work, and mitotic activity" (Moore et al., 1963, pp. 19, 25).

LBM Variation

The data in Figure 5.1 are designed to be a general guide to age and sex differences in body composition. They represent average values for white subjects in the U.S.A. Although such constructs serve to designate—as do the data for the "reference" man—the general features of body composition, they belie the fact that adults exhibit a great deal of individual variation. Just as normal adults vary in height and weight—and in such physiological functions as BMR and muscle strength—so too do they exhibit differences in lean weight and in body fat content.

However, at the outset it should be emphasized that the variability in LBM is much less than that of body fat. This is shown in Figure 5.2, which is based on the author's ^{40}K assays of a large number of women of widely varying body weights, but with a limited range of stature and a specified range of ages. The frequency distribution for LBM is seen to be much narrower than that for body weight, so variations in body fat must account for a large portion of the variation in weight. Burmeister and Bingert (1967) came to the same conclusion after their study of several thousand subjects of both sexes, stating that "the nongaussian distribution of weight (which is seen in most population studies) is thought to be due to the skewness in distribution of fat." The graph of triceps skinfold thickness shown in Chapter 4 (Figure 4.4) portrays the variability in subcutaneous fat thickness and the skewness in the data. There are a number of examples in the literature on body composition that corroborate the fact that body fat content varies more than lean weight among subjects of a given sex, and limited ranges of age and stature. Apparently the constraints, whatever they may be, that limit LBM variability do not operate on body fat. For subjects of a given age and sex stature is similarly constrained.

Having said this, it is nevertheless true that LBM does vary somewhat among normal individuals, and we now consider the sources of such variation. The principal sources of variation for normal individuals are age, sex, stature, and race (the situations for the athlete and the obese are

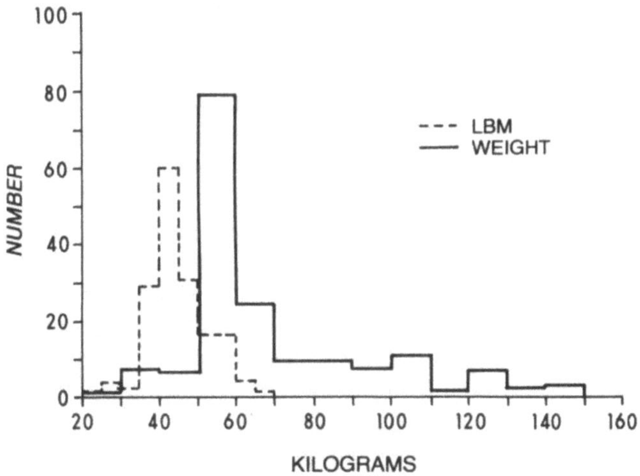

FIGURE 5.2. Frequency distributions of lean body mass and weight for 164 women and girls aged 14–50 years and 156–170 cm in height. Included are patients with anorexia nervosa and obesity, as well as normal subjects. (Author's data.)

described in subsequent chapters). In addition, there is growing evidence for the role of heredity as a factor in the size of the LBM.

Sex

The information in Figure 5.1 demonstrates in a general way the sex difference in LBM, fat, BCM, and total body calcium in adults. As was pointed out in Chapters 3 and 4, boys tend to have a larger LBM than girls (there is even a slight difference in the newborn); however, it is during adolescence that the sex difference becomes pronounced, and once established it persists throughout the adult years. This difference is of such a magnitude that males and females should be considered separately in any analysis of data on lean weight.

Stature

Lean weight is related to stature at all ages thus far examined, from the newborn to the elderly. This statement is based on an extensive review of the literature, and includes results obtained by several techniques for estimating body composition (Forbes, 1974). The correlation coefficients range from 0.21 to 0.95. Although stature accounts for only a portion of the variance in LBM for any given group, its influence is such as to make it necessary for stature to be taken into account in comparing body composition data among individuals or among groups of individuals. This is most important in attempting comparisons between athletes and nonathletes, and between obese and nonobese individuals. The same can be said for evaluating such physiologic functions as pulmonary ventilation, muscle strength, maximum oxygen uptake, and so forth. Although there are obvious exceptions—a short stocky individual may have a larger LBM than one who is tall and lanky—in the main the magnitude of the influence of stature on the size of the LBM is too great to be ignored. Failure to take this into account can lead to faulty conclusions in making comparisons of groups.

The slopes of the regressions of LBM on height gradually increase during childhood, from 0.15 kg/cm in the newborn to 0.46 kg/cm in the adolescent girl and to 0.62 kg/cm in the adolescent boy. During the early and mid adult years the slope is 0.69 kg/cm in males and 0.48 kg/cm in females; then it drops in old age to about 0.35 kg/cm. So in making comparisons between the obese and nonobese, between athletes and nonathletes, between the well and the sick, one must match the subjects by height as well as sex, and as will shortly be shown, also by age.

When the various groups of adults are considered together, it turns out that LBM is related to the third power of height, the equation being

$$\text{LBM (kg)} = 1.03 \times 10^{-5} (H)^3 \tag{1}$$

or, in logarithmic form,

$$\log \text{LBM} = -4.99 + 3.0 \log H \text{ (Forbes, 1974)}$$

These data pertain to presumably normal adults. Later it will be shown that the slope of this line is greater for many groups of athletes (Chapter 8).

Total Body Calcium

Total body calcium is also a function of stature. The data of Ellis and Cohn (1975) indicate that body Ca increases by 20.5 g for each centimeter increment in height. This is one reason why women have less total body Ca than men. Based on these data a short woman (155 cm tall) would be expected to have about 740 g Ca in her body, whereas a tall man (186 cm) would have about 1360 g Ca; thus the short woman has only 54% as much Ca in her body as the tall man.

Borisov and Marei (1974) dissected a number of adult cadavers, and their data on skeletal weight (minus periosteum and ligaments but with marrow intact) also show a relationship to stature. In men skeletal weight increased by 258 g for each cm in stature, and by 273 g/cm in women, the correlation coefficients being 0.77 and 0.72, respectively. The average female skeleton weighed 9.0 kg, the average male skeleton 10.4 kg. Combining the sexes, and plotting skeletal weight and stature on double-logarithmic coordinates generates a linear relationship, the equation being

$$\text{skeletal } W \text{ (kg)} = 0.0326\,(H)^{2.46} \quad r = 0.77 \qquad (2)$$

or, in logarithmic form,

$$\log \text{skeletal } W = -1.487 + 2.46 \log H$$

Total body calcium can also be related to the third power of height. McNeill and Harrison (1982) report an exponent of 3.10 for adults and one of 2.89 for children, and Nelp et al. (1972) one of 3.1 for adults.

It is of some interest to look at the ratio of total body Ca to total body K, since this is also affected by stature. The data of Ellis and Cohn (1975a) show that this ratio averages 8.2 (g/g) in adult men and 10.0 in women, in the age range between 30 and 49 years. A plot of individual data supplied by Stanton Cohn shows that this ratio declines with increasing height, the equation being: Ca/K $= -0.102\,H$ (cm) $+ 26$ ($r = -0.77$). Cadaver analyses conform to this equation as shown in Figure 5.3. Hence it appears that tall individuals have a somewhat greater ratio of muscle mass to bone mass than short individuals. These data pertain of course to presumably normal individuals. Tall people evidently require a bit more muscle to stabilize skeletal function than do short people.

A number of other functions are also related to stature when plots are made on double-logarithmic coordinates. These include vital capacity (exponent 2.8), muscle force (2.9), aerobic power (2.5), maximum oxygen

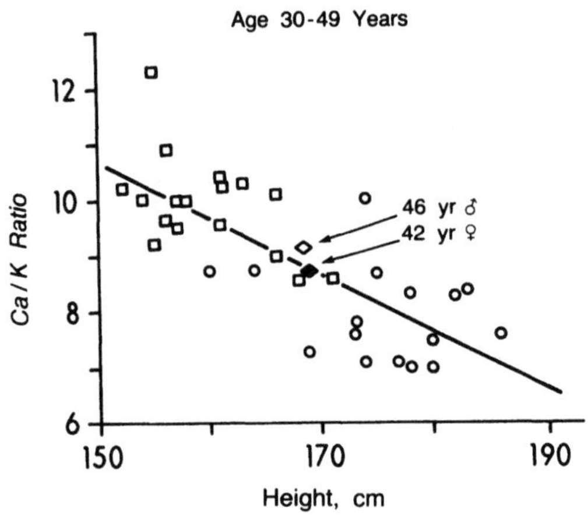

FIGURE 5.3. Plot of total body Ca/K ratio (g basis) against height for 30–49-year-old subjects assayed by Ellis and Cohn (1975; S. Cohn, personal communication). Symbols: (□) women; (○) men. Included are cadaver analysis data for a 46-year-old man (◇) and a 42-year-old woman (◆) (Forbes and Lewis, 1956; R.M. Forbes et al., 1953; Widdowson and Dickerson, 1964).

uptake (2.6 for boys, 2.8 for girls), and handgrip force (average 3.3) as reviewed by Shephard (1982) (pp 327–329). Blood volume and total body water scale to the third power of height (Allen et al., 1956; Nicholson and Zilva, 1964). All of these considerations attest to the importance of stature as a prime physiological parameter.

Race

Orientals are usually shorter and lighter than Caucasians, so it is to be expected that they would have a smaller lean weight. Figure 5.4 is a plot of LBM against stature for Orientals and Caucasians. All of the subjects were young adults; those judged obese were excluded. The general trend is clear: Oriental values for LBM are in keeping with their shorter stature. Novak (1970) also found that Filipino women (who are shorter than Caucasian) excreted less creatinine in their urine than Caucasian women, the average difference being about 12% for creatinine excretion and 9% for height.

North American blacks on the other hand tend to have a slightly larger LBM than whites, together with thicker and denser bones, and hence a larger amount of total body calcium. Throughout most of childhood and during all of the adult years blacks have larger cross-sectional areas of metacarpal cortex than whites, whereas Spanish-Americans have smaller cortical areas (Garn et al., 1976). In their studies of cadaver skeletons Trotter et al. (1960) found blacks' bones to be denser than whites'. Cohn et al. (1977) did extensive body composition assays on North American blacks and whites of both sexes. Throughout the adult years the average black/white ratios were as follows: stature 1.01, total body K 1.15–1.17, total body Ca 1.17–1.22, total body P 1.13–1.12, total body Na 1.09–1.08, and total body Cl 1.15–1.07 (the first figure in each couplet refers to men,

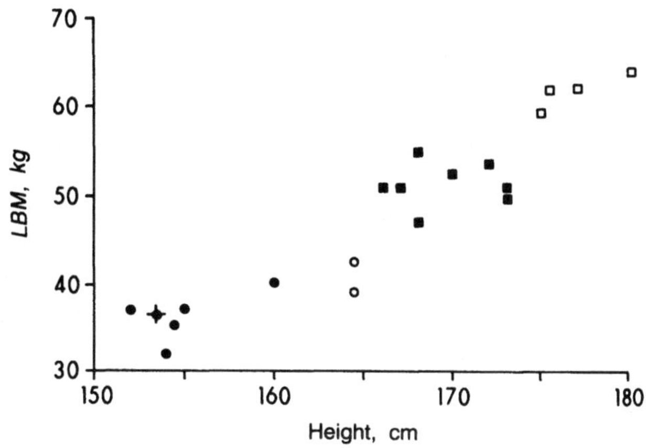

FIGURE 5.4. Plot of LBM against stature. Symbols: (●) Oriental women; (○) Caucasian women; (■) Oriental men; (□) Caucasian men. Data of Nagamine and Suzuki (1964), Novak (1970), Brozek (1952), Kitagawa and Miyashita (1978), Chen et al. (1963), Krzywicki and Chinn (1967), and author's data. LBM calculated from density or ^{40}K counting; young adults only. Orientals include Japanese, Taiwanese, Filipinos, Asiatic Indians. Obese individuals excluded. Vertical and horizontal bars indicate SEM for data of Satwanti et al. (1978) (n = 179).

the second to women). In their densitometric studies of 16-year-old children Huenemann et al. (1974) found that black girls and boys had slightly higher values for LBM than whites. However, Flynn et al. (1970) did not detect a difference in total body K in 4–6-year-old black and white children.

Contrary to popular belief Eskimos (Igloolik tribe of Canada, living at 69° 40′ north latitude) proved to be no fatter than Caucasians when assayed by deuterium dilution (Shephard et al., 1973). Adult men had an average of 13.5% fat, women 24%. Although shorter than the average Caucasian their calculated LBM was about the same, so the people of this tribe have a stocky, nonobese type of body build.

Heredity

A trip to the seashore in summer leaves no doubt as to family resemblances in body build. Although a portion of the likeness in the relative amounts of fat beneath the bared skin of the bathers may be due to a shared environment and eating habits, and the husband–wife similarities the result of mate selection, these cannot account for the parent–child correlation in relative stature, or in the distribution of subcutaneous fat. Studies of twins have shown that monozygous (MZ) pairs are more concordant for height and weight, even though reared apart, than dizygous (DZ) pairs. Most recently Bouchard and his associates (1985) have reported the results of a study of body composition in 871 biological and adopted siblings aged 8–26 years; measurements included body density and skinfold thickness.

When corrected for age and sex the interclass correlations for fat-free weight were as follows: adopted pairs (r = 0.03), unrelated pairs (r = 0.06), cousin pairs (r = 0.28), biological siblings (r = 0.26), DZ twins (r = 0.53), and MZ twins (r = 0.93). The progression of correlation coefficients as the likelihood of shared genes increases from zero to 100% is striking indeed; and the authors rightfully conclude that there is "a clear indication of the fact that body density and fat-free weight are significantly determined by the genotype." Heritability estimates varied from 0.52 to 0.80 of the total phenotype variance within a generation, and skinfold thickness gave heritability estimates of roughly the same magnitude, as did the ratio of extremity to trunk skinfolds. Consequently, both the relative amount of body fat and the distribution of subcutaneous fat appear to be under genetic control. This is not to say that other influences do not play a role. As will be made clear in subsequent chapters both nutrition and physical activity can affect the size of the LBM.

Age

The adult body is not a static system. What was produced during childhood and adolescence soon begins to wither away, but at such a slow pace that the change in LBM is apparent only after a lapse of a decade or more. This makes for difficulties in conducting longitudinal studies: one must be content with waiting; and the technical errors of the assay techniques are often greater than the magnitude of changes to be anticipated. Consequently, most of the extant data represent cross-sectional observations, which carry the risk of being contaminated by secular changes: the 70-year-old subject was born a full half-century before the 20-year-old, and so was the product of a different nutritional environment. Another problem in judging the effect of aging *per se* is the increasing likelihood as the years go by of some disease process—cardiac, renal, vascular, or endocrine—which may operate to compromise body composition and physiologic function. Some cross-sectional studies of "aging" have used healthy free-living subjects for the young adult and middle-aged years and then have chosen hospitalized subjects for the elder years.

Borkan et al. (1983) in an extensive review of the literature were able to partition the adult decline in stature into its biologic and secular components. Over the age span from 22 to 82 years they estimated that 4.3 cm of height was lost as a result of aging, whereas cross-sectional data suggested a loss of 7.3 cm; hence the difference of 3.0 cm represented a secular effect. The conclusion to be drawn is that cross-sectional data on aging are apt to overestimate the true state of affairs.

Another problem in the evaluation of the effect of age on body composition concerns the assumption underlying the various body composition techniques. The density of human bone declines with age (Trotter et al.,

1960); so too does the ratio of body K to total body water (Bruce et al., 1980; Cohn et al., 1980; Pierson et al., 1982). The ratio of estimated muscle mass to visceral mass also declines (Tzankoff and Norris, 1978). The result is that the density of the LBM is probably less in the elderly than in the young adult, and that its K content is also smaller. Skinfolds are more easily compressible in the elderly, so estimates of body fat by anthropometry are apt to be low, and estimates of LBM inappropriately high when this method is used.

As an illustration, a 3% reduction in the density of the LBM from its chosen value of 1.100 g/cm^3 due to age yields a value of 1.067 g/cm^3, so the formula for calculating fraction of fat becomes $5.750/D - 5.389$ instead of $4.95/D - 4.50$ (Chapter 2, Eq. 18). In the case of a thin individual (17% body fat), this new formula yields a value of 3.6% fat; and for an obese person (45% fat) the value becomes 36%. A man weighing 72 kg would be assumed to have 68.4 kg LBM by this new formula and 59.8 kg LBM by the standard formula. One weighing 130 kg would have 83.2 kg LBM by the new formula and 71.5 kg LBM by the standard one.

A reduction in the potassium content of the LBM by 3% due to age would mean that calculated LBM would be 3% higher. For an equivalent change in LBM composition the ^{40}K technique produces a much smaller effect on estimated LBM than does the densitometric technique.

Although the total body water method is not affected by age, the ratio of extracellular to intracellular fluid volume does change, as will be shown later in this chapter. The technique of uptake of fat-soluble gases should not be affected, nor should estimates of total body calcium by neutron activation. Other methods that have been used to estimate skeletal mass, such as metacarpal cortex thickness and photon absorption of the distal radius and ulna, fail to take into account the fact that different portions of the skeleton age at different rates. One must conclude, therefore, that a certain degree of caution is necessary in interpreting changes in body composition that are observed to occur with aging.

The Rise and Fall of Total Body Potassium

The existence of several score whole body counters both in the U.S.A. and abroad, coupled with the nontraumatic nature of the procedure for ^{40}K counting, made possible the assay of large numbers of normal humans. It can be said that there is more information on total body K in man than for any other element contained in the human body. Several thousand individuals ranging in age from the newborn to the elderly have been examined in this manner. In constructing Figure 5.5, the data reported by several investigators were averaged, and from these an age curve for total body potassium extending from age 1 year to age 85 years was developed. The use of double-logarithmic coordinates helps to delineate the age trends.

Total body K is seen to rise during childhood to achieve maximum values

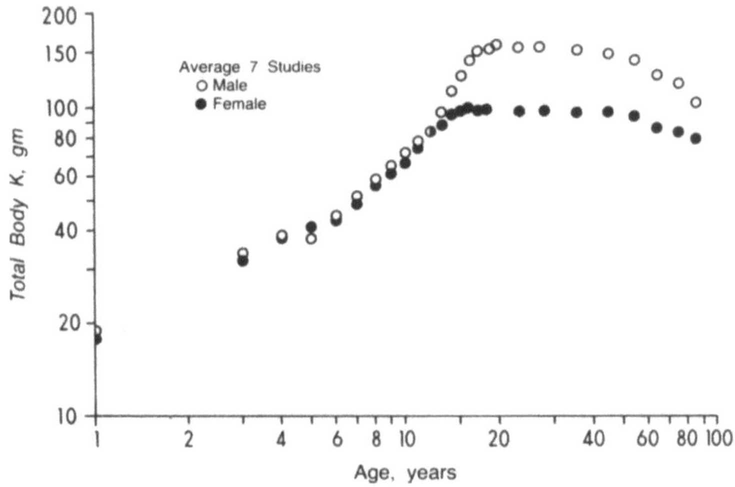

FIGURE 5.5. Plot of total body potassium (g) against age (yr) on double-logarithmic coordinates. Symbols: (○) men; (●) women. Averages of values from several sources: Flynn et al. (1972); Rutledge et al. (1976); Novak (1972); Myhre and Kessler (1966); Woodward et al. (1960); Allen et al. (1960); Burmeister and Bingert (1967); author's data.

by about age 20 years in men and age 17–18 in women. The word "maximum" is used advisedly, for there is a gradual fall as the adult years go by. The sex difference becomes manifest in adolescence, and as mentioned previously males achieve higher values than females, and interestingly enough they tend to lose body K somewhat more quickly than women during the adult years. By the age of 85 years, the average body K content of males is roughly equal to that of the 13–14-year-old boy, and that of females is equal to that of the 11–12-year-old girl. Hence the aging process appears to have somehow dissipated a large portion of the body K that was gained during the adolescent growth spurt. A part of what growth had achieved is taken away again by age. In thermodynamic terms the process of entropy—by which is meant the inexorable tendency for increasing disorder of the universe—was thwarted for a time by force of biological growth, only eventually to manifest itself as the aging processes take over. In a sense the fully grown young adult represents a distortion of the energy balance of nature, and the aging adult a return toward a more primitive orderly state. By age 80 years men have about 110 g K, which is about 45 g less than that of the average young adult man. Women have about 80 g K, or about 20 g less than the average young adult. These values work out to be a loss of about 5% per decade, and 3 ½% per decade, respectively.

Noppa et al. (1979) found a decrease of 202 meq body K in Swedish women between the ages of 44 and 66 years; the average loss was 3.3% per decade.

The rise and fall of body K is approximately a reflection of the rise and fall of the lean body mass. During adult life those individuals who manage to maintain their youthful adult weight status will acquire an increasing amount of body fat.

Creatinine Excretion

The next illustration (Figure 5.6) shows the change in 24-hr urinary creatinine excretion during growth and aging, from age 1 year to age 80 years in men and to age 65 years in women. This curve, which is an approximate reflection of muscle mass, has the same general form as that for total body K as shown in Figure 5.5. The data points are averages from several sources. Not all investigators were careful to exclude meat, which contains creatinine, from the diets of their subjects. Although such precautions would have changed the values somewhat (see Chapter 2), the effect would not have altered the general trends shown in the illustration. There is evidently a "rise and fall" in muscle mass during the course of human life.

It should be emphasized that the data shown in Figures 5.5 and 5.6 are cross-sectional in nature. There is no extensive collection of longitudinal data on infants and children; however there are a few for adults, and these will be shown later in this chapter.

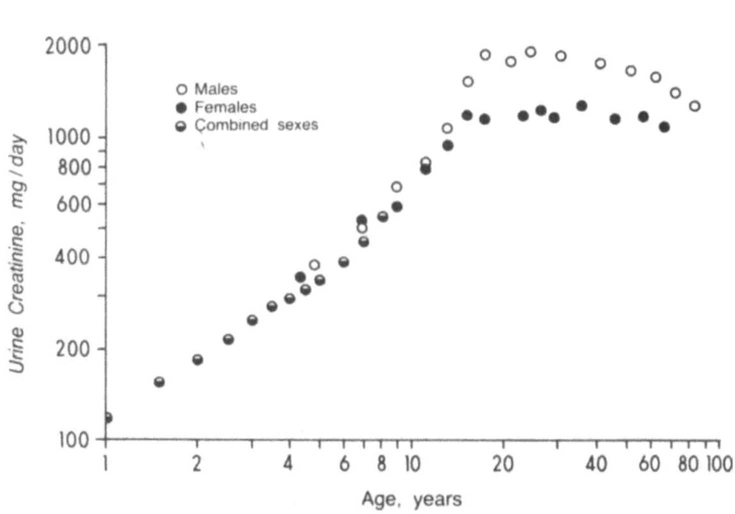

FIGURE 5.6. Urinary creatinine excretion (mg/day) as a function of age (yr) plotted on double-logarithmic coordinates. Symbols: (●) women; (○) men; (◒) combined sexes. Data sources: compilation of Viteri and Alvarado (1970) and data of L.E. Clark et al. (1951) for children; Rowe et al. (1976); Vestergaard and Leverett (1968); C. Young et al. (1963); author's data for adults.

LBM by Other Techniques

These also demonstrate a decline in LBM during the adult years. Figure 5.7 shows the results of cross-sectional studies in which body density and total body water were used to estimate lean weight. It also includes estimates based on the uptake of fat-soluble gases, LBM being calculated as the difference between body weight and body fat. Although there are some variations in the results—due no doubt to differences in technique, and perhaps also to subject selection—the general age trend is clearly evident. Using average values for age 20–30 years and for age 70–80 years in men and 65–73 years in women, the decline in LBM is about 3% per decade in men and about 2% per decade in women. As was the case for body potassium assays and for urinary creatinine excretion, all of these data are cross-sectional.

Stanton Cohn and his associates (Cohn et al., 1976, 1980) have presented two sets of cross-sectional data on adults. Together they show a decline in total body K, N, and P, and total body water with age. In men total body K declines by 36 g (23%), and total body N by 280 g (14%) between the age of 25 and 95 years. For a like period women experience a decline of 24 g total body K (25%), and one of 300 g total body N (21%). Total body phosphorus data are presented for the age range of 35–75 years in women, and 35–85 years in men; the decline for the former is 73 g P (18%), and for the latter it is 53 g P (11%).[1]

Body Water Compartments

Age also affects the relative size of body water compartments. Mention was made in Chapter 3 of the fact that extracellular fluid (ECF) volume exceeds intracellular fluid (ICF) volume in the fetus, and that the ECF/ICF ratio progressively falls during infancy and childhood, to the point where ICF volume occupies the majority position. Now it will be shown that aging tends to reverse this trend, the ECF/ICF ratio reverting towards, but not quite attaining, the infantile status.

Borkan and Norris (1977) assayed 699 men, ranging in age from 30 to 80 years, for total body water by antipyrine dilution and for ECF volume by sodium thiocyanate dilution. Average values for total body water showed a steady decline during these five decades (see Figure 5.7), but this decline was almost entirely due to the loss of calculated ICF volume, for ECF volume remained fairly constant. Consequently, the ratio of ECF volume to total body water steadily rose with age, from an average of 0.42 at age 30 years to one of 0.48 at 80 years. Calculated as ECF/ICF

---

[1] In the paper by Cohn et al. (1976), the 40–49-year-old subjects of both sexes happened to be taller than the 30–39-year-old group, so for purposes of estimating the age decline in total body P the average values for these two age groups was used.

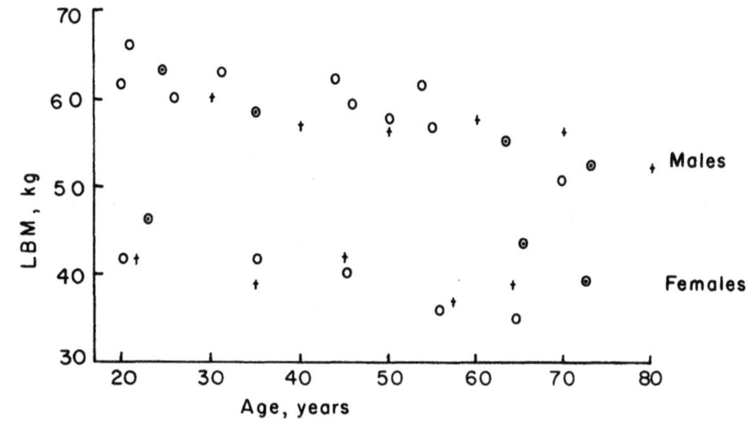

FIGURE 5.7. Lean weight as a function of age estimated from assays of body density, total body water, and uptake of fat soluble gases. Average values for selected age groups. Symbols: (○) densitometry; (+) total body water; (⊙) fat-soluble gases. Data of Myhre and Kessler (1966); Brozek (1952); C. Young et al. (1963); Lesser et al. (1971, 1979); Borkan and Norris (1977).

ratios these are 0.72 and 0.92, respectively. Figure 5.8 shows the changes that take place in the ECF/ICF ratio from mid-fetal life to age 80 years, the graph being drawn on semilogarithmic coordinates. There is a progressive decline in the ratio during the growth period, to reach a nadir in the early adult years, and to be followed by a gradual increase during aging.

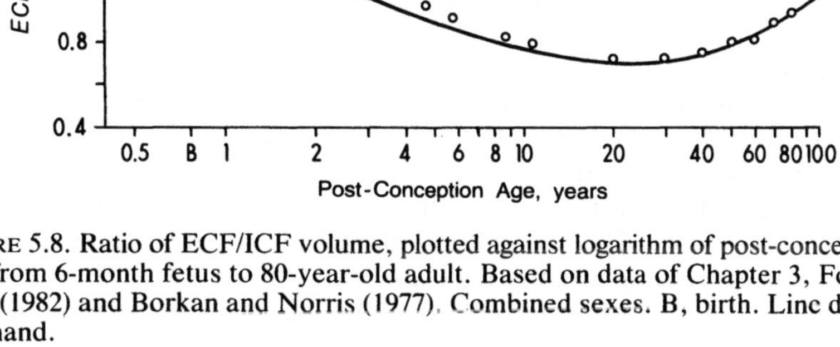

FIGURE 5.8. Ratio of ECF/ICF volume, plotted against logarithm of post-conception age, from 6-month fetus to 80-year-old adult. Based on data of Chapter 3, Fomon et al. (1982) and Borkan and Norris (1977). Combined sexes. B, birth. Line drawn free-hand.

The rather smooth progression of this ratio first in one direction and then in the other as growth is followed by aging suggests that both phenomena are integral parts of the biological processes of life. Growth has as one of its cardinal features a continuing encroachment of body cells on extracellular fluid as the cells hypertrophy; and once the stimuli for growth have been exhausted the body tends to revert towards the compositional status of its earlier years. By age 80 years, the ECF/ICF ratio has attained a value that is roughly comparable to that of the average 5-year-old child. Pierson et al. (1982) found values for Na_e/total body water ratio to be about 20% higher in 80-year-old individuals compared to 20-year-olds.

Longitudinal Observations

As previously mentioned, all of the above data are cross-sectional in nature. These are obviously much easier to collect than longitudinal data, and the anticipated changes in body composition during the adult years occur at such a slow rate that many years of observation are necessary to document a change. However, there are a few longitudinal studies that have been made, and the results of these differ somewhat from those derived from cross-sectional observations.

Shukla et al. (1973) did repeated ^{40}K counts on a group of 10 healthy young and middle-aged adults over a 12-year period. During this time the males lost an average of 6 g body K, and the females lost an average of 4 g body K; these are 3.3% and 3.8% per decade, respectively, of the initial values. However, there was considerable variation: one subject gained a little, two did not change, while the rest lost body K.

Fifteen young and middle-aged males and six similarly aged females had repeat ^{40}K assays after intervals of 4–9 ½ years: 11 lost some LBM during that time, whereas 10 did not (Forbes, 1976).

Albert Behnke was weighed under water on several occasions over a period of 31 years, and his lean weight was calculated from those densitometric measurements (Behnke and Wilmore, 1974). His body weight fluctuated a great deal during this period, but except for one 11-month period when he was taking a low-energy diet and temporarily lost 18 kg, his calculated LBM declined at a fairly constant rate, achieving a total loss of about 5 kg in 31 years. This works out to be an average loss of 2.3% of the initial value per decade (Figure 5.9, subject C).

Two males have had repeated assays of ^{40}K content in the author's laboratory over a period of many years, one for 26 years, the other for 27 years. One had several significant shifts in body weight, the maximum spread being 12 kg. The second maintained his weight within ±2 kg for many years (including a sabbatical year abroad!), and then showed a tendency to lose weight after age 67. All weighings were made with the sub-

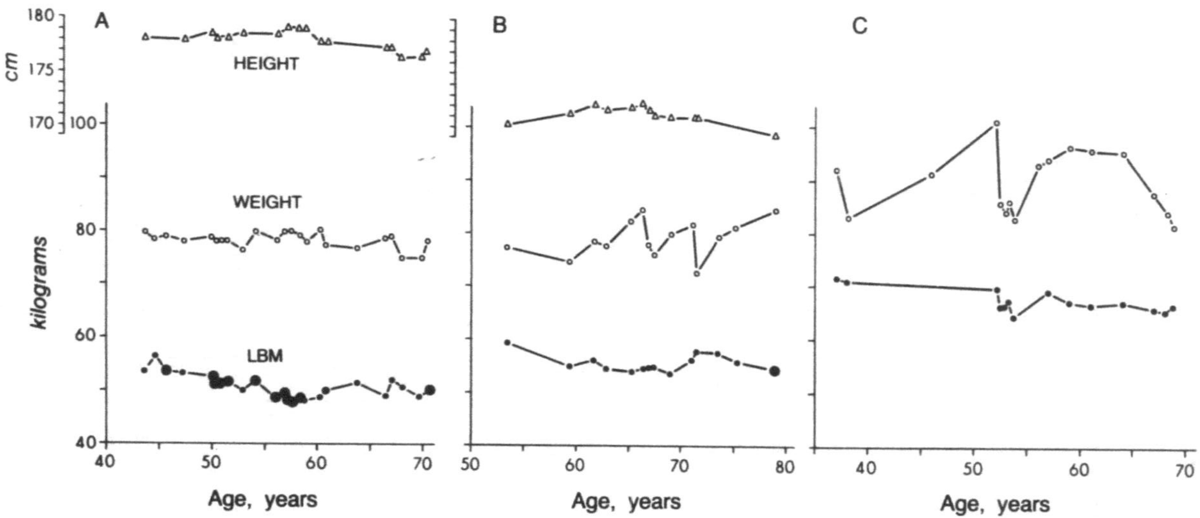

FIGURE 5.9. Longitudinal data on height, weight, and LBM for three adult men assayed repeatedly over a period of many years. Subjects A and B had ^{40}K counts in the author's laboratory; subject C had assays of body density (Behnke and Wilmore, 1974, p. 207.) All weighings of subjects A and B made with subjects dressed in cotton pajamas. The large dots for LBM represent averages of two or more consecutive assays.

jects dressed in cotton pajamas. Neither was in the habit of monitoring his weight at regular intervals (Figure 5.9). The best estimate is that subject A lost 3% of initial LBM per decade and that subject B lost 0.4% per decade.

The weight loss of 9 kg sustained by subject B at age 71 was the result of a vigorous exercise program, and it is of interest that his calculated LBM showed a tendency to rise at the same time.

Some of the variations in calculated LBM for subjects A and B were due to changes in the counting apparatus—replacement of the sodium iodide crystal, installation of a new multichannel analyzer. The data of Shukla et al. (1973) also show some variation in ^{40}K counts in their subjects during the 12-year period of observation which coincided with changes in the counter apparatus.

Even the supposedly easy technique of measuring stature is not immune to technical problems. The recorded increase in stature for subject A at age 56 and for subject B at 63 coincided with the installation of a Holtain stadiometer[2] in the author's laboratory, an instrument that has a direct reading dial, a proper headboard, and is wall mounted. It is capable of excellent precision provided standing posture does not vary. Using only those measurements made from age 56 years for subject A and age 63 years for subject B, the declines in stature were 2.3 cm and 2.1 cm, respectively, or 0.8% per decade.[3]

Thus changes in assay equipment can make for problems in conducting longitudinal observations in adults, in whom the anticipated changes with time are only a few percent. Technical errors in estimating body composition must also be considered. Nevertheless, it can be said that these longitudinal assays generally show a smaller drop in LBM, and in stature, than do cross-sectional assays, which are confounded by secular effects. Of significance is the fact that LBM exhibits much less variation over time than body weight, indicating that weight variations are principally the result of changes in body fat.

Basal metabolic rate bears a relationship to LBM, so it is of interest to look at longitudinal changes in this physiologic function. Tzankoff and Norris (1978) studied a large group of men of varying age, many of whom were observed over an 11-year period, and concluded that BMR declined at a rate of 3.7% per decade. Based on urinary creatinine excretion they estimated that this fall was due to the decline in muscle mass, and that portion of the metabolic rate assigned to the nonmuscle components of the body did not change with advancing age, at least during the range of 30–80 years.

---

[2]Holtain Limited, Crosswell, Crymmych, Pembrokeshire, Wales.

[3]Parizkova and Eiselt (1980) observed a group of 65-year-old men until they reached the age of 82 years. During this interval the average loss in height was 1.5 cm, with a standard deviation of 1.4 cm. This indicates a considerable degree of variation in the decline of stature.

Keys et al. (1973) did repeat assays of body density in 58 young men after a lapse of 19 years, and found an average loss of 1.0 kg LBM, or 0.8% per decade. Basal metabolic rate also fell, at a rate of 3.2% per decade. However, in another group of men whose ages averaged 49 years, and who were measured again after a lapse of 22 years, BMR fell more slowly, at a rate of 0.7% per decade. The authors conclude that the rate of decline in BMR as determined longitudinally is about one-half that recorded for cross-sectional observations. Unfortunately, individual data are not given so the variability cannot be determined.

Two individuals who had repeat BMR determinations, one after 20 years, the other after 50 years, showed declines of 4.3% and 4.8% per decade, respectively. Another had a second estimate of blood volume 32 years later: plasma volume had dropped at a rate of 5.0% per decade, and total red cell mass by 2.7% per decade (see Forbes, 1976 for references).

Although these longitudinal data are not extensive, and although most of the cited work does not contain a clear indication as to the health of the individuals who were assayed (as will be shown later, illness is accompanied by a loss of body K) they firmly suggest that the changes in body composition, and in BMR, represent a true aging phenomenon. It is also apparent that not everyone ages at the same rate. As shown in Figure 5.9, one subject had an almost imperceptible decline in LBM, whereas the other two had a definite drop. Healthy infants and children grow at different rates; it is to be expected that ostensibly healthy adults would show some variation in the rate of aging.

However, the studies reported by Lesser et al. (1973) show that the male rat appears to lack this age effect. These investigators made longitudinal studies of body fat by cyclopropane uptake, and by difference fat-free weight. They could not detect a significant change in calculated fat-free weight between 350 and 900 days of age in this species, although body fat increased. Cross-sectional observations, however, did show a 4–5% decline in fat-free weight over this same age range. They ascribe this difference to the fact that the older animals in the cross-sectional study—and they imply that such is the case for all cross-sectional studies—did not represent a random sample of older animals, but rather the survivors. Their comments are worth quoting: "Average colony FFB (fat-free body) decreased during senescence. However, the longitudinal data demonstrated this to result from selective longevity for rats of smaller FFB, rather than to loss of lean tissue by healthy aged animals."

It behooves the human biologist to conduct more long-term longitudinal observations on man, with careful attention to technique and to the health of the subjects. If our species is at all comparable to the rat—the only species so far studied in longitudinal fashion—it may well be that the process of aging in regard to body composition as shown by cross-sectional studies has been exaggerated. The fact that there appear to be differences in the rate of loss of LBM in the longitudinal data quoted above raises the distinct possibility of inherent individual differences in the rate of aging.

Granting the possibility that cross-sectional data may exaggerate the true rate of aging, such data are not without merit. They do define the differences in body K, in LBM, in urinary creatinine (and hence muscle mass), and, as will shortly be seen, differences in body Ca, between the elderly and the young adult at the time the data were collected. From the practical standpoint it matters not whether these represent secular effects or true aging; their very existence is of importance to the physiologist, the pharmacist, the nutritionist, and to those who administer fluids by vein.

Age Changes in Bone, Muscle, and Fat

Aging and Total Body Calcium

Cohn et al. (1976, 1980) present two sets of cross-sectional data on total body Ca as determined by neutron activation. In men, total body Ca averages 1220 g at age 20–29 years, and 1006 g at age 80–89 years, a drop of 36 g per decade. As presented in their 1976 paper, the women aged 30–39 years had an average of 785 g Ca and those 70–79 years old had an average of 634 g Ca; the 1980 paper gives an average of 924 g Ca at age 20–29 years and 671 g Ca at 70–79 years. So the amount lost per decade is 38–51 g. Women enter adult life with less body Ca than men, and lose it at a more rapid rate.

Age Changes in Bone

Earlier mention was made of Trotter's finding (Trotter et al., 1960) of the fall in bone density with age in cadaver specimens. However, the samples that they analyzed included bone marrow, which makes it difficult to interpret the magnitude of the change in density of bone per se.

Osteoporosis is recognized as a common problem among older adults in modern society, and the information just presented documents the age decline in total body Ca. It is not limited to the human species, for 11–18-year-old mongrel dogs show an increase in bone porosity, histologic evidence of bone resorption, and occluded Haversian canals compared to younger animals (Detenbeck and Jowsey, 1969). Nor is it a new disease: ancient skeletons of older aged Nubians (350 B.C.–550 A.D.) show evidence of osteoporosis (Dewey et al., 1969).

A number of studies have been made on separate regions of the skeleton by ordinary roentgenograms, photon absorptiometry, CAT scans, and measurements of bone cortex thickness both in children and adults. Some techniques measure only cortex thickness, whereas others give estimates of changes in trabecular bone as well. All demonstrate a decline in skeletal size during adult life, but the loss rate varies in different regions of the skeleton. This is shown in the cross-sectional data of Genant et al. (1982), who made measurements of the spine by quantitative computed tomography, of the thickness of the second metacarpal cortex on plain roentgenograms, and of photon density of the radial diaphysis in their subjects (Figure 5.10A and B). All three regions of the skeleton show a decline with age, but the changes in the spine greatly exceed those that occur at the other two sites. In the females, the density of the spine at age 80 is only 40% that of the 20-year-old woman, whereas that for the male spine is about 55%. The graphs also contain data for an osteoporotic woman

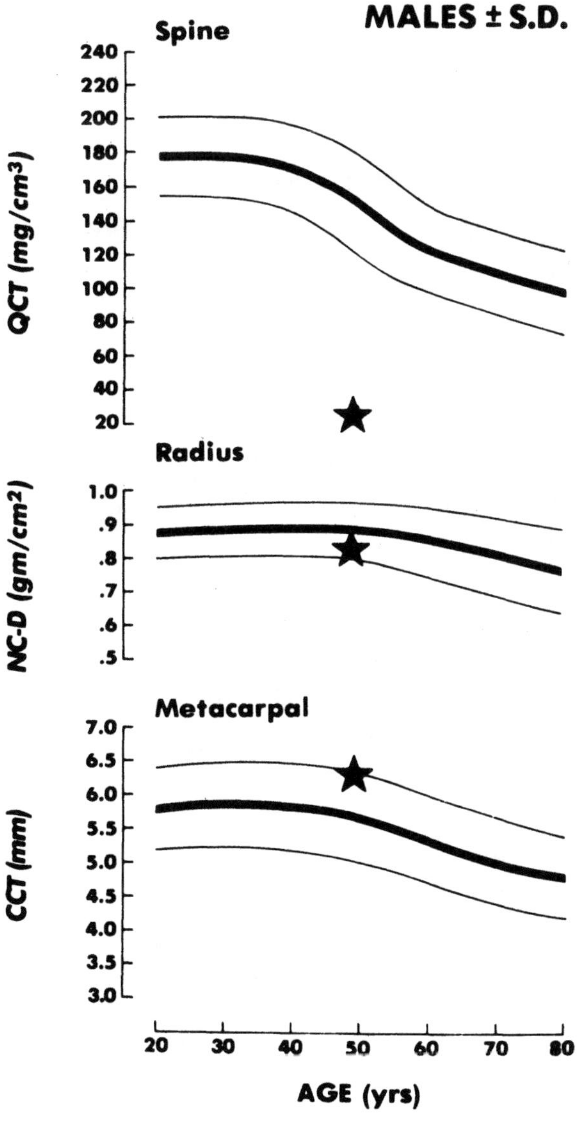

FIGURE 5.10.A: Age changes in skeletal size for males; cross-sectional data, means ± SD. Quantitative CT of spine (QCT), photon density of radial diaphysis (NC-D), combined cortical thickness of second metacarpal bone (CCT). Stars indicate values for 50-year-old man with Cushing's disease. [From Genant (1982), with permission.]

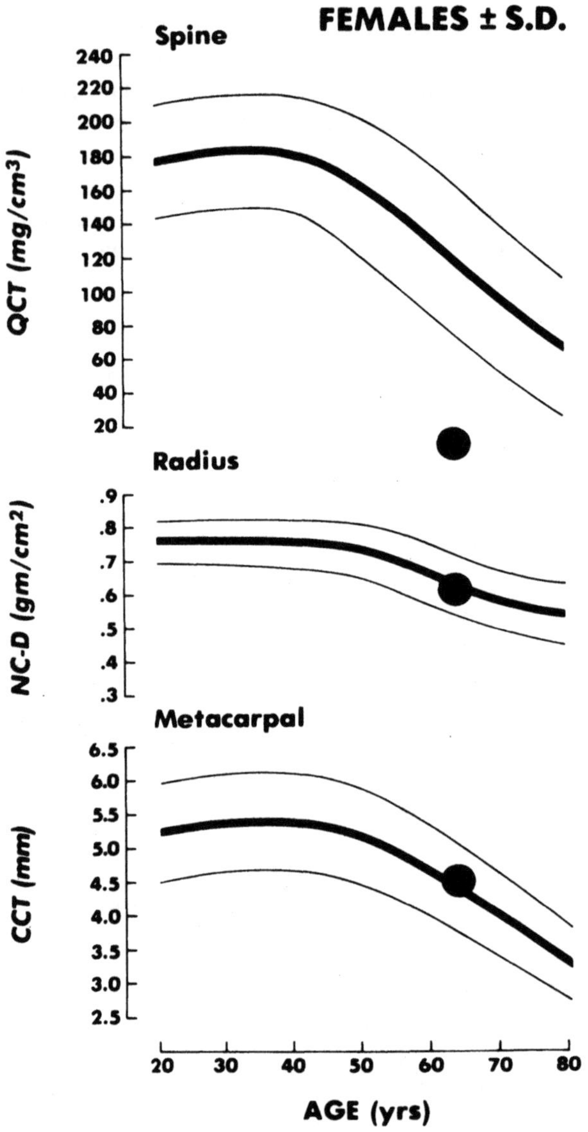

FIGURE 5.10.B: Similar data for women. Solid circles indicate values for 59-year-old woman with postmenopausal osteoporosis. [From Genant et al. (1982), with permission.]

and for a man with Cushing's disease. Here, too, the patients exhibit a decrease in spine density while the other two sites remain within the normal range.

There have been a number of cross-sectional studies of metacarpal cortex thickness area on individuals in the U.S.A. and abroad. Finland is represented by Virtama and Helelä (1969), Germany by Ringe and Rehpenning (1977), England by Exton-Smith et al. (1969), and the U.S.A. by Garn et al. (1976). All show a decline in cortex thickness with age. Figure

5.11 shows the computation of the cross-sectional area of the second metacarpal cortex made by Garn et al. (1976) from roentgenograms obtained during a country-wide survey of several thousand U.S.A. whites. This has the same general configuration as the data for total body K and for urinary creatinine excretion shown in Figures 5.5 and 5.6. However, there are some quantitative differences: maximum values are not attained until the third decade of life, and the subsequent decline with advancing age is not as marked as with the other two parameters. Similar data for Finnish children and adults have been published by Virtama and Helalä (1969). These data are more representative of the axial skeleton than the spine, a point clearly shown in Figure 5.10. Data for the "reference man" compiled by the International Committee on Radiation Protection (1975) show that trabecular bone makes up only about 20% of the total mineralized adult skeleton (see Table 5.2), so a selective loss of, say, vertebral bone would have only a modest effect on total body calcium content.

Cross-sectional studies show a decline in bone mineral content of the forearm (photon absorptiometry) in Danish adults (Christiansen and Rödbro, 1975) and in Alaskan Eskimos (Mazess and Mather, 1974).

A few investigators have collected longitudinal data on skeletal mass. Adams et al. (1970) remeasured 55–64-year-old adults 11 years later, and recorded a 10% loss of bone cortex thickness in men and a 15% loss in women during this interval. However, there was considerable individual variation, inasmuch as 38% of the men and 22% of the women showed no change. Milne (1985) failed to find a change when elderly adults had second assays 5 years later, whereas cross-sectional observations did show

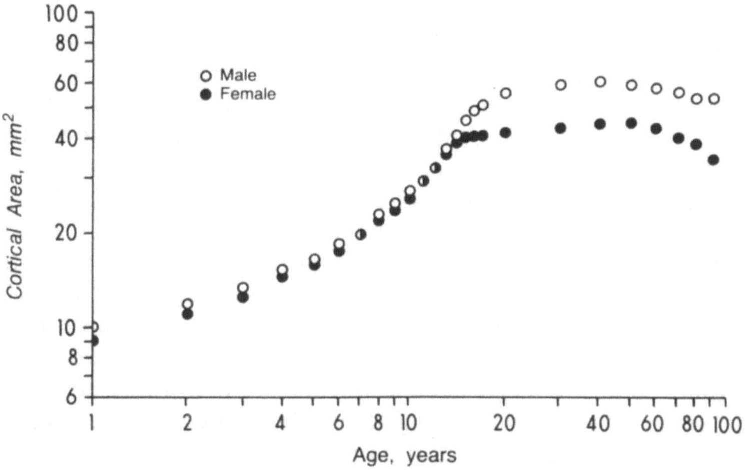

FIGURE 5.11. Cross-sectional area of the mid-cortex of the second metacarpal bone plotted against age on double-logarithmic coordinates. Symbols: (○) men; (●) women. Drawn from data of Garn et al. (1976).

a loss. D. M. Smith et al. (1976) made repeated photon absorption measurements of the mid-radius area during a 4.5-year period on a group of elderly women. The average loss in bone mineral content was 1.32% per year for those subjects who were 51–65 years old, but only 0.02% per year for those who were 70–91 years old. Cross-sectional observations of a larger population of women in the same age range revealed a greater loss rate.

Age Changes in Muscle

Lowry and Hastings (1942) assayed muscle from rats who ranged in age from 30 to 1010 days. On a fat-free wet tissue basis there was little change in water content or ECF volume until after age 668 days, when water content rose by about 4% and ECF volume by about 70%. The concentration of potassium in ICF remained at 170 meq/liter throughout. Hence these animals had to become very old (over 2 years) before changes in fluid volume occurred. In the very old animals the ratio of ECF volume to total water tended to revert towards infantile values (see Chapter 3).

Inokuchi and his associates (1975) assayed samples of rectus abdominus muscle from humans aged 20–80 years who had died in accidents. They determined the areas occupied by various tissue elements on microscopic sections of tissue. The cross-sectional area occupied by connective tissue rose progressively from 3% of the total at age 20 years to 20% by age 50 in nulliparous women, after which there was no further increase. Values for men were about the same, whereas those for parous women ranged as high as 30%. Areas occupied by fat were similar in all three groups, rising from 1–5% of the total at age 20 years to 40–50% at age 80 years.

Granted that cross-sectional surveys may overestimate losses due to aging, why does the aging process produce a decline in lean weight, in muscle mass, and in skeletal size? The loss of height is easy to understand: as the years go by thoracic kyphosis is accentuated, and there is compression of intervertebral discs and an increasing angulation of the neck of the femur. Robert Heaney and his associates (1982) discuss the hypotheses that have been put forward to explain the loss of skeletal mass. It is known the gastrointestinal absorption of calcium decreases with age, and that patients with osteoporosis absorb less Ca than do age-matched controls, and further that they tend to have lower serum levels of calcitriol (1,25-hydroxycholecalciferol, the active form of vitamin D). One hypothesis states that the age-induced decline in renal function acts to reduce the formation of calcitriol, and this together with the decrease in gonadal steroids diminishes intestinal absorption of calcium; the result is an increase in parathyroid hormone secretion, which leads to increased bone resorption. Another takes for its starting point the decrease in physical activity and in estrogen production with age, with the result that bone resorption exceeds bone formation; this leads to diminished parathyroid hormone secretion, resulting in a decrease in the synthesis of calcitriol by the kidney and thence to diminished intestinal absorption of calcium. In this view clinically significant osteoporosis is not a separate entity—underlying disease processes excluded—but rather an extreme version of the universal phenomenon of aging. Although there is some evidence that can be offered in support of both hypotheses neither has been proven. Certainly the role of reduced physical activity has to be seriously considered, since it has been shown that only a few weeks of bed rest is enough to produce detectable changes in the density of the lumbar spine (Krølner and Toft, 1983). Some attention is now being given to the possible role of the generous amounts of protein in the modern diet as a factor, for high protein intakes do lead to hypercalcuria.

There are a number of factors that might possibly explain the loss of muscle mass, and of LBM, during aging. The age-associated decline in physical activity, which is aided by the ubiquitous automobile, is one such. Earlier mention was made of the effect of bed rest on nitrogen balance. There is some disagreement as to just how much aging affects the secretion of the various hormones that possess anabolic activity. The problem in deciding this question revolves around the health of the subjects from whom the blood samples are obtained. Hospitalized elderly patients who are often used as subjects are not necessarily representative of the normal elderly population. A recent review of the endocrinology of aging contains information to the effect that both somatomedin-C and dehydroepiandosterone levels fall by 50% or more during adult life, and that nocturnal surges in plasma hormone levels are reduced both in amplitude and frequency. Some report a decline in serum testosterone levels, others no change. However, the levels of testosterone binding globulin tend to increase, so the level of the free hormone must decline somewhat. There is also disagreement about the effect of age on serum thyroxine and triiodothyronine, and as to whether there is a change in insulin resistance. However, there can be no doubt about the fall in serum estrogen after the menopause (Wilson and Foster, 1985).

Although this book is not concerned with function *per se,* mention should be made of the decline in maximal oxygen consumption, and of muscle strength and endurance. (The decline in BMR was noted earlier.) Athletic performance suffers with age, and injury repair is delayed. Shephard's book (1982) presents these aging phenomena in considerable detail, and the recent book by Whitbourne (1985) discusses the many facets of aging.

One of the challenges of today is whether the decline in muscle mass and in physiological functions can be lessened by well-designed programs of regular exercise. A proper study of this question may well require a longer period of longitudinal observation than most investigators are willing to undertake, and a degree of compliance that most subjects lack.

Age Changes in Body Fat

The great variability in skinfold thickness, in calculated cross-sectional area of upper arm subcutaneous fat, and in estimated total body fat make for great difficulty in establishing norms at any age. The problem is compounded by the fact that most frequency distributions of body fat are skewed to the right, so that standard deviations are not suitable for estimating the extent of the variability. Some authors have attempted to normalize the data by means of logarithmic transformations.

It is unlikely that lean weight will increase after age 25, except under unusual circumstances (androgen administration or vigorous and prolonged exercise programs), so that adults who gain weight after that age are going to accumulate body fat. For example, Dill et al. (1967) measured body density and maximum oxygen consumption on a group of champion runners 20–50 years after their record performances. The eight individuals who had either lost a little weight or had gained less than 5 kg during that interval had an average body fat content of 15%; the six who had gained 6–12 kg had 20% fat, and the two who had gained 15–22 kg had 30% fat. Unfortunately, none of these individuals had had density assays at an earlier age; however, earlier estimates of maximum oxygen consumption were available, and repeat estimates showed that all of them had sustained a decline regardless of whether they had continued to exercise.

Sex is an obvious factor. Girls acquire more body fat than boys during adolescence, and this difference persists throughout the adult years. During the large-scale nutrition survey of 1971–74 data on triceps skinfold thickness and mid-arm circumference were collected on 19,067 white subjects aged 1 to 74 years; these have been analyzed by Frisancho (1981), who constructed percentile values for all age groups. The 50th percentile value for 20-year-old males is 10 mm. By age 50 this has reached 12 mm; then it falls to 11 mm by age 70. Twenty-year-old women have a value of 18 mm at the 50th percentile; this rises to 25 mm at age 50, and then falls to 24 mm by age 70 years.

Calculated values for cross-sectional arm fat area from triceps skinfold and arm circumference show the same age trends but to a more pronounced degree.[4] It takes more fat to cover a larger limb with a given thickness of subcutaneous fat than it does to cover a smaller limb with a comparable layer. For men the 50th percentile values are 14.1, 17.4, and 16.2 cm^2 for ages 20, 50, and 70 years, respectively; and for women they are 21.7, 32.4, and 30.6 cm^2, respectively.

The variability recorded for this large group of individuals is evident from the spread between the 10th and 90th percentile values. Using age 20 years as an example, the 10th percentile value is about one-half of the

---

[4]The equation for arm fat area (Chapter 2, p. 160) reduces to $CT/2 - \pi T^2/4$, where C is arm circumference and T is triceps skinfold thickness.

50th percentile value, and the 90th percentile value is about twice the 50th in both sexes. This also documents the fact that the distributions are strongly skewed to the right.

Mention was made earlier of the fact that body fat content is more variable than LBM, and that the variability in body weight is largely—but not entirely—due to the variation in body fat content. The effect of nutrition on body fat content will be discussed in Chapter 7: underfed individuals lose body fat, and overfed individuals gain fat. There is good evidence that obese individuals require more food to maintain body weight and that they have a larger energy expenditure than thin people. A major factor in the variability of body fat is the plane of nutrition.

Heredity

The role of heredity, first described in twin studies a half-century ago, has recently been confirmed. Using body mass index (W/H^2) as a criterion of thinness and obesity, Stunkard et al. (1986) have shown that adoptees resemble their biological parents more than their adoptive parents. Bouchard et al. (1985) measured body density and skinfold thickness in a large number of individuals. Interclass correlations for skinfold thickness were close to zero for adopted and unrelated pairs of individuals, then progressively increased as cousin pairs, biological siblings, DZ twins, and MZ twins were considered; the correlation coefficient for this last group was 0.83.

Variability also exists for the distribution of body fat. The Swedish studies show that the ratio of abdominal to buttocks circumference varies from 0.59 to 1.0 in adult women and from 0.75 to 1.10 in adult men (Lapidus et al., 1984; Larsson et al., 1984). Bouchard et al. (1985) estimated the heritability of the ratio of trunk to extremity fat to be 0.7.

However, the factor(s) that operate at the gene level have thus far escaped detection. High on the list of possibilities are those genes that have to do with behavior, and that program those areas of the brain that control appetite. In a society such as ours which is so abundantly supplied with food, such genes can find full expression.

6

Pregnancy

There all the Learn'd shall at the labour stand,
And Douglas lend his soft, obstetric hand.

Alexander Pope

Weight Gain

Nutritionists have long been interested in this state of vigorous anabolic activity, for the feto–maternal unit gains weight at a relatively rapid rate, a gain that amounts to about 20% of the mother's prepregnant weight in a period of 40 weeks. Estimates of the average weight gain during the first trimester of pregnancy (13 weeks) are not very accurate, although it is generally held that it is only of the order of 1 kg. Thereafter there is an acceleration, so that by the 20th week about 4 kg of the average total gain of 12.5 kg has been accomplished. For the last half of pregnancy the average gain is 0.43 kg per week, or 60 g per day. However, the variations in recorded weight gain encompass a range of 100–1000 g per week. The fetus and amniotic fluid account for a little more than a third of the total weight gain, the remainder being the mother's body *per se*.

The total energy cost of the average pregnancy has been estimated to be some 80,000 kcal (Hytten and Chamberlain, 1980), or about 6.5 kcal per gram gained. In their studies of women in the third trimester of pregnancy Johnstone et al. (1981) found an average weight gain of 16.2 g for each additional 100 kcal consumed, or an energy cost of 6.2 kcal per gram of weight gain. This value is not too far from the average for nonpregnant women who were deliberately overfed for a period of 3 weeks under controlled conditions, namely 7.2 kcal per gram of weight gain (Forbes et al., 1986). Seitchik (1967) studied the energy cost of quiet sitting in pregnant women; the average was 24 cal/min per kg fat-free weight, a value comparable to that for nonpregnant women. Data such as these support the contention that overall energy balance is not abnormal in pregnancy. The rise in basal metabolic rate, estimated to be about 15% by Hytten and Chamberlain (1980), is not far from the percent increase in body weight.

Composition of Weight Gain

There has been for many years a great deal of interest about the composition of the weight gain. Numerous nitrogen balances have been done on pregnant women, beginning early in the present century. A review of these by Macy and Hunscher (1934) showed that the average increment in body nitrogen from the 4th to the 10th (lunar) month of pregnancy amounted to 515 g of nitrogen. This suggests an increment of 3200 g of protein, an enormous quantity most of which cannot be assigned to any known site in the mother or the fetus. In accepting the results of the only technique for assessing changes in body composition available to them at the time these early workers were forced to the conclusion that the excess nitrogen was somehow stored in the body. These hypothetical "stores" could then be drawn upon to take care of the losses incurred during parturition and to act as a reservoir for meeting the needs of lactation. The end result was the recommendation by nutritionists of generous intakes of protein during pregnancy, and their opinion that such intakes could contribute to a successful outcome.

It has been only in recent years that the shortcomings of the nitrogen balance method have been clearly appreciated, and reasonable results produced (Johnstone et al., 1972; King et al., 1973), but by this time other methods for estimating body composition were in place. Incidentally, King et al. (1981) have used an ingenious maneuver in dealing with this problem. They measured nitrogen balance in a group of normal nonpregnant women, and found an apparent positive balance of 0.72 g/day +0.9 (SD), which in the absence of weight gain obviously represented an overestimation of the true state of affairs. Then, in studying pregnant women they subtracted 0.72 g N/day from the observed balance, and so derived a value for retention of 1.0 g N/day, which is close to the expected value for late pregnancy.

A number of balance studies have also been done in which the retention of other elements—Ca, P, Na, Cl—has been determined, and these too have produced apparent retentions that are excessive (Macy and Hunscher, 1934). I dwell on this matter because the (spurious) results of the numerous metabolic studies done in the past are so firmly entrenched in the minds of some nutritionists. Indeed, it was with some fear of criticism that Rush and his colleagues (1980) decided to have a control group as they embarked on their study of the effects of protein supplements in underprivileged pregnant women in Harlem; such supplements did not improve pregnancy outcome, as judged by newborn weights.

Rat

Direct analyses are possible for other species, and Widdowson (1981) has provided some interspecies comparisons of the relative size of newborn and maternal weights. Total litter weight as percent of mother's weight ranges from 68% in the guinea pig and 23–25% in rabbit and rat to as low as 3.6% in the elephant and 2.5% in the blue whale. The human newborn constitutes only 5–6% of the mother's weight, roughly comparable to the situation in the cow and pig.

Widdowson has also provided data on the increment in the rat dam's own body together with changes in her body composition, and has compared this to the weight and composition of the litter of pups (Table 6.1). In this species the dam must acquire relatively large amounts of protein and minerals in order to provide for her own body and the pups she eventually delivers. She of course delivers a litter the total weight of which is a far larger proportion of her prepregnant weight than is the case for the singleton newborn human. It is of interest that the rat dam acquires some additional body fat during pregnancy, amounting to about one-fourth of her own gain, i.e., exclusive of the pups. Such maternal fat stores have been regarded as a source of energy for subsequent lactation, and in this regard it should be mentioned that some mammals—notably the elephant seal—are known to fast during lactation (Laws, 1985).

Widdowson (1981) also presents data for the rat to show that during the course of pregnancy, maternal cardiac and kidney size increase by 30%, liver size by 44%, and gastrointestinal tract by 42%.

The data shown in Table 6.1 are of interest from the standpoint of other aspects of body composition. In considering the gain in body constituents other than fat that take place in the mother's body (exclusive of the fetuses), the relative increments in water, sodium, potassium, and protein contents are 34, 18, 32, and 22%, respectively; and that for fat-free weight is 30%. The ratio of K to fat-free weight in the tissue gained by the mother is 89 meq/kg, a reasonable number, whereas that for Na is only 37 meq/kg, which is lower than the ratio of 61 meq/kg that obtains at the start of

TABLE 6.1. Pregnancy in the rat.[a]

Prepregnant status		Gain in maternal body (%)		Young at birth[b]	Total increment (%)	
Weight, g	230	75	(33)	53	128	(56)
Water, ml	130	44	(34)	46	90	(69)
Protein, g	40.4	8.7	(22)	5.7	14.4	(36)
Fat, g	51	21	(42)	0.6	22	(43)
Fat-free wt, g	179	54	(30)	52	106	(59)
Ca, mg	2460	30	(1)	159	189	(8)
P, mg	1450	140	(10)	185	325	(22)
Mg, mg	76	17	(22)	13	30	(39)
Fe, mg	12.8	-0-	(-0-)	3.1	3.1	(24)
Cu, mg	0.27	—	—	0.22	0.22	(81)
Na, meq	10.9	2.0	(18)	5.3	7.3	(67)
K, meq	15.1	4.8	(32)	2.8	7.6	(50)

[a]From Widdowson (1981).
[b]Nine pups

pregnancy. Another strange aspect of this table is lack of increase in maternal body iron during pregnancy. The data shown in Table 6.1 obviously had to be constructed from two sets of rats, one prepregnant, the other at the end of gestation; and although the results do not correspond in all respects to the human state, they are of importance in assessing the relative contributions of the mother and the fetuses to the total changes in body composition during pregnancy.

Human

The human mother also acquires additional constituents for her own body during the course of pregnancy: the uterus increases in size, the mammary tissues enlarge, and it is known that the increase in body Na_e and K_e are greater than can be ascribed to the fetus alone. The well-known increase in blood volume is evidence that she has acquired extra protein and iron; but aside from this measurement there is no other presently available body composition technique that can accurately partition the increase in body composition between mother and fetus. The problem is that materials used to estimate total body water, ECF volume, and exchangeable electrolyte readily cross the placenta.

Cardiac size is known to increase by about 12% during human pregnancy; and although the size of the kidney as determined from its length on radiographs increases by only 8–10%, creatinine clearance (a measure of glomerular filtration rate) is augmented by about 40%.

Oxygen consumption also increases, the increment being about 13 ml O_2/min at 10 weeks and 38 ml O_2/min at term. Of the latter it is estimated that 11 ml O_2/min is accounted for by the fetus. The total oxygen consumption at term represents about a 15% increase over the prepregnant state.

Frank Hytten and his associates have compiled a large amount of data concerning body composition changes during human pregnancy. The aspect studied most frequently is plasma volume and red cell mass (Letsky, 1980). Although blood hemoglobin concentration tends to fall during pregnancy, there is an absolute rise in total red cell mass which however is exceeded in magnitude by the increase in plasma volume. The increase in the latter is about 1250 ml, an increment of almost 50% above the average prepregnant value. Although there is considerable variation in reported results, the red cell mass increases about 240 ml (18% of prepregnant value) in women not given iron medication, and about 400 ml (30%) in those who are given iron. Some investigators have determined that plasma volume and red cell mass tend to fall towards the end of pregnancy, but in such instances it is likely that the assays were done with the subject in the supine position. In this position the gravid uterus may compress the inferior vena cava and so compromise the return of blood from the lower extremities. When studies are done with the subject in the left lateral position this late fall in blood volume is not observed.

Figures 6.1 and 6.2 show the increments in plasma volume and red cell mass as the pregnancy advances. Note the influence of multiple births, and that of iron medication. Letsky (1980) emphasizes that these are total increments, not percentage increases, and concludes that it is unlikely that they are related to prepregnancy values, although they are related to the size of the fetus. Both plasma volume and red cell mass are functions of body weight, so the percentage increase during pregnancy will depend on their magnitude at the start of pregnancy.

Figure 6.3 shows a scheme for the components of weight gain. About 40% of the total gain is represented by the fetus plus placenta and amniotic fluid. It is to be expected, therefore, that there would be a relationship between weight gain during pregnancy and the weight of the newborn infant. Although there is considerable variation, the slope of the plot of birth weight against pregnancy weight gain shows that the average birth weight of white infants increases by 37 g for each additional kilogram of weight gained by the mother above a baseline of 3 kg (Institute of Neurological Diseases and Stroke, 1972). This same trend holds for infants born to black mothers, with the line shifted downward in accordance with their slightly lower birth weights.

A number of studies of body composition have been made in pregnant women in addition to plasma volume and total red cell mass. These include assays of body density, ECF volume, total exchangeable electrolyte, ^{40}K content, anthropometry, and urinary creatinine excretion.

MacGillivray and Buchanan (1958) studied eight women early in preg-

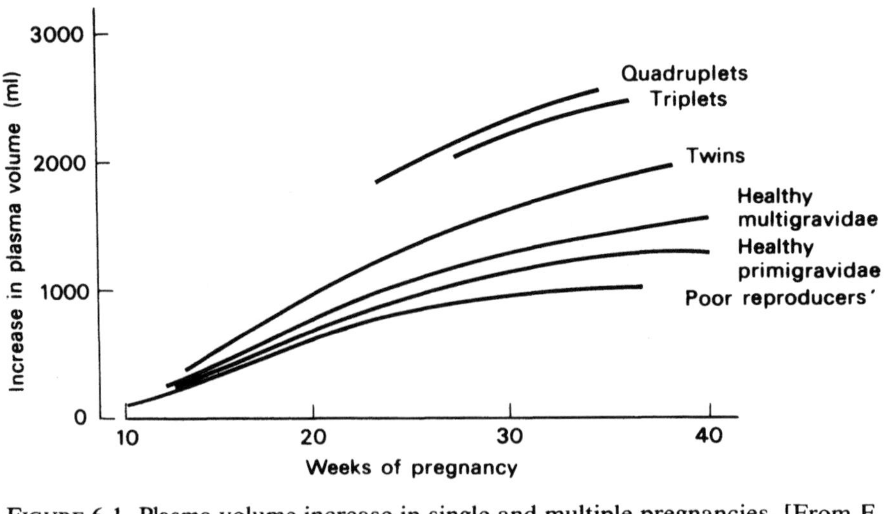

FIGURE 6.1. Plasma volume increase in single and multiple pregnancies. [From E Letsky (1980): The haematological system. In: F Hytten and G Chamberlain (eds), Clinical Physiology in Obstetrics. London: Blackwell Scientific Publications, with permission.]

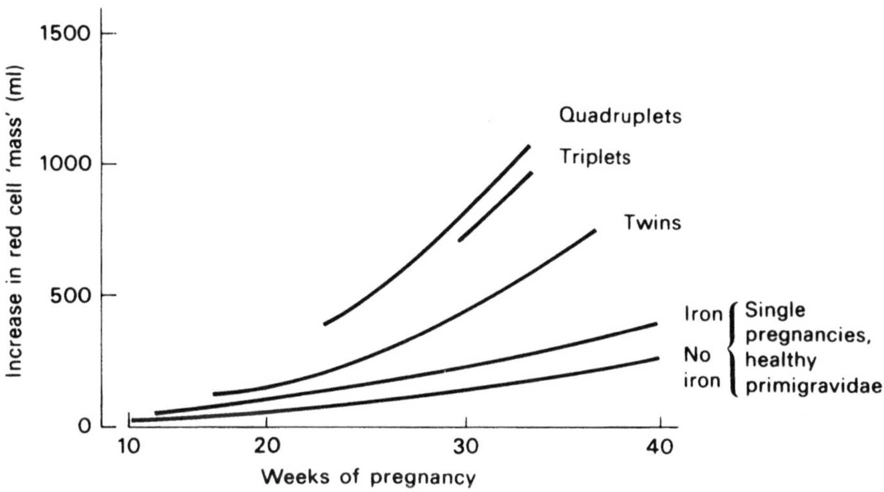

FIGURE 6.2. Increase in red mass in pregnancy. [From Letsky E (1980) (see legend to Figure 6.1), with permission.]

nancy and again in late pregnancy. Total exchangeable sodium (Na_e) increased by 773 meq, and K_e by 559 meq. Weight gain was 14 kg, so the increase in Na_e was 55 meq per kg of weight gain, and the increase in K_e was 40 meq/kg. McCartney et al. (1959) found a decrease in body density (from 1.043 g/cm^3 to 1.024) during pregnancy, an increase in sodium "space" and in Na_e, and in total body water. On average water accounted for about 60% of the total weight gain of 11.9 kg, and Na "space" for about 50%; hence intracellular water accounted for only 10% of the increase in body weight.

Pipe et al. (1979) measured total body water, and total body K by ^{40}K counting early and again late in pregnancy in 27 women. The average gain in K was 312 meq and in weight 10.4 kg, the ratio of K gained to weight gained being 30 meq K/kg. The gain in total body water averaged 7.2 liters; such a large value, which leaves only 3.2 kg of solids, suggests that all of the gain was lean weight in their subjects. Seitchik et al. (1963) measured body density in 21 women early in pregnancy and again at the 37th–40th week. On average body density fell from 1.026 g/cm^3 to 1.020. For those who gained less than 8.5 kg the calculated density of the tissue gained was 1.050, and for those who gained more it was 0.990. The authors concluded that a large share of the weight gained by this latter group was fat (density 0.90 g/cm^3). An analysis of their published data shows a reasonable correlation (r = 0.64) between the increase in total body water and the increase in body weight which ranged from 6 to 16 kg in their group of women. The slope of the regression line is 0.59 liters of water per kilogram body weight. The increase in body K was also correlated (r = 0.56) with the increase in body weight, the regression slope being 27

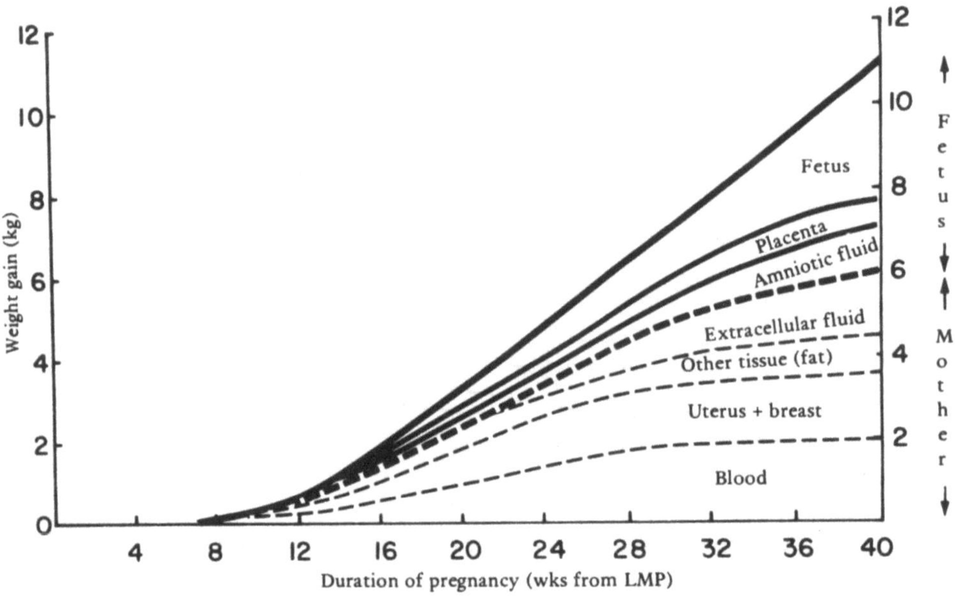

FIGURE 6.3. Components of weight gain during pregnancy. [From RM Pitkin (1977). In: Nutritional Support of Medical Practice (HA Schneider, CE Anderson, and DB Coursin (eds). New York: Harper & Row, p 408, with permission.]

mmol K/kg weight. The ratio of the two regression slopes (27 mmol K/ 0.59 liter H_2O) suggests that 46 mmol K is gained for each liter of body water gain. This is the same ratio as that of the newborn infant plus amniotic fluid, namely 150 mmol K and 3200 ml water, or 46 mmol K/liter.

We studied 50 pregnant women by ^{40}K counting on two occasions during pregnancy, at about the 16th week and again at about the 38th week. The average weekly gain in weight was 520 g, and that of potassium 22 meq. In their studies of pregnant teenagers during the last trimester King et al. (1973) found an average potassium increment of 24 meq/week. They also found an increase in urinary creatinine excretion during the 10-week period of their study; the average increment was 98 mg/24 h, an amount that corresponds to 2.8 kg LBM according to the formula listed in Chapter 2, p. 58). An objection might be raised here that such an interpretation is faulty in that the rise in creatinine excretion merely reflects the known increase in creatinine clearance by the pregnant kidney. Against this argument is the demonstrated finding of a good relationship between urinary creatinine excretion and LBM in normal individuals over a wide range of body sizes and who therefore must exhibit a range of creatinine clearance values (Forbes and Bruining, 1976).

Cheek and co-workers (1985) did biopsies of the rectus abdominis muscle in women undergoing Caesarean section at the 39th–40th week of pregnancy, and compared the results to those obtained on muscle samples from nonpregnant women undergoing elective surgery. Muscle from the pregnant women contained 2% more water (fat-free fresh weight basis), 42% more ECF, and 12% less ICF than muscle from nonpregnant women. However, there were no differences in RNA or DNA concentrations. These authors conclude that the increase in total body water that occurs during pregnancy is mainly extracellular in nature. Unfortunately, they failed to provide information on the type and amount of parenteral fluids administered to these women at the time of surgery.

The interpretation of the results of body composition assays done on pregnant women is difficult. The estimated increment in ECF volume exceeds that of ICF volume, and about 60% of the total gain of the feto–maternal unit is water. The estimated density of the fat-free body of the newborn is less than that of the adult, as is its potassium content (Chapter 4, Table 4.5). Tracer substances used to estimate ECF and ICF volumes and total exchangeable electrolytes traverse the placenta. Retentions of elements determined by metabolic balance include both mother and fetus, and the ^{40}K assay method detects K in both. Maternal skinfold thickness increases during pregnancy, but much of this increase, which amounts to about 11 mm for the sum of the triceps, biceps, subscapular and suprailiac sites, is rapidly lost postpartum (Pipe et al., 1979), so that these measures cannot provide a valid estimate of the gain in body fat. The thigh and suprailiac regions show the greatest increase in skinfold thickness during pregnancy.

The following calculations are an attempt to partition body composition changes between mother and fetus, using body potassium assays on pregnant women. MacGillivray and Buchanan (1958) reported an increment of 559 meq K by ^{42}K dilution, Pipe et al. (1979) one of 312 meq K by ^{40}K counting; we found one of 540 meq K (G. Forbes, unpublished data), and extrapolation of the data of King et al. (1973) yields a value of 600 meq K (24 meq per week times 25 weeks). The average of these four values is a gain of 500 meq K during pregnancy. The placenta at term contains about 25 meq K, and the fetus 150 meq; hence the gain for the mother *per se* is 325 meq K. This includes the amounts of K in the gravid uterus (65 meq) and in the expanded red cell mass (25 meq) and plasma volume (6 meq), so that the amount represented by other maternal tissues is 229 meq K. One can speculate that this amount represents expanded cardiac and kidney size (the liver does not enlarge during pregnancy), an enlarged muscle mass, and perhaps an increase in intestine mass.

Calculations such as these suggest that a portion of the pregnancy gain represents maternal lean tissue apart from ECF volume expansion, a conclusion supported by the observed increase in urinary creatinine excretion.

Hytten and Chamberlain (1980) have made an exhaustive review of the literature on body composition changes during human pregnancy, and Table 6.2 presents the results of their analysis. Admittedly, the values are not as precise as they appear to be from the number of listed digits for each; nonetheless they do represent a concerted attempt to partition the

TABLE 6.2. Pregnancy increments, human.[a]

Site	Total water (ml)	ECF (ml)	ICF (ml)	Protein (g)	Fat (g)	Total weight (g)
Fetus	2414	1400	1014	440	440	3294
Placenta	540	260	280	100	4	644
Amniotic fluid	792	792	0	3	0	795
Uterus	800	528	272	166	4	970
Breasts	304	148	156	81	12	397
Plasma	920	920	0	135	20	1238
RBC mass	163	0	163			
Maternal stores					3345	3345
Total	5933	4048	1885	925	3825	10,683
					Additional interstitial fluid	1817
					Total	12,500

[a]Adapted from Hytten and Chamberlain (1980).

feto–maternal weight gain among its various components. Accordingly, the total gain in protein is 925 g, 543 g of which, or 59%, represents fetus and placenta; gain in total body fat is 3825 g, 444 g of which, or 12%, represents fetus and placenta. The value for the increment in blood volume (1238 ml) is somewhat less than the total of the increase in plasma volume (1250 ml) and red cell mass (240 ml) listed by Letsky (1980) in the Hytten and Chamberlain (1980) book.

The estimated increase in interstitial fluid volume (1817 ml) is in keeping with the observation that many pregnant women have mild edema. Women with generalized edema have up to 5 liters of additional interstitial fluid.

The estimate of body fat increase is subject to the same degree of uncertainty mentioned earlier. The fact that the existing body composition methods, save for assays of total red cell mass, measure the entire feto–maternal unit means that estimates of maternal fat accumulation are subject to considerable error. In this respect neither anthropometric techniques nor the metabolic balance method can offer additional help. The reader should understand that the data presented in Tables 6.2 and 6.3 are gross estimates at best. Perhaps the nuclear magnetic resonance instrument—which can identify body fat depots—will be able to provide better information.

Hytten and Chamberlain (1980) have also estimated the time course of changes in body composition. As one could have concluded from the total weight gain during the first 10–13 weeks (Figure 6.3) the estimated increases in water, protein, and fat represent only a few percent of the total gains achieved at term. Consequently, the need for extra food for mother does not become appreciable until later in pregnancy.

The data in Table 6.3 pertain to women who gain 12.5 kg during pregnancy. Although this is considered to be a desirable goal, in practice there is a great deal of variation, ranging from zero to 16+ kg as recorded in the national collaborative study (Institute of Neurological Diseases and Stroke, 1972). Although it is likely that some women who have excessive

TABLE 6.3. Components of weight gain (g).[a]

	Weeks of pregnancy				
	10	20	30	40	Total
Protein					
Fetus + placenta	2	43	220	540	925
Mother	33	122	278	385	
Fat					
Fetus + placenta	—	3	83	444	3824
Mother	328	2060	3510	3380	

[a]From Hytten and Chamberlain (1980).

gains have put on large amounts of body fat, it is possible that the gain in others represents edema fluid. Hence the "standard" pregnant woman who gains 12.5 kg represents merely an idealized construct, analogous to the "50th percentile" child.

Although the data in Table 6.2 suggest that the gain in mother's body protein during pregnancy is due solely to increases in blood volume and in size of uterus and breasts, other data show that heart and kidney sizes are augmented; and as noted on p. 203 potassium retention exceeds the amount contained in uterus and expanded blood volume. Moreover, the increase in urinary creatinine excretion, an increment greater than can be assigned to the fetus, suggests that maternal muscle mass also increases during pregnancy. What are the mechanisms responsible for bringing about these changes?

The increase in body fat is easy to explain: women usually eat more during pregnancy [overfeeding increases body fat—and LBM—in women (Forbes et al., 1986)], and estrogen and cortisol levels both rise (both are lipogenic). Women in underprivileged areas gain more weight during pregnancy (and deliver bigger infants) when they are given food supplements (Lechtig et al., 1975). The increase in ECF volume can be explained by the higher levels of aldosterone and desoxycorticosterone.

Pregnant women also have higher serum levels of a number of other hormones that have anabolic properties. These are placental lactogen, progesterone, insulin, and testosterone. The latter two hormones are especially potent in this respect.

It is entirely likely, therefore, that the changes in body composition during pregnancy are the result of hormonal changes in concert with an augmented intake of food.

Adolescent Pregnancy

There is great concern today about the welfare of the girl who becomes pregnant in her "teen" years. Several studies have suggested that their infants tend to be smaller and somewhat less viable than those born to mature women, and there is now a debate as to whether this is due to biological immaturity, or to socioeconomic factors. In reviewing data from the National Collaborative Study Garn and Petzhold (1983) showed that although very young mothers delivered smaller infants they also tended to weigh less at the beginning of pregnancy. They found a good relationship between birth weight and prepregnant weight over a range from 40 to 90 kg, the slope of the regression line indicating a 10-g increase in birth weight for each kilogram increase in prepregnant weight. Of interest is the finding that this relationship is about the same for white girls 13–19 years old as it is for white women 20–29 years old. The regression slopes for blacks are slightly less, but here too the two age groups had similar regression slopes. On this basis the teenager performs as well during pregnancy as the older woman. However, Garn and Petzhold included all mothers 13–19 years old in their teenage grouping, so their analysis fails to provide a clear answer about the fate of infants born to young teenagers.

Generally speaking, young teenagers who become pregnant tend to be poor, black, socially disadvantaged, and less likely to receive adequate

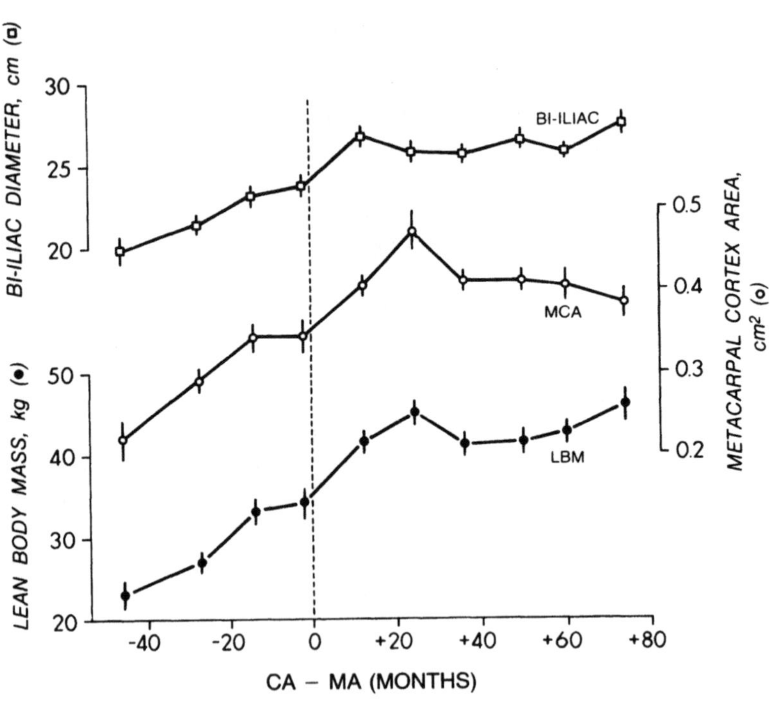

FIGURE 6.4. Data on biiliac diameter, transverse area of the second metacarpal cortex, and LBM for 113 normal girls, plotted according to difference between age at measurement (CA) and menarchal age (MA). Vertical bars are ± SEM. [From Forbes (1981), with permission.]

prenatal medical care. All of these factors may compromise the welfare of the fetus.

As mentioned earlier, girls who have an early menarche tend to be taller and heavier than their premenarchal age peers, so in these respects they are more mature. We have made ^{40}K assays and other measurements on 113 adolescent girls whose age of menarche was known. This ranged from 10.5 to 15.8 years (average 12.9 years), and the ^{40}K assays were made at times ranging from 48 months before to 80 months after menarche. Figure 6.4 shows plots of LBM, biiliac diameter, and cross-sectional area of the second metacarpal cortex for these subjects, the abscissa scale representing the difference between chronological age when the assays were done and age of menarche. Minus values on the abscissa indicate girls who were premenarchal when the assays were done, and positive values those who were postmenarchal. The graph shows that all three measurements were not far from mature values by about a year post-menarche.

On the other hand, Moerman (1982) points out that an early menarche does not provide for a mature pelvic inlet, a dimension of obvious importance at the time of obstetric delivery. In reviewing a series of pelvic roentgenograms made serially in adolescent girls she found that those who had an early menarche had smaller pelvic inlet dimensions than those who

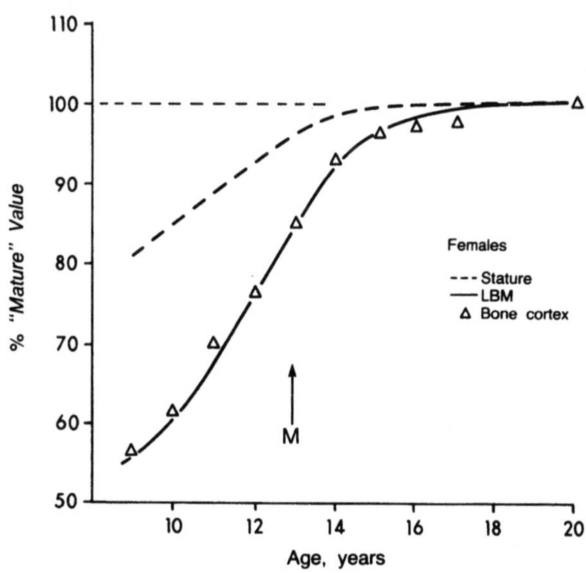

FIGURE 6.5. Percent of "mature" value for stature, LBM, and metacarpal cortex cross-sectional area for adolescent girls. Symbols: (---) 50th percentile height (National Center for Health Statistics, 1979); (—) average LBM, from Burmeister and Bingert (1976) and author's data; (△) cortical area, from Garn et al. (1976); "M" signifies average age at menarche.

had a later menarche. In this respect the growth of the adolescent female pelvis does not keep pace with other features of sexual development, and it appears that the growth of the outer pelvic dimensions, as illustrated by the biiliac diameter in Figure 6.4, occurs somewhat more rapidly than does the pelvic inlet.

The data shown in Figure 6.5 are plotted according to chronological age, without regard to sexual maturity. These are all cross-sectional in nature. The ordinate scale is in terms of percent of the values achieved by age 18–20 years. The 50th percentile value for stature is seen to reach this goal by age 16 years. Lean body weight as estimated by ^{40}K counting has attained about 91% of this goal by age 14 years, and so too has metacarpal cortex thickness. However, both of these measurements show a further increase of 2–4% during the third decade of life.

From the standpoint of these biologic aspects the girl of 16–18 years is practically mature and should be able under favorable circumstances to carry a fetus successfully to term. That is not to say that there may not be other obstetrical risks for the young pregnant teenager. There are at present no data on body composition of these young girls early in pregnancy.

There has been some concern about the possibility of added nutritional risk for the young pregnant woman because she is still in her growth phase. However, the data shown in Figure 6.5 make it clear that the increments in LBM and in bone cortex expected from the mother's growth *per se* during the 9-month gestation period are rather small in comparison to the demands imposed by the pregnancy.

7

Influence of Nutrition

Talking of a man who was grown very fat, so as to be incommoded with corpulency:
Johnson: 'He eats too much, Sir.'
Boswell: 'I don't know, Sir; you will see one man fat, who eats moderately, and another lean, who eats a great deal.'
Johnson: 'Nay, Sir, whatever may be the quantity that a man eats, it is plain that if he is too fat, he has eaten more than he should have done. One man may have a digestion that consumes food better than common; but it is certain that solidity is encreased by putting something to it.'
Boswell: 'But may not solids swell and be distended?'
Johnson: 'Yes, Sir, they may swell and be distended, but that is not fat.'

Samuel Johnson on Obesity
(Monday, April 28, 1783)

The availability of body composition techniques has made it possible to study the long-term changes in body composition without having to deal with the problems inherent in the metabolic balance method.

There are many reported studies on the composition of the weight loss in obese individuals subjected to short-term energy deficiency, and there is some discussion about the relative merits of various diets in attempting to preserve body nitrogen (and hence lean weight) under such circumstances (Bistrian et al., 1977b; Yang et al., 1981; Fisler et al., 1982). A far more important consideration, however, is the composition of the long-term loss, for this is the goal of all weight reduction programs. Is it possible, for instance, to lose only body fat during significant weight reduction over extended periods? Is the composition of the weight loss the same in the nonobese as in the obese?

Many years ago Voit (1901) reported that obese animals withstood starvation better than thin animals, and there is recent information that suggests that the same is true for man. Large numbers of obese individuals have tolerated fasts of 100 days or longer (Drenick et al., 1964), whereas the Irish hunger strikers (who presumably were not obese) died in 57–73 days. A few obese patients have died while taking very low-energy diets, and in analyzing the time elapsed from the start of the diet to death van Itallie and Yang (1984) found that this was inversely related to the degree of obesity. The question of interest, then, is whether the composition of the weight loss differs between the obese and the nonobese.

In considering the opposite situation, do people who gain weight put on fat, or lean, or both? How do changes in body composition induced by nutritional intervention correspond to the compositional status of individuals with established obesity on the one hand, or with undernutrition on the other?

An important question is whether the lean and fat components of the body are independent entities: can one component experience a significant change without a change in the other? Or is there a link between the two? There is now an abundant amount of information on human subjects that can be brought to bear on these questions. In this chapter we shall first discuss the compositional status of individuals with established obesity and undernutrition, and then proceed to examine the composition of induced weight gain and weight loss in man.

Obesity

This appears to be a very common condition in Western society, affecting both children and adults, and occasionally infants. Indeed it is second only to dental caries in frequency among the various nutritional abnormalities. Although the prevalence varies somewhat among various ethnic groups and nationalities, and by sex, it is a common condition in all societies that possess an adequate food supply; hence one can justify designating obesity as a nutritional disease. Although it is not our purpose to discuss the cause(s) of obesity, it can develop only in the face of a positive energy balance, namely an intake of energy that exceeds energy expenditure. Publications that purport to show that obese individuals eat no more, or even less, than the nonobese must be regarded with suspicion. Such conclusions are based on dietary history, and as Mahalko and Johnson (1980) have shown, obese individuals are apt to underestimate their food intake, and thin individuals are prone to overestimate their intake. Observation of obese children in the home show they usually eat more than their nonobese siblings (Waxman and Stunkard, 1980). Bessard et al. (1983), Jéquier and Schutz (1983), and Van Es et al. (1984) all report that energy expenditure as measured in the whole body calorimeter is proportional to body weight in adults.

Prentice et al. (1986) studied 21 women (50–102 kg weight) with doubly labelled water (^{2}H and ^{18}O) over a 14–31-day period.[1] They found a relationship between energy expenditure and body weight by this method; the regression slope is 15.7 kcal/kg/day ($r = 0.72$). Based on the amount of food needed to maintain body weight in elderly men Calloway and Zanni (1980) found a regression slope of 36 kcal/kg/day. Using a similar approach we found a regression slope of 20.5 kcal/kg/day in 24 women and one of

[1]This technique involves the estimation of body water turnover and the turnover of water plus carbon dioxide; carbon dioxide production is calculated by difference, and from this an estimate of energy expenditure can be made.

21.1 kcal/kg/day in four men (G. B. Forbes and M. Brown, unpublished data.)

The weighted average for these seven groups of subjects is 20.0 kcal/kg/day, which means that an individual who weighs 1 kg more than another will on average need to eat 20 kcal more each day to maintain his/her weight.

A further analysis of our subjects, which included a range of body fat from 2 to 74 kg, showed that the amount of food needed to maintain body weight could be described by the following equation:

$$\text{kcal/day} = 35.7 \text{ LBM (kg)} + 15.3 \text{ fat (kg)} + 198. \tag{1}$$

The correlation coefficient is 0.93 and the standard error of estimate is 242 kcal/day. At constant body weight energy intake must equal energy expenditure, so for these subjects, all of whom were engaged in light physical activity, one can deduce the latter from the former. From these data it would appear that energy intake required to maintain body weight is a function of both lean body mass (LBM) and body fat, with the former accounting for a larger fraction of the total. The coefficients of Equation 1 would undoubtedly be different for individuals engaged in moderate or strenuous physical activity.

The observed value of about 20 kcal/day for each kilogram of weight differential that is needed for weight maintenance suggests that weight control is a finely regulated process. The fact that many adults maintain their weight within a kilogram or two over long periods means that they are able to achieve an energy balance of ±20–40 kcal/day, or 0.8–1.6% of an average intake of 2500 kcal/day. The works of Neumann (1902) and Gulick (1922) are often quoted as showing that these two individuals, who conducted long-term feeding experiments on themselves, could achieve relatively constant weight despite variations in energy intake. However, a review of their reports showed that they always gained weight when they ate more food, and always lost when they ate less (Forbes, 1984).

It has long been known that obese children of both sexes tend to be a little taller than their nonobese peers, and to have some advancement in skeletal maturation. To answer the claim that these features could be due to constitutional or genetic factors, we studied 17 children who had become obese during childhood and for whom longitudinal height data were available throughout childhood, including the period before they had become obese (Forbes, 1977). It turned out that most of them had had an increase in height percentile status once they had become obese, and that this change in height status occurred either coincident with the increase in weight or sometime thereafter, but never before. Indeed, by the time they had been obese for 5 years there was a modest, though significant, correlation ($r = 0.47$) between the change in relative height status and change in relative weight status. These data strongly support the contention that the acceleration in stature growth is the result of nutritional surfeit, just as faltering in growth is a consequence of food deprivation.

The above comments are offered in support of the concept of the usual form of human obesity as a nutritonal disease. The rare exceptions will be mentioned later. The current wave of interest in adipocyte size and number need not concern us here.

Body Composition in Obesity

Although it has long been known that obese individuals tend to have an elevated basal metabolic rate (BMR) when compared to their age and height peers, it is not generally appreciated that most also have an increase in LBM. Some years ago Forbes (1964) and Cheek et al. (1970) reported that a portion of the excess weight of obese children consisted of LBM; prior to these reports Passmore et al. (1958) had said, "The excess weight of obese people is not all attributable to fat . . . there may be hypertrophy of skeletal and cardiac muscle, and perhaps other organs, which have to support and move the increased mass of fat." Since that time additional studies of obese children and adults have confirmed these observations. In view of the fact that LBM is a function of stature at all ages thus far examined (Forbes, 1974), and that it is also a function of age, comparisons of LBM between the obese and the nonobese are best done when these two variables, together with sex, are controlled.

Figure 7.1 is a plot of the LBM/height ratio for obese subjects assayed by the author by ^{40}K counting, compared to that for normal subjects studied in our laboratory. The 61 obese subjects ranged in size from 117% to over 200% of ideal body weight; only three were less than 130%. None had evidence of endocrinopathy, edema, or hypertension, and all had been obese for a number of years.

Three-quarters (47/61) of the obese subjects had an LBM/height ratio that exceeded one standard deviation of the mean for normals, and more than half of them (32/61) exceeded 2 SD. Only three had a value below the mean, and none fell below -1 SD. On the basis of chance alone [LBM/height distributions for nonobese subjects are approximately normal (G. Forbes, unpublished)], one would expect only 10 subjects to exceed $+1$ SD, and only two to exceed $+2$ SD. Hence the majority of the obese subjects had a larger LBM than normal subjects of the same age, height, and sex.

However, the data of Figure 7.1 show that some obese individuals possess a normal LBM. How do they differ from the majority who have a supranormal LBM? In studies of children both Forbes (1964) and Cheek et al. (1970) found some distinguishing features. Compared to obese children with normal lean weights, those with increased LBM tended to be tall for their age, to have increased bone age, and to have been obese for a longer period of time. Obviously, this is a matter deserving of more research.

Data from the literature are confirmatory. Table 7.1 shows that when

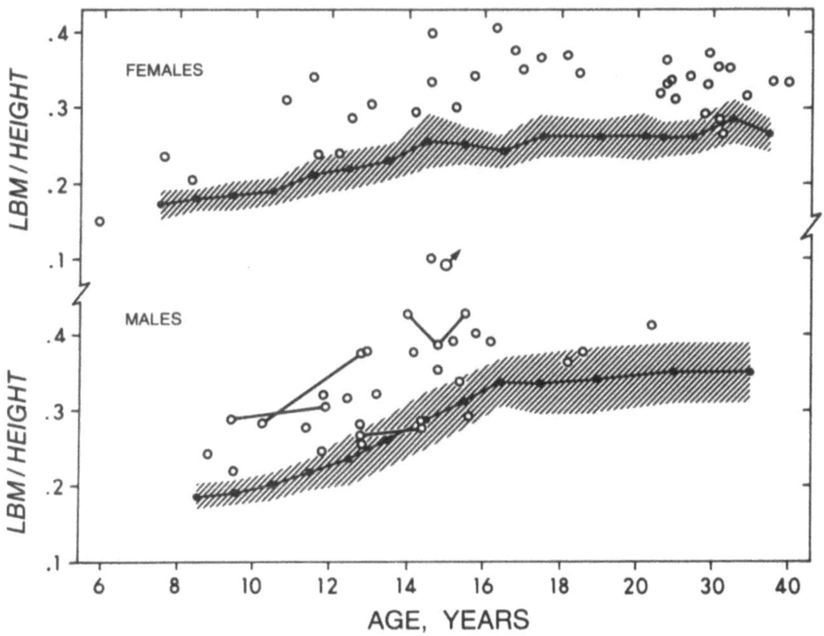

FIGURE 7.1. LBM/height ratios (kg/cm) against age for 61 obese subjects (open circles) assayed in the author's laboratory by ^{40}K counting compared to 484 normal females and 571 normal males (Forbes, 1978; unpublished). Cross-hatched areas define ±SD and solid dots the means for normal subjects. Repeat assays on four obese subjects shown by connecting lines. [From Forbes and Welle (1983), with permission.]

the obese are matched for height, age, and sex with the nonobese, and the same assay technique is used for both, the obese on average have a larger LBM. The lean component accounts for 20–40% of the excess weight; the average for the 10 groups of subjects is 29%. Recently Webster et al. (1984) reported their observations on 104 females of widely varying fat content, from which they concluded that the excess weight in obesity consists of 70–78% fat and 22–30% lean.

The increased LBM in obesity may explain the increased BMR (James et al., 1978; Ravussin et al., 1982; Prentice et al., 1986), the increased glomerular filtration rate (Stokholm et al., 1980), and the greater urinary excretion of creatinine (Tager and Kirsch, 1942; Hankin et al., 1972) and of adrenal steroids (Schteingart and Cohn, 1956; Prezio et al., 1964; Stokholm et al., 1980) found in such patients. In a study of lean and obese subjects Streeten et al. (1969) found that both cortisol secretion rate and urinary 17-hydroxycorticoid excretion were correlated with urinary creatinine excretion. The autopsy data of Naeye and Rovde (1970) show that the obese have larger hearts, livers, kidneys, spleens, and pancreases than

TABLE 7.1. Body composition in obese and normal subjects.

Subjects	n	Height (cm)	Weight (kg)	LBM (kg)	Δ Weight	Δ LBM	Method	Reference
Adults								
Female	9	161	60	40			D_2O dilution	Linquette et al. (1969)
Female	13	162	124	56	64	16	D_2O dilution	Linquette et al. (1969)
Male	12	169	70	56			D_2O dilution	Linquette et al. (1969)
Male	8	169	117	66	47	10	D_2O dilution	Linquette et al. (1969)
Female	16	164	77	53			^{40}K, THO dilution	Krotkiewski et al. (1977)
Female	18	164	92	59	15	6	THO dilution	Krotkiewski et al. (1977)
Female	34	165	104	61	27	8	THO dilution	Krotkiewski et al. (1977)
Female	15	162	56	40			^{40}K counting	James et al. (1978)
Female	61	163	96	53	40	13	^{40}K counting	James et al. (1978)
Male	11	176	67	52			^{40}K counting	James et al. (1978)
Male	11	175	115	70	48	18	^{40}K counting	James et al. (1978)
Female	13	163	54	41				Franklin et al. (1979)
Female	23	163	76	47	22	6	Density	Franklin et al. (1979)
Male	10	178	75	62			D_2O dilution,	Morse and Soeldner (1963)
Male	9	178	102	67	27	5	density	Morse and Soeldner (1963)
Adolescents								
Male	25	162	50	44			Density	Parizkova et al. (1971)
Male	12	161	69	49	19	5	Density	Parizkova et al. (1971)
Female	25	157	50	41			Density	Parizkova et al. (1971)
Female	9	158	69	47	19	6	Density	Parizkova et al. (1971)

the nonobese. In reviewing data obtained during the national nutrition survey, Garn and Ryan (1982) conclude that obese individuals have slightly higher blood hemoglobin levels (average difference 0.2 g/dl) than the nonobese. The data of Dalén et al. (1975) show that obese adults have a thicker metacarpal bone cortex, by about 15%, than the nonobese.

Some massively obese individuals have edema of the lower extremities, suggesting an increase in extracellular fluid (ECF) volume. The question whether there is generally an increase in the ratio of ECF volume to total body water has been addressed by several groups of investigators.

Shizgal et al. (1979) compared a group of obese adults (average weight 137 kg) to a group of nonobese adults (average weight 70 kg). They found no significant differences in the ratio of ECF volume to total body water, in the ratio of total exchangeable Na to total exchangeable K, or in the ratio of total exchangeable K to total body water between the two groups of subjects. Egusa et al. (1985) found comparable values for plasma volume per unit fat-free weight in obese and nonobese adults. However, Wang and Pierson (1976) report a higher ECF/total H_2O ratio in obese than in normal adults. The problem in interpreting their data is that the reported ratio for their obese subjects is almost the same (0.42) as that found by others for normal individuals, whereas the ratio for their normal subjects (0.30) is abnormally low.

Our own data on bromide space and total body water combined with those of others show that the regression slope of the former on the latter is much the same in the obese as in the nonobese (Chapter 4, p. 140). The preponderance of the evidence, therefore, speaks against an abnormal ECF/body H_2O ratio in human obesity.

However, there are two—possibly three—types of human obesity that are associated with a decrease in LBM. Ellis and Cohn (1975a) did ^{40}K assays on eight women and six men with Cushing's syndrome. Body K content (an index of LBM) was only 89% of that predicted for normal males, and only 76% of that for normal females. Ernest (1967) assayed total body water and total exchangeable K in 12 patients with Cushing's syndrome; the majority had values lower than those expected for normal adults, and in 10 who were assayed again after treatment total exchangeable K increased and body fat declined. Ross et al. (1966) also found an increase in total exchangeable K and a loss of weight following operative treatment in three patients with Cushing's syndrome.

Observations on adults with hypothalamic obesity, though few in number, show that these individuals do not exhibit the increase in LBM that characterizes ordinary obesity. Bray and Gallagher (1975) measured total exchangeable K and total body water in three women with obesity secondary to hypothalamic lesions and in five with "essential" obesity. When calculated per unit height the average exchangeable body K of the former group was only 80% of the latter, and total body water was only 90%. Nelson et al. (1981) studied seven children with the Prader-Willi syndrome and compared them to 10 children with ordinary obesity. Calculated LBM in the former group was less than that of the latter, as was BMR, but since the authors failed to specify the sex of their subjects or to list individual ages, it is not possible to interpret their findings. As will be noted later (p. 218) animals rendered obese by hypothalamic injury have a reduced LBM.

With regard to LBM in obesity, some years ago Pitts (1962) described a linear relationship between LBM and fat in the guinea pig; over a range of 50–300 g fat, the calculated regression slope was 0.71 g LBM/g fat.

(Unfortunately, the animals were eviscerated prior to analysis.) With this in mind, we surveyed our data on female subjects with varying amounts of body fat as determined by ^{40}K counting. Included were normal adolescents and adult women who were within the range of ideal body weight for height, those with anorexia nervosa, and those with varying degrees of obesity. They ranged in age from 14 to 50 years. Since LBM is known to be a function of stature, only those subjects in the range of 156–170 cm are included. None had complicating disease or evidence of edema. None of the patients with anorexia nervosa had a history of vomiting or laxative abuse, and ^{40}K assays were done after a week of hospital diet at which time serum proteins and serum K were normal. None of the normal subjects were engaged in athletics.

When these 164 subjects are grouped according to body fat content, the relationship between LBM and body fat is seen to be a curvilinear one (Figure 7.2), and of great interest is the fact that the curve is semilogarithmic in character. The regression equation calculated from the grouped data is

$$\text{LBM (kg)} = 23.9 \log \text{fat (kg)} + 14.2. \quad (2)$$

The correlation coefficient as calculated from individual subjects is 0.79, and standard error of estimate is 4.7 kg. In differential form,

$$d(\text{LBM})/d\,(\text{fat}) = 10.4/\text{fat}. \quad (3)$$

Accordingly, a doubling, or halving of body fat content is accompanied on average by a $10.4 \times \ln 2 = 7.2$ kg change in LBM regardless of the initial body fat content, at least over the 14-fold range shown in Figure 7.2.

It should be noted that the data presented in Figure 7.2 pertain only to females. Although there are suggestions that the same general phenomenon holds true for males, the available data are not sufficient to permit the type of mathematical analysis used for females. There are some animal data that are in keeping with the relationship shown in Figure 7.2. Recalculation of the data on both adult mice (Dawson et al., 1972) and squirrel monkeys (Ausman et al., 1981) shows a good relationship between chemically determined fat-free weight and log body fat.

The derived relationship between change in LBM and change in body fat depicted in equation 3 permits one to make certain predictions about the composition of weight losses and gains in man. It suggests that significant changes in body weight the result of alterations in food intake will involve a larger proportion of lean tissue in individuals who are thin, and a larger proportion of fat in those who are obese. Information that bears on the accuracy of these predictions will be presented later in this chapter.

Jen et al. (1985) found that the majority of male monkeys in their colony became obese during their adult years, with values of 30–50% body fat as determined by the total body water technique. As in humans, the obese animals had higher values for plasma insulin, and for adipocyte size and

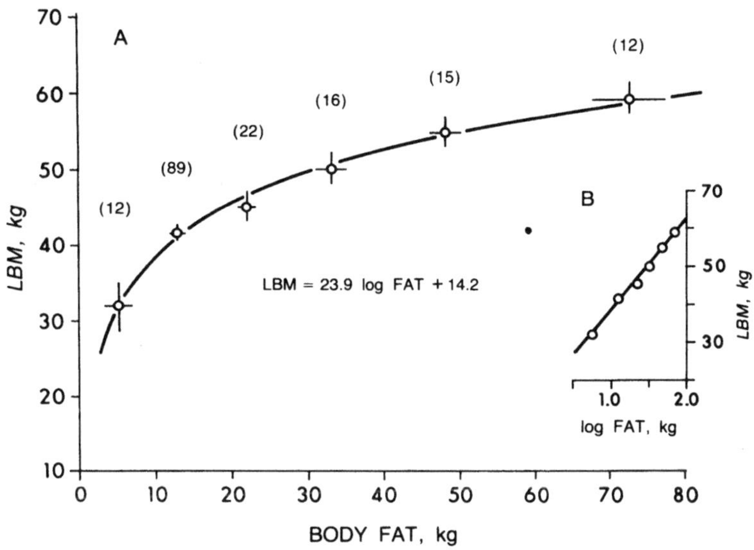

FIGURE 7.2. (A) Plot of LBM (kg) against body fat (kg) for females grouped according to body fat content. The thinnest group consists of individuals with anorexia nervosa, the next group represents normal subjects who were in the range of ideal body weight for height, and the rest represent those with progressive degrees of obesity. The vertical and horizontal bars show ±2 SEM. Numbers of subjects in parentheses. The calculated regression line is included. (B) Plot of LBM against log body fat.

number. Those who had more than 25% fat had slightly higher values for LBM than those who were thinner: the former weighed on average 5.3 kg more than the latter, and the difference in LBM was 2.1 kg. Although this latter difference is said not to be statistically significant, the wide variation in age of the entire group of monkeys (12–27 years) makes comparisons difficult, for if monkeys are like humans, one would expect that the older members would have a lower LBM than the younger ones.

Experimental and Genetic Obesity in Animals

There are a number of reports on body composition of such animals. Rats who are fed high-fat diets, or palatable diets, will usually gain weight if they are not allowed to exercise, and become moderately obese. Strains of mice and rats are known in which obesity seems to be inherited, either as a recessive or dominant trait. Obesity has been produced in rats by injuring the hypothalamus by surgery or the administration of gold thioglucose. As illustrated in Table 7.2 the effect on body composition is variable, some showing an increase in LBM, others a decrease in comparison to controls. Among the genetic forms the yellow-obese (A^y/a) mouse has more carcass protein and is longer than controls; and likewise the "*Ad*"

TABLE 7.2. LBM in experimental and genetic obesity.

Species	Cause of obesity	Change in LBM[a]	References
Rat	High-fat diet	↑	Schemmel et al. (1970), Oscai (1982), Applegate et al. (1984)
Rat	High-fat diet	-0-	Peckham et al. (1962)
Rat	Small litters	↑	Harris (1980)
Rat	Force-fed	↑	Rothwell and Stock (1979), McCracken and McNiven (1983)
Rat	Force-fed	↓[b]	C. Cohn et al. (1957)
Monkey	Overfed	↑	Jen and Hansen (1983)
Rat	Hypothalamic lesion	↓	Han (1967), J.K. Goldman et al. (1974), Frohman and Bernardis (1968)
Mouse	ACTH-secreting tumor	↓	Mayer et al. (1956)
	Genetic		
Mouse	Yellow-obese (A^y/a)	↑	Hausberger and Hausberger (1966) Plocher and Powley (1976)
Mouse	*Ad*	↑	Trayhurn et al. (1979)
Rat	Zucker	↓	Pullar and Webster (1974), Deb et al. (1976)
Mouse	*ob/ob*	↓	Bergen et al. (1975), Thurlby and Trayhurn (1978)

[a]Compared to controls.
[b]One group weighed the same as the controls, another actually weighed less; both had excessive fat and less lean.

mouse exhibits a modest increase in carcass protein. On the other hand the Zucker rat has a reduced lean weight, and the *ob/ob* mouse has less carcass protein than controls; and Bergen et al. (1975) found the reduced muscle weight to be present early in life, prior to the development of obesity in this latter species.

In considering the influence of nutrition *per se*, rats nursed in small litters (three pups per mother) acquire 10–20% more carcass protein, and are fatter than those nursed in litters of nine. Most investigators have found that rats fed high-fat diets become obese and acquire more carcass protein than those fed grain diets. Rothwell and Stock (1979) measured total body water in force-fed and control rats and report that body fat accounted for 83% of the excess weight gain in the former group. The lack of increase in LBM reported by C. Cohn et al. (1957) may have been due to the fact that their animals were tube-fed only twice daily and weighed no more (some actually weighed less) than the controls.

Obesity produced by lesions in ventromedian nucleus of the hypothalamus is accompanied by a *reduced* lean weight and length. In the series of rats studied by Han (1967) the lesioned rats had 3.6 times as much body fat as controls but only 60% as much lean; in those studied by J. K. Goldman et al. (1974) the lesioned animals had 5.6 times as much fat, but only 58% as much lean. Femur length and body length were respectively 81% and 67% those of controls. The lesioned rats studied by Frohman and Bernardis (1968) had 3.2 times as much fat, but only 85% as much lean as controls; and they were 10% shorter. Mice with adrenocorticotropin

(ACTH) secreting tumors have a reduced lean weight. Hence it would appear that animals with endocrine abnormalities behave differently than those with food-induced obesity, in regard to body composition.

Generally speaking, those animals who become obese through overfeeding show an increase in lean weight, whereas those with hypothalamic injury suffer a decline in LBM as they gain body fat. Of the four hereditary obesities listed, two have an increased LBM compared to controls, and two a decrease. Dietary induced obesity resembles the human state more than some of the other types in this respect. On the basis of LBM response, obesity induced by injuring the hypothalamus cannot be used as a model for human obesity. These animals as well as the Zucker rat exhibit abnormalities in pituitary function, carbohydrate metabolism, and tolerance to cold in addition to their excess store of body fat. The lesioned rats studied by Frohman and Bernardis (1968) had smaller pituitary glands and lower plasma growth hormone levels than controls.

However, rats made obese by hypothalamic injury can still respond to overfeeding. Walgren and Powley (1985) tube-fed normal and lesioned animals: both gained weight and both gained some LBM as a result of this overfeeding procedure, and in both the gain in weight could be related to the amount of excess energy consumed. However, it should be noted that in contrast to the findings of others (see Table 7.2) these lesioned rats did not have a reduced LBM at the initiation of the overfeeding procedure, although they were massively obese.

The reason(s) for the increased LBM that occurs in most patients with obesity is not known with certainty. As mentioned previously, it is known that organs such as heart, kidney, liver, and muscle are involved. Possibilities include a work hypertrophy of skeletal muscle in response to the greater bulk of the obese, the increased number of adipocytes each with its component of intracellular fluid, and perhaps hypertrophy of the gastrointestinal tract in response to the greater food intake. Many obese individuals have elevated serum insulin levels, and some have an increase in urinary excretion of adrenal androgens, both of which are anabolic. As has already been noted, overfed animals acquire a larger LBM than controls, and as will be noted subsequently the same is true for humans who are deliberately overfed. One cannot escape the conclusion, therefore, that human obesity develops in response to food in excess of need, i.e., a positive energy balance.

Undernutrition

In many areas of the world food is in short supply, so that large numbers of people do not have enough to eat. The result is growth faltering in children and reduced body weight in adults. The situation is intensified by frequent intercurrent infections that deplete the body of protein and essential nutrients. Although such is true for certain segments of Western society—the poor and the mentally disturbed—the most striking examples are the result of disease. Malignancy; psychosomatic disorders; intestinal malabsorption; and serious cardiac, hepatic, and renal disease all can have an impact on nutritional health. The changes in body composition vary somewhat, depending on the nature of the nutritional insult.

The post-war years saw the application of body composition techniques to the study of malnutrition. Kerpel-Fronius and Kovach (1948) estimated ECF volume in malnourished infants and found that ECF volume as a percent of body weight was inversely related to the degree of body weight deficit. Severely malnourished infants had values as high as 45% of body weight in contrast to those of about 25% in normal infants. In their studies of weight loss in young adults Keys et al. (1950) found that plasma volume and ECF volume remained the same in the face of a weight loss of about 15 kg.

In contrast to the situation in obesity, where the ratio of ECF volume to total body water is similar to that in normal weight individuals (see p. 215), this relationship is disturbed in malnutrition. Numerous studies of malnutrition, both in animals and man, whether induced by underfeeding or occurring under conditions of reduced food intake, have consistently shown an increase in the ratio of ECF volume to total body water. It would appear that the shrunken cells of the body retain their fluid covering; in a sense the ratio of ECF to ICF reverts to its infantile status.

Waterlow et al. (1960) and Widdowson and Dickerson (1964) have reviewed the earlier work on this topic. Severely undernourished humans and rats may have ECF volumes equal to controls despite the difference in weight. The ratio of ECF volume to total body water is higher, as is the body Cl/K ratio, and with nutritional rehabilitation the increase in cell mass proceeds to the point where these ratios revert to normal. Since the mass of bone does not suffer as much as soft tissue during undernutrition, the undernourished animals have more Ca per unit of fat-free weight than the controls, though they have less total body Ca.

Recent work has confirmed these trends. Flynn et al. (1967a,b) found an increased ECF/total H_2O ratio in malnourished children, and Barac-Nieto et al. (1978) reported a similar finding in malnourished adults. Shizgal (1981) compared a group of undernourished patients to controls; although the patients had a weight deficit of 11 kg, they had a somewhat greater ECF volume. A striking feature was the ratio of total exchangeable sodium (Na_e) to total exchangeable potassium (K_e): this ratio was 1.95 (meq/meq) for the wasted subjects, or about twice that of the controls (ratio 0.98). The ratio of K_e to total body water was reduced. Beddoe et al. (1985) studied protein-depleted patients by neutron activation and by tritium dilution. Total fat-free weight was calculated from total body water and total nitrogen. The calculated water content of the fat-free body mass was 750 ml/kg for the female patients and 732 ml/kg for the male patients, compared to 726 and 711 ml/kg, respectively, for the normal subjects.

LeMaho and her associates (1981) made an extensive series of observations on fasting geese. Nitrogen excretion paralleled the fall in body weight, and total red cell mass declined, but ECF volume and plasma volume remained unchanged throughout most of the fast. Of interest is their finding that the mass of the pectoral muscles was maintained while other muscle groups diminished in size.

Small animals, such as the marmot, attempt to conserve body tissues during hibernation by dropping their metabolic rate to about 5% of its usual value, and dropping their body temperature to about 5°C (Morgulis, 1923). By the end of the hibernation period the marmot has lost 35% of its initial weight: 16% of this loss is body fat, 7.6% is muscle, 5.6% is skin, 2% is skeleton, and 1.9% is liver. From the standpoint of body composition, about 54% of the loss is lean tissue and 46% is fat. Weinland (1925) reported similar findings in the hibernating hedgehog. The bear is a clear exception, for as will be mentioned later this animal loses only body fat during hibernation.

Emperor penguins normally fast—and somehow survive the Antarctic winter!—while incubating their eggs. At the end of this long period they weigh only half as much, their body fat stores (which are about 30% of body weight at the start) are almost depleted, and their fat-free weight has been reduced by 25–30%. Despite this severe insult, they are able to waddle off to the nearest open water to hunt for food! (Dewasmes et al., 1980).

Young miniature pigs gained very little weight during 12 weeks on a low-protein diet, and at the end of this time they had a higher ECF/total water ratio (0.57) than the controls (0.51), and a higher ratio of ECF volume to total body potassium (Tumbleson et al., 1969). Five-week-old rats fed either a high-protein diet in limited amounts or a low-protein diet *ad libitum* lost weight and had a much higher ECF/total body water ratio (0.58) than the controls (0.38) (Čabak et al., 1963a). Other findings included an increase in the ratio of collagen N to total N and a drop in the K/N ratio in the whole body. Analyses of liver, skin, and quadriceps muscle showed that these organs, which together accounted for a large share of the weight differential, also had an abnormal ECF/total water ratio. However, the Ca/N ratio in the femur was unchanged.

These animals represented an extreme circumstance, for the ones who were malnourished weighed only about one-third as much as the controls at the end of the 26-day experiment. Of some interest is the finding that the changes in body composition in those animals fed the high-protein diet in limited amounts were roughly similar to those in the animals fed the low-protein diet *ad libitum*. The former group ate less food than the latter—although neither ate as much as the controls—and had less body fat at the end of the experiment (0.6% vs. 2.2%); however total body weight was about the same. It is evident that under conditions of reduced food intake the provision of large amounts of protein does not alleviate the effects of malnutrition on body composition; one must conclude that it is energy intake which is important.

In studying adult rats maintained on a protein-deficient diet Angeleli et al. (1978) found that total body collagen N declined to a lesser extent than total body N. Individual organ analyses found that this was true for skin but not for muscle. Anasuya and Narasinga Rao (1970) had earlier reported that the same was true for total rat carcass minus skin, and that urinary excretion of hydroxyproline—an amino acid unique to collagen—was reduced; however, in contrast to Angeleli et al. (1978) they found that skin collagen N was reduced more than noncollagen N. They suggested that the various collagen pools in the body have different turnover rates.

Picou et al. (1966) analyzed the bodies of infants dying of malnutrition and compared these results to those who were clinically well nourished and who died from other causes. The bodies of the malnourished infants had only about half as much noncollagen protein as the others whereas the amount of collagen protein was 85–129% of the expected value. This tendency for body collagen to resist the ravages of malnutrition poses a problem in interpreting the results of *in vivo* body composition assays. Tendon and skin have a lower K/N ratio than muscle and viscera, so significant weight loss will act to reduce the total body K/N ratio, with the result that the use of total body K as an index of loss of LBM may provide a larger calculated LBM loss than will assays for total body N. Alleyne et al. (1970) measured total body K and urinary creatinine excretion in infants with severe malnutrition. When referred to height, both measurements were reduced to about two-thirds of the expected values.

Metcoff et al. (1960) did muscle biopsies on children with kwashiorkor. The Na/K ratio was far higher in the malnourished children [1.41 compared to 0.59 (meq/meq) in normal subjects]; intracellular K concentration was lower and intracellular Na concentration was much higher. A number of carbohydrate intermediates were also assayed by these authors, and they postulate that the disturbed ionic content in malnourished muscle is the result of interference with Krebs cycle metabolism. Nichols et al. (1973) found an increase in muscle water and Na in malnourished children, and a decrease in K; liver and brain were also partially depleted of potassium. Leukocytes from malnourished children also exhibit an abnormally high Na/K ratio (Patrick and Golden, 1977).

Samples of quadriceps muscle from patients with Crohn's disease have less K and more water than controls, and the calculated ECF/total water ratio is increased (Nyhlin et al. 1985).

Alleyne (1975) has reviewed the electrolyte abnormalities that occur in infants with severe malnutrition: hyponatremia, potassium and magnesium deficits, a relative increase in total body sodium, and a reduction in the K/N ratio in brain and muscle. Nichols et al. (1969) found that muscle K accounted for a much smaller proportion of total body K than in normal subjects. Magnesium deficiency is largely due to losses in stool, and there is evidence that this can accentuate K deficiency.

Thus information from a number of studies in both animals and man reveals a rather consistent picture of altered body composition in states of malnutrition: cells shrink and ECF volume tends to be preserved; and the shrunken body cells have lost some of their K, Mg, and P while gaining Na. With recovery from malnutrition body fluids return to their normal status. However, it should be noted here that osmotic equilibrium is preserved, for Edelman et al. (1958) have shown that the ratio of $Na_e + K_e$ to serum Na + K is equal to total body water in a variety of clinical conditions including malnutrition.

By what mechanisms are these changes in body fluid and electrolyte distributions brought about? The hypothesis that body cells merely shrink inside their ECF envelope is not sufficient: what keeps the ECF volume from shrinking too? Although a complete answer is not possible, some of the hormonal and functional changes that have been described in severe malnutrition are worthy of consideration. In addition, the frequent occurrence of diarrhea in malnourished infants and children results in dehydration and depletion of body potassium and magnesium, and the ensuing acidosis compromises the Na-K pump of the cell membrane. It is known that cardiac output and glomerular filtration rate are reduced, and that salt and water loads are not excreted promptly (Alleyne, 1975). Some children with kwashiorkor have elevated levels of serum aldosterone (Migeon et al., 1973), which revert to normal on recovery. Srikantia and Mohanran (1970) report elevated plasma and urine levels of antidiuretic hormone, and Kritzinger et al. (1972) found elevated levels of plasma renin. Hence there are several factors that operate in the direction of preserving ECF volume, and one (aldosterone) that promotes Na retention and K loss.

The finding of some investigators that plasma volume tends to be preserved in the face of malnutrition (Keys et al., 1950; Le Maho et al., 1981) suggests the existence of factors that operate to equalize the ratio of interstitial fluid (ISF) volume to plasma volume. If such factors do exist, this could explain the finding of normal ECF volume in malnutrition.

It is known that animals made K-deficient will retain excessive amounts of Na and acquire additional amounts of intracellular Na when they are provided with generous amounts of dietary Na, and that this process is augmented when they are given adrenal mineralocorticoids. It might be of interest to look at the level of the newly described natriuretic hormone "auriculin" which is produced by the cardiac atria (Laragh, 1985); if perchance the shrunken hearts of individuals with severe malnutrition curtail the production and/or secretion of this hormone, this could compromise the organism's ability to excrete sodium. It is unusual to see mention of salt intake in reports of patients with malnutrition; such information would be helpful in understanding the pathogenesis of the abnormal body fluid distribution.

Acute malnutrition has been produced in fetal rats by clamping one uterine artery late in pregnancy (Hohenauer and Oh, 1969; Roux et al., 1970). Such fetuses are considerably smaller than those residing in the unclamped uterine horn. Body nitrogen content is lower and water content slightly higher in the stunted fetuses; brain weight is somewhat reduced, but the most striking change is in liver, whose weight is half that of the controls, and whose glycogen content is a mere one-third. RNA and DNA contents are reduced. Strangely, body fluid distribution was not altered appreciably.

Arara and Rochester (1982) did careful dissections of the diaphragm in patients who came to autopsy. Some had been ill prior to death; others died suddenly. They found a good correlation between diaphragm weight and body weight in those who had died suddenly. Patients who were underweight had lighter diaphragm muscles (average 150 g) than those who were of normal weight (average 218 g). Keys et al. (1950) found a decrease in the size of muscle fibers in their subjects who lost weight during underfeeding. CAT scans of the upper arm of patients with malnutrition show an irregular, "moth-eaten" appearance of the muscle (Heymsfield et al., 1979).

In their studies of adult rats fed low-protein diets Uezu et al. (1983) found that about half of the weight loss, which amounted to 14% of initial body weight, consisted of muscle and bone (the two were not separated), 19% was skin, and 6% was liver; brain weight did not change. Of the total loss of nitrogen, 71% was accounted for by muscle and bone, 17% by skin, 6% by liver, and 3% by gastrointestinal tract.

Patients with anorexia nervosa have had assays for total body potassium and nitrogen (Fohlin 1977; Russell et al., 1983; Forbes et al., 1984). Total body N, total body K, and blood volume are all subnormal; glomerular filtration rate is reduced in the face of a normal renal plasma flow, so the filtration fraction is less. Using values obtained 2 weeks after admission to the hospital the ratio of body K (meq) to body N (g) was 1.45, which is lower than values for normal individuals in the laboratory of Russell et al. (1983).

It should be remembered that most patients with anorexia nervosa fail to exhibit some of the cardinal features of other forms of severe malnutrition. They usually have normal values for blood hemoglobin and serum albumin, they rarely have edema or signs of vitamin deficiency, and they do not have diarrhea. One has the impression that their diet has been adequate except for energy. However, Ljunggren et al. (1961) have documented an excess of ECF volume relative to cell mass in these patients.

James et al. (1984) compared wasted adult patients to normal subjects of the same height and shoulder width, the latter being taken as an index of skeletal size. The patients had 40% less total body K and 27% less total body N than the controls; their total body K/N ratio averaged 1.64 meq/g compared to the normal value of 2.0. Burkinshaw and Morgan (1985) found an average total body K/N ratio of about 1.6 meq/g in patients with heart disease and those with malnutrition compared to one of 1.8 meq/g in normal individuals. This low total body K/N ratio in malnutrition reflects the fact that body collagen does not diminish during weight loss to the same extent as the noncollagen proteins of muscle and viscera.

All of these data are offered in support of the well established principle that not all organs participate to the same degree during weight loss. This fact when combined with the above-mentioned distortions in body fluid distributions make for some difficulty in interpreting the results of body composition assays in situations involving severe malnutrition. Soft tissues suffer more than bone, so the density of the LBM is increased; the fat-free body mass has a somewhat larger complement of water, so tritium or deuterium dilution will yield a falsely high value for LBM; intracellular K concentration is subnormal, at least in muscle, and the total body K/N ratio is subnormal. What one would like to have is an easily applicable method for directly estimating body fat, but as was mentioned in Chapter 2 the only one available (namely, uptake of fat-soluble gases) is extremely difficult to use. Despite valiant efforts on the part of many investigators, anthropomorphic techniques have not shown themselves to be equal to this task. The compressibility of skinfolds is altered in states of dehydration and in patients with marked weight loss; the usual sites chosen for the measurements are not representative of the entire subcutaneous fat mantle, and there is no uniformity of opinion as to the relative amounts of body fat contained in the subcutaneous and internal sites.

So the techniques for studying body composition *in vivo* provide estimates, and one cannot claim for them a high degree of precision. They work well in individuals who are normal, and although they have their faults in subjects who have gross abnormalities of nutritional status, it is likely that they are not too far off the mark in situations where the abnormalities are moderate in degree.

Composition of Weight Gains and Losses

Weight Gain

Many investigators have shown that the rehabilitation of malnourished individuals is accompanied by a positive nitrogen balance as body weight is gained. Such a phenomenon is to be expected, but does the same occur when normal people are overfed? Do they put on only fat, or a mixture of fat and lean? The early experiments of Cuthbertson et al. (1937) and of Wiley and Newburgh (1931) showed that the feeding of excess food to adult men resulted in positive nitrogen balance, and Keys et al. (1955) spoke of "obesity tissue" as containing water and protein as well as fat. Passmore et al. (1963) then reported that nitrogen retention also occurred in overfed women.

In this section we will first consider compositional changes in normal humans and animals who gain weight, and later the changes that occur during recovery from malnutrition.

A number of overfeeding experiments have now been done in which body composition techniques have been used, and in which the subjects were carefully observed so that energy intake in excess of maintenance needs could be estimated. In the review to follow only those subjects who consumed at least 20,000 excess kilocalories during the experimental period were included, since conclusions based on short-term experiments could be compromised by changes in fluid or glycogen stores. Some were fed excess energy as fat, others as carbohydrate, and others as a mixture of carbohydrate and fat; all had an adequate protein intake; and they either had body composition assays prior to and at the end of the overfeeding period or complete nitrogen balance during the course of the study. All were said to be in good health; and a further prerequisite was a fore-period of weight maintenance.

We have studied two adult males and 13 adult females in this manner, and an additional 31 subjects have been taken from the literature. They ranged in body weight from 44 to 118 kg; the overfeeding periods ranged from 14 to 83 days. Figure 7.3 is a plot of weight gain against total excess energy for these 46 subjects, 16 of whom were females. Although there is a certain amount of scatter in the data, the correlation coefficient is significant at the 5% level, and the *y*-axis intercept is close to zero. This suggests that the relationship between weight gain and excess energy intake is indeed linear. There is no obvious sex difference in the response. The slope of the regression line indicates that 0.124 kg were gained for each additional 1000 kcal consumed; hence on average the energy cost of the weight gain was 1/0.124 = 8.05 kcal/g (Forbes et al., 1986).

Figure 7.3 also shows the changes in LBM recorded for the 39 subjects who had estimates of body composition. For those subjects who had determinations of nitrogen balance, the change in LBM was calculated on the basis that this component of the body contains 33 g N/kg. Although the correlation coefficient is not as high as for the data on weight change, the *y*-axis intercept is close to zero. Whereas all of the subjects without exception gained weight when overfed, three failed to experience a change in LBM. The reason for the more variable response in LBM is probably the relative lack of precision of body composition assays compared to the scales used for measuring body weight. The average increment in LBM was 2.1 kg, which represents an increment of about 4%, whereas the technical error of the assay procedure is at least 2%. The ratio of the two regression slopes suggests that 0.048/0.124, or 38% of the weight gain, consisted of LBM. Ravussin et al. (1985) found that LBM comprised 44% of the weight gain in men given an extra 17,250 kcal during a 9-day period.

As one contemplates these data, it should be realized that there is considerable room for error in the various procedures employed. These include errors in food preparation and consumption, in body composition assays, in estimating the maintenance requirement, and variations in

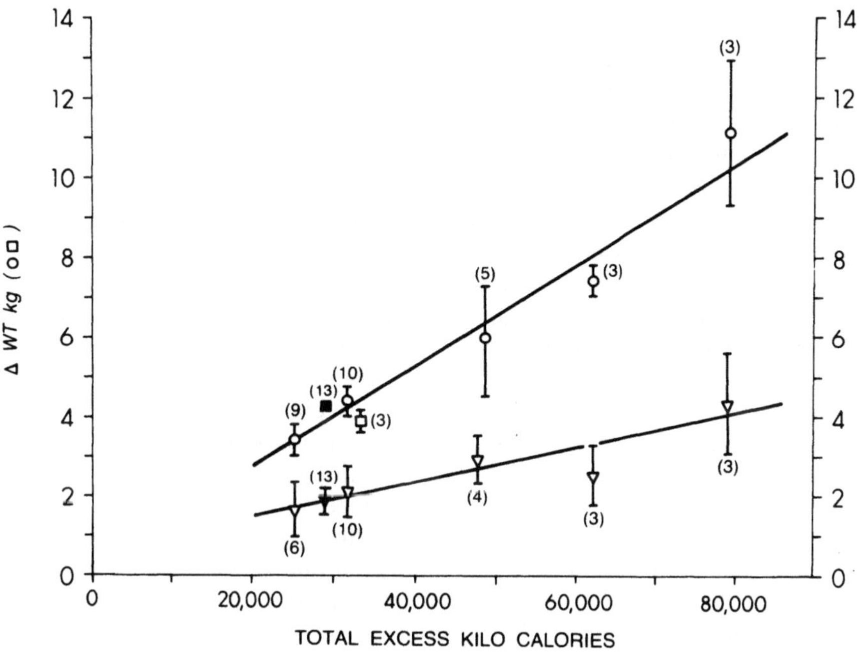

FIGURE 7.3. Upper line is a plot of increment in body weight against total excess energy consumed during overfeeding in man. Subjects grouped in energy categories. Symbols: males (○), females (□) (author's subjects (■)); vertical bars are SEM. Regression equation: y(kg) = 0.29 + 0.124x (Mcal), r = 0.77. Lower line is a plot of increment in LBM against total excess energy. Symbols: males (▽), females (▼). Regression equation: y = 0.28 + 0.048x, r = 0.49. Data sources: Forbes et al. (1986) (^{40}K counting); Passmore et al. (1955), Wiley and Newburgh (1931) (nitrogen balance); R. F. Goldman et al. (1975), Norgan and Durnin (1980), Webb and Annis (1983) (densitometry); Miller and Mumford (1967) (weight only).

physical activity. Errors of this type are difficult to avoid in experiments of this type.

There are two recent reports showing that weight gain in overfed rats is directly proportional to energy intake (McCracken and McNiven, 1983; Walgren and Powley, 1985). Furthermore, both investigators found an increase in LBM as well as fat; the ΔLBM/ΔW ratio was in the range of 0.15 to 0.26.

In their studies of squirrel monkeys fed various diets Ausman et al. (1981) were able to monitor food intake as well as body weight. In 24 instances involving 11 animals there were periods of 4 weeks during which body weight was either maintained, declined progressively, or rose steadily. When change in body weight was plotted against energy excess, or deficit, compared to maintenance requirement, there evolved a linear relationship (r = 0.83). Of great interest to the present argument is the fact that the slope of the regression line is 0.13 g/kcal (= 7.7 kcal/g) which is surprisingly close to that shown in Figure 7.3. The range covered was from −40 g to +70 g/week and from −300 to +500 kcal/week. The intercept on the y-axis of 3.8 g/week, which is not far removed from zero, suggests that the relationship between change in body weight and energy intake holds for both weight gain and weight loss (see p. 230 for a further analysis of their data).

The longest period of controlled overfeeding in human subjects recorded to date is 83 days; it is impossible to say whether the linear relationships shown in Figure 7.3 would persist for longer periods. In animal experiments Blaxter (1971) has shown a curvilinear relationship between energy balance and intake. However, his data represent a wide range of intakes (4–6-fold) and include both positive and negative energy balance.

The observed relationship between body weight and the energy need for weight maintenance (see pp. 210, 211) suggests that weight gain on excess energy diets would indeed be curvilinear for man over the long term. As body weight increases during overfeeding maintenance energy requirement also increases, at the rate of 20 kcal/day for each kilogram of weight gained. Hence if the new energy intake persists, body weight will eventually increase to the point where this energy intake represents a maintenance level; and at that point weight gain will cease. Hence, it is likely that the weight gain during very prolonged overfeeding would be curvilinear, with an asymptote at this new level. Based on the average energy cost of weight gain of 8050 kcal/kg gain, it can be estimated that this new equilibrium would be approached with a half-time of 278 days.[2]

The fact that the relationship between weight gain and excess energy consumed appears to be linear in Figure 7.3 is not incompatible with this formulation, simply because the longest observation period (83 days) is

---

[2]The maximum weight gain will be Δkcal/day ÷ 20 kcal/kg/day, and the fractional rate of weight increase will be $1 - \exp(-\lambda t)$, where λ is 20 kcal/kg/day divided by the energy cost of the weight gain (8050 kcal/kg), and t is time in days. Hence the formula is

$$\Delta W\,(\text{kg}) = \Delta\text{kcal/day} \div 20\,[1 - \exp(-0.0025t)] \qquad (4)$$

The final weight, after many days, is initial weight plus Δkcal/day/20, and this is approached with a half-time of $\ln 2/\lambda = 278$ days. For example, were an extra 400 kcal to be consumed each day, 10 kg would be gained in 278 days, i.e., one half-time, 15 kg at the end of 556 days, 17.5 kg at the end of 834 days, and eventually the maximum gain would approach 20 kg. This assumes, of course, that energy expenditure remains the same and that the composition of the tissue gained does not change.

One can also use Equation 4 to estimate the energy excess necessary to achieve a given increase in body weight at any time t:

$$\Delta\text{kcal/day} = 20 \times \Delta W/[1 - \exp(-\lambda t)]$$

When t is very large, the denominator on the right hand side approaches one, so that as in the example just cited, $\Delta\text{kcal/day} = 20 \times 20 = 400$.

I am indebted to Dr. William Simon for this formulation.

so short in relation to the theoretical half-time of 278 days. Assuming a constant excess of 1000 kcal/day for the entire 83 days, the estimated weight gain calculated from Equation 4 would be 1000/20 [1 − exp (0.0025 × 83)] = 9.5 kg, which is so close to the observed value of 0.29 + 0.124 × 83 = 10.6 kg (see legend to Figure 7.3) that the difference between the observed linear formulation and the exponential formulation cannot be resolved. The only way to test the validity of Equation 4 is to study subjects for long periods of time (at least 278 days, which is one half-time) under strictly controlled conditions, a feat most unlikely to be accomplished given the vagaries of human subjects.

The data shown in Figure 7.3 suggest that the energy cost of the weight gain was 8.05 kcal/g gain, i.e., the reciprocal of the regression slope. A question of some importance is how well this value corresponds to the theoretical value derived from the composition of the gain.

According to Spady et al. (1976), the cost of depositing a gram of fat is 12 kcal, and for a gram of protein it is 8.66 kcal; and since the LBM contains 20.6% protein the energy cost is 1.78 kcal per gram LBM. Based on the observed ratio of the two regression slopes (0.38), the energy cost of the weight gain is:

$$1.78 \times 0.38 \text{ g LBM} + 12 \times (1 - 0.38) \text{ g fat} = 8.12 \text{ kcal/g } W, \quad (5)$$

which is not too far from the value of 8.05 kcal/g derived from the regression line of Figure 7.3.

It may be of interest to compare the composition of the weight gain induced by overfeeding (namely 38% LBM) with those recorded for other situations involving increased body weight. Spontaneous weight gain also is accompanied by an increase in LBM: my observations (G. Forbes, unpublished) on three obese adolescents (average body fat 31 kg) who gained an average of 15 kg over a period of several months showed that LBM accounted for 40% of the weight gain; Sjöström (1980) restudied five obese adult women (average body fat 61 kg) who had gained an average of 10 kg over a period of several years, and found that LBM accounted for 31% of the weight gain. According to Table 7.1, an average of 29% of the excess weight in individuals with established obesity consists of LBM (range 19–40% in 10 reported series). Since most obese individuals usually will have acquired their added weight over rather long periods of time, one could speculate that the ΔLBM/ΔW ratio would eventually diminish were the overfeeding period to be continued beyond the 83-day maximum recorded for the subjects in this study.

A further point should be made about overfeeding experiments. All of the subjects listed in Figure 7.3 were fed adequate, but not excessive, amounts of protein, and all were in good health at the start of the experiment. However, when very high protein intakes are given, as in Müller's (1911) early experiment, nitrogen balance is strongly positive, to the point of suggesting that almost all of the weight gain consisted of lean tissue. On the other hand, when low-protein diets are used there is evidence that although body fat is gained some LBM is actually lost. For example, the subjects studied by Miller and Mumford (1967) who were fed low-protein–excess-energy diets gained weight but lost body K, whereas those fed normal protein-excess energy diets gained body K as they gained weight. The experiments of Barac-Nieto et al. (1979) showed that although undernourished men can gain weight on low-protein diets some LBM is lost at the same time that body fat is gained. When adequate protein was given to these same subjects without a change in energy intake, they gained LBM very readily. A moment's reflection on Equation 5 makes it obvious that low-protein–high-energy diets would be inefficient, in that the energy cost of the weight gain is higher than that for diets containing adequate amounts of protein.

Another point concerns the initial status of the subjects who are overfed. The data shown in Figure 7.2 and the nature of Equation 3 (see p. 216) predict that weight gain in thin individuals would consist, at least initially, of a large proportion of lean tissue since the right-hand denominator is small; and that individuals with more normal stores of body fat would have a somewhat smaller $\Delta LBM/\Delta W$ ratio as they put on weight. Such is indeed the case: the weight gain of patients with anorexia nervosa during nutritional rehabilitation consists of 64% LBM (Forbes et al., 1984), and Barac-Nieto et al. (1979) found that LBM made up 68% of the weight gained by undernourished men during nutritional repletion. Motil et al. (1982) supplemented the diet of undernourished adolescents who had Crohn's disease. Energy intake was increased from 67 to 96 kcal/kg body weight per day, and protein intake from 2.3 to 3.2 g/kg/day. During the 3-week period of study these six patients gained an average of 3.3 kg in weight and 7.7 g K (by ^{40}K counting); and they retained about 60 g of nitrogen. Based on body K the gain in LBM was 2.9 kg and based on N retention it was 2.0 kg; hence well over half of the weight gain consisted of LBM. All of these values are far higher than those reported for overfed normal subjects (38% LBM) or for obese individuals who gain weight (31–40% LBM).

The body composition data presented by Ausman et al. (1981) for the squirrel monkey can be offered in support of this principle. Figure 7.4 which is taken from their publication shows the composition of weight increments as the animals progress from thinness (body fat 7%) to obesity. The initial weight increment is only 33.5% fat—and hence 66.5% lean—while the final increment is 77.6% fat and 22.4% lean. This trend is quite in keeping with the prediction contained in Equation 3 (p. 216), which states that the change in LBM with respect to fat is inversely related to body fat content: the fatter the animal the larger the proportion of weight gain that is represented by fat.

A final consideration has to do with the age of the subjects. When young animals are supplied with a surfeit of food they grow faster than controls, and a large fraction of the weight differential consists of lean tissue. Adult

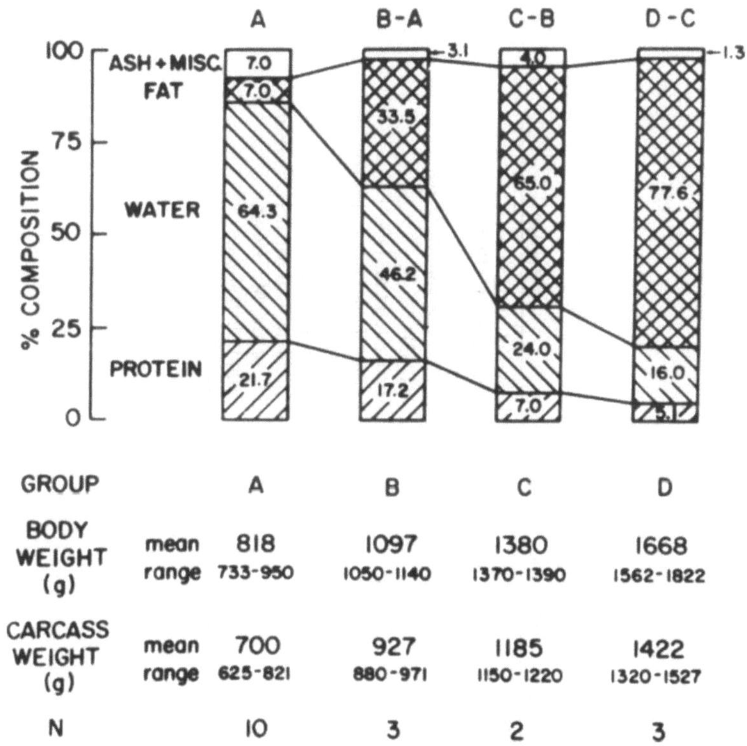

FIGURE 7.4. Carcass analysis of 18 male squirrel monkeys recalculated to show composition of carcasses of 10 lean control monkeys (body weight 733–950) (group A) and composition of incremental tissue for each of the next three groups: fatted control (B), and obese (C and D). [From Ausman et al. (1981), with permission.]

animals also gain weight under such conditions, but the majority of their weight gain consists of fat, and the same is true for those in whom dietary manipulation was begun after weaning. The data listed in Table 7.3 illustrate the effect of age on the nature of the response. Here the difference in weight and in LBM between the experimental animals and the controls are given, together with the composition of this difference, namely the ratio of LBM difference to the weight difference.

It should be noted here that for all of the animals listed in Table 7.3 in whom body composition was determined by carcass analysis the entire body (minus gastrointestinal tract contents) was assayed. There are a number of reports on animals that have been eviscerated prior to analysis; such analyses are obviously not representative of the entire animal, and hence the data are not analogous to those obtained by *in vivo* techniques, such as isotopic dilution.

It appears that the ability to increase LBM in response to nutritional surfeit is somehow blunted by the aging process. Young organisms possess

TABLE 7.3. Effect of feeding on body composition.

Species	Age, duration	Procedure	Weight difference[a]	LBM difference[a]	Ratio	Reference
Baboon	Birth, 18 wk	High-energy diet	+575 g	+447 g	0.78	Lewis et al, (1983)
Rabbit	Birth, 21 d	Double-fed	+169 g	+141 g	0.83	Spencer et al. (1985)
Rat	Birth, 21 d	Overfed	+16 g	+13 g	0.81	Heggeness (1962)
Rat	Adult, 30 d	Tube-fed	+254 g	+39 g	0.15	McCracken and McNiven (1983)
Rat	5 wk, 10 wk	High-fat diet	+44 g	+13 g	0.30	Applegate et al. (1984)
Rat	Adult, 3 wk	Tube-fed	+58 g	+11 g	0.19	Rothwell and Stock (1979)
Monkey	Adult, 2 mo	Gastric feeds	+3.3 kg	+1.13 kg	0.34	Jen and Hansen (1983)
Rat	Adult, 3 wk	"Cafeteria diet"	+49 g	-0-	-0-	Rothwell and Stock (1979)
Rat	3 wk, 20 wk	High-fat diet			0.20[b]	Schemmel and Michelsen (1970)
Rat	Adult, 6–10 wk	"Cafeteria diet"			0.10[c]	G. Forbes, M. Brown, unpublished data
Rat	Adult, 6–10 wk	"Cafeteria diet" + exercise			0.23[c]	

[a]Experimental − control.
[b]Regression of Δ LBM on Δ W for seven groups of females and seven groups of males.
[c]Regression of LBM on body weight.

the ability to lay down significant amounts of lean tissue as they gain weight in response to nutritional surfeit; the adult having lost much of its ability to do so responds by laying down proportionally larger amounts of fat. The observations of Winick and Noble (1967) provide a possible explanation: they found an increase in DNA content of young rats who were overfed, but no increase in protein/DNA ratio. Hence cell number increased in response to overfeeding but not cell size. Since later growth in this species is largely accomplished by cell hypertrophy rather than hyperplasia, the older animal has lost much of its ability to manufacture additional amounts of lean tissue.

Is there a similar age effect in man? Although a number of adults have had body composition assays during deliberate overfeeding experiments, there are no such studies on normal infants and children. However, it is known that children who become obese during childhood exhibit a slight but definite acceleration in statural growth (Forbes, 1977). Children whose obesity became manifest in infancy tend to have a somewhat larger LBM than those who became obese later in childhood (Forbes, 1964; Cheek et al., 1970).

These data are in keeping with those on young animals who have been overfed (Table 7.3). An interesting question for speculation is whether these nutritionally induced increases in LBM persist in later life. The study reported by Widdowson and McCance (1960) suggests that they do. These authors studied rats who were nursed in small litters (3 pups/mother) and in large litters (17 pups/mother); after weaning they were fed *ad libitum*. The former group of rats weighed more at weaning and had a larger LBM (and more body fat) than the latter group. Of significance is the fact that these weight and body composition differences tended to persist, albeit in lesser degree, into adult life. The question of whether such early nutritional influences would have long lasting effects in man is impossible to answer.

The results of the "cafeteria diet" experiments shown in Table 7.3 are less striking than those in which the animals were fed high-fat diets or were tube-fed normal diets. The diet offered to our female rats consisted of chocolate chip cookies, lard mixed with rat chow, and sucrose added to the drinking water; fresh water and rat chow were also freely available. The control animals had chow *ad libitum*. Some of the animals were killed after 5 weeks, others after 10 weeks on diet. Those fed the "cafeteria diet" gained more weight (average 198 g) than those fed chow (average 83 g), but as shown in the table most of the weight differential was fat. The weight differential of the animals studied by Rothwell and Stock (1979) on a comparable diet was entirely fat; however their experiment was briefer in duration.

Caged animals are forced to be sedentary. When we provided a second, and similarly fed, group of animals with a running wheel that they could use as desired, the weight gain was less—average of 150 g for the "cafeteria diet'' animals, and 63 g for those fed chow. However, as shown in Table 7.3, these animals had a slightly higher ratio of LBM gain to total weight gain.

There are also several experiments in which normal adults have been overfed lesser amounts of food than those depicted in Figure 7.3. V.R. Young et al. (1984) fed young adult males 35–57 kcal/kg/day for 84 days on a metabolic ward. Of the six subjects who took more than 45 kcal/kg five gained weight, and of the eight subjects who took less than 45 kcal/kg six lost weight. The change in weight ranged from −3700 to +3500 g, and the cumulative nitrogen balance ranged from −90 to +40 g. When all of the subjects are considered, nitrogen balance turns out to be a function of weight change: the slope of the regression line is 12.6 g N per kilogram ($r = 0.86$), which suggests that LBM made up $12.6 \div 33 = 38\%$ of the tissue lost or gained in this experiment (LBM contains 33 g N/kg).

In experiments involving the overfeeding of young adult males on diets providing an extra 400–800 kilocal/day, Butterfield and Calloway (1984) found a retention of nitrogen, in the amount of 1–2 mg per excess kilocalorie fed. This works out to be about 0.05 g LBM per excess kilocalorie, which is close to the value of 0.048 derived from the regression slope shown in Figure 7.3. The remarkable thing about this study is that nitrogen retention occurred in the face of a very modest intake of protein, namely 0.57 g/kg body weight per day. The average ratio of LBM gain to weight gain was 0.41.

Chien et al. (1975) studied a group of Taiwanese adults in 1954 and the same subjects again in 1966, by means of body density. During this 12-year period the diet of the general population improved, and some of the subjects gained weight. The following analysis pertains only to those who were less than 50 years of age at the time of the last measurement, so as to exclude the effects of aging. Of the 16 males and three females who gained weight during these 12 years the average gain was 10.1 kg, and the average increment in LBM was 2.58 kg, so the $\Delta LBM/\Delta W$ ratio was 0.26. On the other hand, for the five males and two females who either lost a little weight or who gained less than 1 kg, the average change in weight was −0.2 kg, and the average change in LBM was −1.0 kg. It would seem likely, therefore, that the weight gain in the first group was due to better nutrition since there was a concomitant increase in LBM. The magnitude of the loss of LBM in the second group, amounting to perhaps 2% of LBM in 12 years, is about the same rate of loss as others have found in normal adults (see Chapter 5).

Nancy Butte and her co-workers (1985) have published some interesting observations on lactating women who had monthly assays of body density from the first to the fourth month postpartum. There were 37 women whose body density assays did not show unexpected fluctuations. Some gained weight during the 3-month period of lactation, some lost weight, whereas others had no change; the change in body weight ranged from a loss of 7 kg to a gain of 7 kg. Although there is a fair amount of scatter in the individual data, which were kindly provided by Dr. Butte, the trend is clear: those women who lost weight tended to lose both LBM and fat, and those who gained weight tended to gain both LBM and fat. As calculated by least-squares regression analysis, the equation is $y = +0.54 + 0.29x$ ($r = 0.74$), where y is change in LBM and x is change in body weight (both in kg). Thus on average 29% of the weight change consisted of LBM and 71% was fat. These observed effects, which took place in "free-living" subjects whose diets were not prescribed, are similar in magnitude to those seen in subjects whose diets were imposed.

Earlier mention was made of the fact that thin and/or undernourished individuals acquire relatively large amounts of LBM during nutritional rehabilitation, at least in the initial stages. Alleyne et al. (1970) reported very low values (about two-thirds normal) for urinary creatinine/height index and for total body K/height index in malnourished infants, and that both indices achieved normal values after about a month of nutritional repletion. Brooke and Cocks (1974) showed that malnourished infants who were not edematous gained relatively more total body K than body weight during recovery.

Repeated assays of body density were made of the subjects studied by Keys et al. (1950) as they recovered from their imposed undernutrition, which amounted to a loss of about 23% of their initial weight. For the first 12 weeks the weight gain consisted of about equal portions of LBM and fat, but thereafter body fat content increased rapidly to the point where it clearly exceeded the initial value by about 4.5 kg, then fell back to the starting value of 10 kg as LBM progressively rose to normal by the end of the 58th week of rehabilitation. In this particular experiment, body fat content overshot the mark, so to speak, during recovery. The experiments of Keys et al. (1950) are the most detailed and the longest in duration of any thus far reported so it is impossible to say whether these trends in body composition during recovery from undernutrition were unique.

Since the energy cost of laying down a gram of lean tissue (about 1.8 kcal/g) is far less than cost of depositing a gram of fat (12 kcal/g) (see Equation 5, p. 229), the energy cost of weight gain should vary according to the composition of that gain. Jackson et al. (1977) indeed found an inverse relationship between the energy cost and the relative proportion of muscle gained in children recovering from malnutrition. The energy cost of the weight gain in our patients with anorexia nervosa, two-thirds of whose weight gain was LBM (Forbes et al., 1984), was estimated to be 5.3 kcal/g, whereas in normal individuals induced to gain weight, in whom 38% of the gain was LBM (see p. 226), this value was 8.05 kcal/g. As was mentioned in Chapter 4, p. 138, prematurely born infants acquire a large proportion of LBM (some 68–77% of weight gain) during growth, and it has been estimated that their energy cost is 2.8–4.0 kcal/g. For children recovering from malnutrition Spady et al. (1976) found an average value of 4.4 kcal/g.

When the energy cost of weight gain for these groups of subjects is

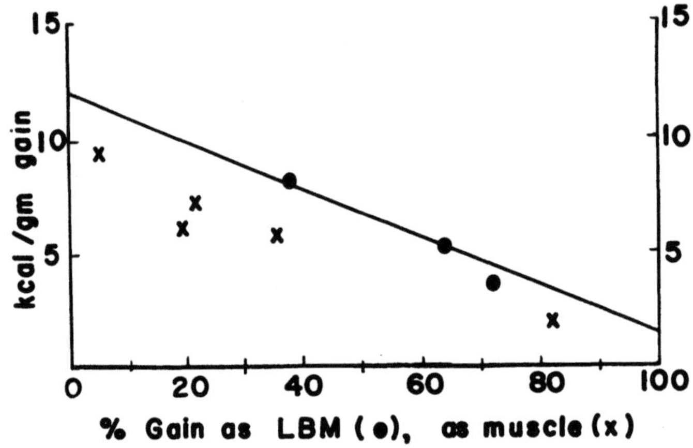

FIGURE 7.5. Plot of energy cost of weight gain against % of gain as LBM (•): overfed normal adults (38% LBM) (Forbes et al., 1986); patients with anorexia nervosa (64% LBM) (Forbes et al., 1984); prematurely born infants (72% LBM) (see Chapter 4). Percent of gain as muscle (x), calculated from [^{15}N]-creatine kinetics in malnourished children, data of Jackson et al. (1977). Solid line is based on estimates by Spady et al. (1976) of energy cost of fat deposition (12 kcal/g) and of protein deposition (8.66 kcal/g = 1.8 kcal LBM/g).

plotted against the composition of the gain the result is as shown in Figure 7.5. The higher the proportion of lean the lower is the energy cost, a result that could have been anticipated from the estimated costs (Spady et al., 1976) of depositing a gram of fat and a gram of protein. The solid line in the figure has been drawn between these two points, namely zero LBM and 100% LBM. The good correspondence between the observed data and the theoretical values suggests that the estimates of Spady et al. (1976) were very close to the mark.

The data reported by Jackson et al. (1977), which are represented by the (X) symbols, refer to the proportion of the weight gain represented by muscle tissue in children who were recovering from malnutrition. The trend is the same as that for proportion of LBM, the lower absolute numbers being due to the fact that muscle comprises only a portion of the LBM.

It seems fair to conclude, therefore, that the observed variation in the energy cost of weight gain can be ascribed to the composition of the tissue gained.

To summarize briefly, significant weight gains both in animals and man, whether induced by overfeeding, observed during rehabilitation of malnourished individuals, or occurring under natural circumstances, almost always consist of a mixture of LBM and fat. The relative contribution of these two body components to the total weight gain depends on the initial body composition status of the subjects, and their age. Obviously, protein intake must be adequate, and other dietary essentials as well, but given those prerequisites the type of food fed makes little difference. A detailed analysis of the data shown in Figure 7.3 shows that variations in the response to overfeeding of normal adults could not be ascribed to sex, type of food fed, or smoking behavior (Forbes et al., 1986). Our patients with anorexia nervosa gained lean weight as well when fed modest (10% of calories) amounts of protein as they did on high-protein intakes (Forbes et al., 1984).

The need for a diet adequate in all respects is made clear by the report of Rudman et al. (1975). These investigators fed undernourished adult patients by the intravenous route. Use of a complete feeding high in energy produced satisfactory weight gain and adequate retentions of N, P, K, Na, Cl, and Ca. However, the withdrawal of either N, P, Na, or K impaired or abolished the retention of the other elements although some weight was gained. Just as satisfactory growth in children demands an intake of all dietary essentials so too does the rehabilitation of undernourished adults and the promotion of weight gain in normal subjects.

Weight Loss

It has been known for many years that people who fast continue to excrete nitrogen in their urine, and so must be losing LBM as well as fat (Benedict, 1915). Despite Voit's (1901) early demonstration that obese animals lose proportionally less body nitrogen than thin animals during fasting, it is not generally appreciated that the same is true for fasting humans.

An analysis (Forbes and Drenick, 1979) of individuals who had fasted showed that the obese lost an average of 10 g N for each kilogram of weight lost, whereas the nonobese lost twice as much, or 20 g N/kg. Hence the weight lost by the latter contains a larger proportion of lean tissue. This undoubtedly accounts for the greater tolerance for fasting by the obese. Of interest is the finding that the decline in body N content in fasting individuals behaves as a two-component exponential function (Forbes and Drenick, 1979); in the obese about 6% of body N is lost with a half-time of 10 days, and the remainder with one of 433 days. Values for the nonobese yield half-times of 2.4 and 116 days.[3]

This means that the fractional loss of body nitrogen per day is about 3½ times greater in thin individuals as it is in the obese during fasting; since their body stores of nitrogen are usually less (Table 7.1) the thin person is at much greater risk. In those parts of the world where famine is apt to occur, obesity constitutes a survival factor.

Figure 7.6 shows the average rates of nitrogen loss and total weight

---

[3]The equations as derived from least-squares regression analysis are $N_t = N_o [0.06 \exp(-0.066t) + 0.94 \exp(-0.0016t)]$ for the obese and $N_t = N_o [0.01 \exp(-0.284t) + 0.99 \exp(-0.0060t)]$ for the nonobese. Initial body N was calculated from urinary creatinine excretion. The standard deviations from regression are 0.1 g and 0.08 g, respectively.

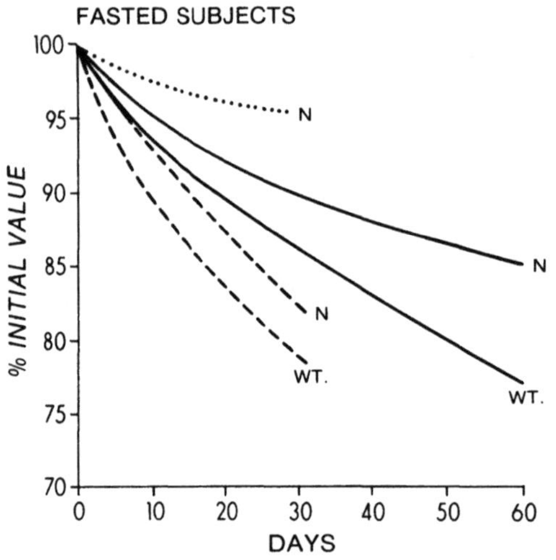

FIGURE 7.6. Effect of fasting on body weight (W) and body nitrogen (N) in human subjects: obese (—) ($n = 9$), nonobese (---) ($n = 4$). Dotted line (• • •) shows reduction in body nitrogen (N) for subject on protein-free, energy-adequate diet. [From Forbes and Drenick (1979), with permission.]

loss during fasting. Included is the calculated change in body N for an individual who consumed a protein-free adequate-energy diet for 30 days, and whose body weight showed very little change.

The relative loss of body nitrogen is less than that of body weight in both the obese and the nonobese individuals. It may be mentioned here that the decline in body weight is also exponential in nature during fasting (Forbes, 1970). After the first few days the obese lose weight at the rate of 0.32% per day, the nonobese at the rate of 0.55% per day. The respective half-times are 220 and 127 days. The exponential nature of the weight loss process means that the loss rate, and hence the energy expended, body tissues being the only source of energy, is proportional to body weight.

However, animal studies have shown that these exponential relationships do not last indefinitely: there is an antemortem increase in both weight and nitrogen loss.

Of more practical importance are the observations on individuals who are given low-energy diets. Nitrogen balances have also been recorded for such individuals, and a number have had assays of body composition. Generally speaking, N balance is strongly negative for the first few days, and then becomes less negative as time goes on.

Benedict et al. (1919) recorded negative nitrogen balance in a group of nonobese young men who were fed a 1900-kcal diet for 7 weeks. During this period they lost on average 7 kg weight and 105 g N; hence LBM contributed 3.2 kg, or 46% of the total weight loss. Later Keys et al. (1950) repeated the experiment, combining a 1600-kcal diet with a vigorous exercise program, and using densitometry to assess body composition. They found that LBM constituted 63% of the weight loss during the first 12 weeks, and 56.6% for the entire 24 weeks of the experiment. By this time some of the subjects had visible edema.

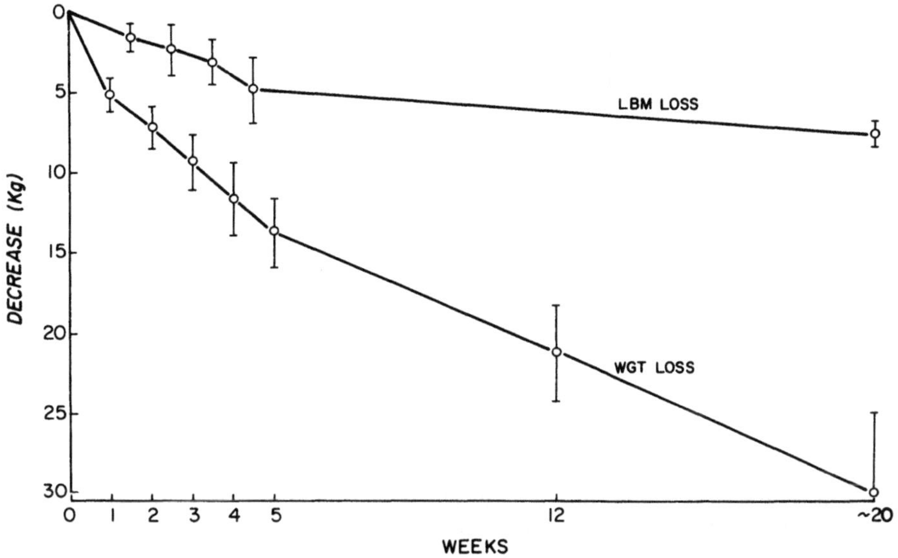

FIGURE 7.7. Decrease in weight and LBM (^{40}K counting) in obese adolescents consuming a 500–700-kcal diet (protein 1.5 g/kg/day). Vertical bars are SD. n = 8 for the first 5 weeks, n = 5 subsequently. [From Brown et al. (1983), with permission.]

Some years ago Albert Behnke decided to lose weight and so put himself on a diet. During the ensuing 10 months his weight dropped from 101.5 kg to 86.5 kg and his LBM (by densitometry) fell from 69.9 to 67.6 kg. The ratio of LBM loss to weight loss was 2.3/15.0 = 0.15 in response to what must have been a very modest energy deficit since he lost an average of only about 50 g/day (Behnke and Wilmore, 1974, p 207).

The construction in recent years of special weight-reducing diets for the obese has brought with it a renewed interest in the composition of the weight loss in such individuals. Claims have been made that low-energy diets composed largely of protein will "spare" protein as weight is lost. There are now a number of reports in the literature having to do with the composition of the weight that is lost in response to various diets, and to the intestinal bypass and gastric stapling procedures. Since it is known that low-protein diets lead to negative nitrogen balance in the face of adequate calories (see Figure 7.6), the analysis to follow will concern itself solely with experiments employing diets that provide at least 45 g of protein daily (fasting excepted), and that were at least 4 weeks in duration. Short-term observations may be difficult to interpret due to fluid shifts and changes in liver glycogen, and after all the goal of all weight reduction programs is what happens in the long run.

Figure 7.7 shows the course of weight loss and calculated LBM loss in a group of obese adolescents (four boys and four girls, aged 11–18 years)

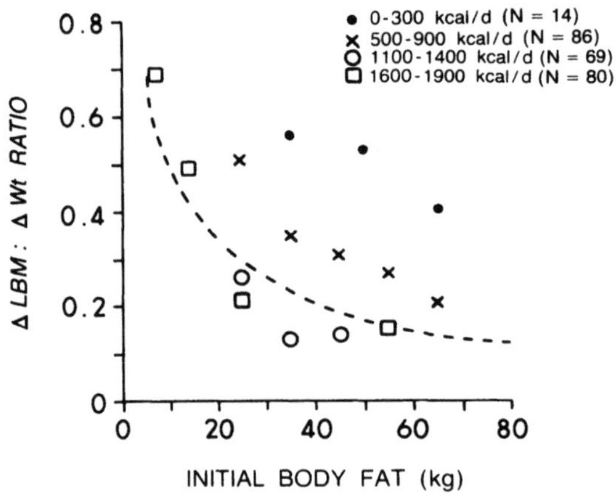

FIGURE 7.8. Plots of fraction of weight loss due to LBM in underfeeding experiments of at least 4 weeks' duration, grouped according to initial body fat content and energy intake. Symbols: 0–300 kcal/day (●); 500–900 kcal/day (x); 1100–1400 kcal/day (○); 1600–1900 kcal/day (□). Data points are averages of three or more subjects in each category. The dotted line is based on data from stable individuals of varying fat content (Equation 3); see text for explanation. Data sources arranged according to technique used. Densitometry: Parizkova (1977), van Seters et al. (1983), Belko et al. (1985), Bogardus et al. (1981), Jacobson et al. (1964), Keys et al. (1950), C Young et al. (1971), C Young and DiGiacomo (1965), Phinney et al. (1980), Ditschuneit et al. (1985). ^{40}K counting: Brown et al. (1983), Warnold et al. (1978)*, Lantigua et al. (1980), Archibald et al. (1983), Doré et al. (1982), Barnard et al. (1979)**, Vaswani et al. (1983)***, Björntorp et al. (1975). Total body water: Woo et al. (1982). *Also H_2O method; **Ke, H_2O; ***also body N by neutron activation, H_2O.

who were given a 500–700-kcal "protein sparing" diet over a period of several weeks (Brown et al., 1983). They lost significant amounts of weight, and despite adequate amounts of protein, potassium, and other dietary essentials they also lost lean weight. At the end of 5 weeks the ratio of LBM loss to weight loss was 0.36, and by 20 weeks it was 0.26. Although the changes in body composition are not as profound as those seen in people who take no food at all, they are nonetheless appreciable.

To return once more to Figure 7.2 and Equation 3 (pp. 216, 217): the latter predicts that the composition of the tissue lost during weight reduction will depend, all other things being equal, on the initial body fat content of the subject. The thinner the person, the smaller will be the right-hand denominator of Equation 3 and hence the higher the ΔLBM/Δfat ratio. In the analysis to follow we shall use the $\Delta LBM/\Delta W$ ratio, since the error in determining body fat is larger than that for weight and for LBM by the usual body composition techniques.

We have already seen that obese individuals lose proportionally less body nitrogen during fasting than do nonobese persons. We shall now consider the question as to whether initial body fat has an influence on the composition of the weight lost by people who are given food, but in submaintenance amounts.

Figure 7.8 is a plot of the fraction of the total weight loss due to LBM against initial body fat recorded by body composition assays for individuals on various energy intakes. A total of 249 subjects were available for analysis; all had lost significant amounts of weight. For ease of illustration these are grouped into various body fat and energy intake categories. It is at once apparent that the ΔLBM/ΔW ratio is indeed a function both of initial body fat content and the magnitude of the energy deficit. For a given range of energy intake, the ΔLBM/ΔW ratio progressively falls as initial body fat increases; and furthermore, for a given range of body fat content, the ΔLBM/ΔW ratio is inversely related to energy intake.

The dotted line in this figure is based on a modification of Equation 3 (p. 216), which describes the relation between LBM and body fat in individuals of varying body fat content. The rearranged equation is

$$d\,(\text{LBM})/d(W) = 10.4/(\text{fat} + 10.4)^4 \qquad (6)$$

This equation says that for a given (small) change in weight [$d(W)$], the ratio ΔLBM/ΔW is an inverse function of body fat. It is derived, as stated earlier, from body composition studies of individuals of varying body fat content and whose weights were reasonably stable. The dotted line thus represents what would happen if the composition of an individual's weight loss on a given diet were to mimic the difference in body composition between himself and the person who is normally just a bit thinner. It is of great interest that the points in Figure 7.8 representing diets of 1100 kcal or more are not greatly removed from this line, suggesting that intakes of this magnitude will result in approximately the expected decline in LBM.

Bortz's (1969) famous patient, who lost 213 kg of his initial 304 kg on an 800-kcal diet over a period of 700 days, had 38 kg less LBM at the end as judged by urinary creatinine excretion. The calculated ΔLBM/ΔW ratio for this extended period of weight loss was 0.18. This patient's initial body fat content of over 200 kg puts him beyond the range covered in Figure 7.8.

There is some variability in the results reported by the various investigators, due no doubt to variations in physical activity, compliance with the prescribed diets, and to technical errors in assessing body composition. The results might have been more uniform had it been possible to estimate the actual energy deficit; obviously a 1000-kcal diet, for instance, represents a larger energy deficit for the average man than for the average

---

[4]Starting with Equation 3, d(LBM/d(fat) = 10.4/fat; substitute $d(W) - d$(LBM) for d(fat), and rearrange:
fat = [10.4 $d(W)$ − 10.4 d(LBM)]/d(LBM),
whence fat = 10.4 $d(W)/d$(LBM) − 10.4,
and 10.4 $d(W/d$(LBM) = fat + 10.4.
Inverting, we have Equation 6.

woman. Nevertheless, the general trends shown in Figure 7.8 are clearly evident. Unfortunately, the data in many of the published reports are presented in a manner which does not permit a calculation of the variability of the ΔLBM/ΔW ratios.

It would appear that there is no level of reduced energy intake that will completely spare LBM when significant amounts of body weight are lost. Nevertheless, the obese have a distinct advantage over lean individuals in that their weight loss consists of a smaller proportion of LBM; in addition, as mentioned earlier, most obese individuals have a larger store of lean tissue at the start. Even modest reductions in energy intake can lead to some loss of LBM. Welman et al. (1980) arranged for a group of adult men to consume 500 kcal/day less energy than their usual diet; over a period of 70 days they lost 6 kg in weight, and about a third of the loss consisted of LBM.

Data from animal experiments show that dietary-induced weight loss involves a number of organs, brain and bone being the least affected. Human subjects fed low-energy diets excrete less creatinine, so muscle mass suffers, and Keys et al. (1950) found a diminution in heart size in their subjects who had lost 16 kg (23% of initial weight) on a 1600-kcal diet. This occurred despite the superimposition of a regular exercise program. Now MacMahon et al. (1986) have found echocardiographic changes in individuals who sustained a smaller weight loss (8.3 kg, or 9% of initial weight) in response to a 1000-kcal diet over a period of 21 weeks. In these moderately obese hypertensive adults, left ventricular mass declined by 20%, posterior ventricular wall thickness by 11%, and septum thickness by 13%. Blood pressure also fell, but unfortunately body composition assays were not done.

Subtle changes in brain size and configuration have been noted on CAT scans of patients with anorexia nervosa (Datlof et al., 1986).

A number of investigators have determined nitrogen balance in subjects given low-energy diets. Earlier mention was made of the experiments of Benedict et al. (1919) on nonobese young men fed a 1900-kcal diet. Later Benedict et al. (1923) fed two steers on reduced rations (about one-half their usual intake) for 140 days. These animals lost an average of 150 kg in weight, or about 25% of their initial weight, and 1630 g nitrogen during this period. The ratio of N loss to weight loss was 11 g N/kg which is not too different from that which he and his associates had found for underfed men.

Neumann (1902) did a series of feeding experiments on himself over a period of many months. Whenever he reduced his energy intake to 1600–1900 kcal/day he lost weight and went into negative nitrogen balance.

Recently there has been considerable talk about the possibility of constructing a "protein-sparing" diet, namely a low-energy diet composed completely or almost completely of high-quality protein. Some investigators who have claimed that nitrogen equilibrium eventually occurs on such diets have failed to estimate unmeasured N losses (skin and body secretions), have omitted any reference to sweating, and have recorded positive N balances for some of their subjects—the last a most unlikely circumstance indeed. Then there is the matter of the rate of weight loss, which tends to decline as time goes on. Take for example a ratio of 5 g N/kg weight loss: if a subject on a low-energy diet continues after a time to lose 200 g in weight per day—a not unreasonable figure—the expected negative balance would be only 1 g N per day. Lacking meticulous attention to detail, the possibility that this could be recorded as zero balance is easy to foresee. As mentioned earlier, one very careful study, which included losses through skin and via body secretions, of normal women who were at stable body weight recorded an average nitrogen balance of +0.7 g/day (King, 1981); so an error of this magnitude in a subject who is losing weight at the rate of 140 g/day could easily be construed as zero balance.

With regard to the possibility of eventual nitrogen equilibrium on low-energy diets, during the 6th week of their experiment Durrant's (1980) subjects lost an average of 183 g/day, and were said to be in slight positive balance (0.09 g N/day). If this latter figure is corrected for unmeasured losses (amounting to about 0.5 g/day) the result is actually a negative N balance of about 0.4 g N/day, so that the $\Delta N/\Delta W$ ratio was about 2.2 g N/kg at this time. Looked at in this way, true N equilibrium was actually never achieved.

The following table (Table 7.4) lists the ratios of nitrogen loss to weight loss recorded for individuals given low-energy diets providing at least 45 g protein/day. To correct for unmeasured losses, five mg N/kg/day was added to the recorded N loss if not done by author.

Using the value of 33 g N/kg LBM, the ratio of $\Delta LBM/\Delta W$ turns out to be 0.27 for the obese subjects and 0.46 for the nonobese.

Archibald et al. (1983) gave 10 adolescents a diet providing 920 kcal and 150 g protein (= 24 g N) for periods ranging from 70 to 133 days. Although they all lost body K as they lost weight, only eight lost body N as assayed by neutron activation. The fact that two actually gained body

TABLE 7.4. $\Delta N/\Delta W$ for subjects on weight reduction diets.

n, sex	Subjects	Intake/day g N	Intake/day kcal	Duration	W loss (kg)	$\Delta N/\Delta W$ (g N/kg)	Reference
6 M	Obese	11–21	600–700	64 d	22	8.6	Yang et al. (1981)
9 M	Obese	8	700–800	68 d	19	11.9	Wynn et al. (1985)
16 F	Obese	8	700–800	68 d	12	9.6	Wynn et al. (1985)
17 F	Obese	11.2	800	84 d	12.5	5.0[a]	Vaswani et al. (1983)
1 M	Obese	8.8	500	30 d	13.6	14.7	Drenick (1967)
5 F	Obese	7.2	850	42 d	9.8	7.7	Durrant et al. (1980)
11 M	Nonobese	10	1900	49 d	7	15	Benedict et al. (1919)
10 M	Obese	16	400	40 d	19.4	8.2	Fisler et al. (1982)
5 F	Mod. obese[b]	10	512	42 d	12.7	4.4	Bistrian et al. (1977)

[a]N by neutron activation.
[b]Includes 1 week of fasting: values for energy and N intake are averages for entire 42-day period.

N suggests that there may have been some problems in the calibration of their neutron activation apparatus. Hence their results are not included in Table 7.4.

A number of investigators have recorded negative nitrogen balances on subjects given low-energy diets that provide less than 7 g N per day. Since these represent subnormal protein intakes in addition to inadequate energy intakes they have not been included in the table.

To return to the concept of protein sparing, Fisler et al. (1982) actually studied two groups of obese subjects, one given 400 kcal/day and 16 g N/day for 40 days (Table 7.4), whereas the others were fasted for 40 days. As would be expected, the latter lost more weight (23.4 kg vs. 19.4 kg) and more nitrogen (277 g vs. 160 g) than the former. The $\Delta N/\Delta W$ ratios were 11.7 and 8.2 g N/kg. The former ratio is almost identical to that reported by Forbes and Drenick (1979) for fasting adults. It is of interest to compare the nitrogen balance of these two groups during the final week of the experiment. The average for those who fasted was -4.8 g N/day, and for those on 400-kcal diets it was -1.6 g/day. Hence the provision of some food does reduce the wastage of body nitrogen that occurs in a fast, but it does not "spare" protein completely.

In general terms, it appears that the relative losses of K are apt to be somewhat greater than those for N during weight reduction. For example, Vaswani et al. (1983) found that the K/N ratio of the tissue lost by their subjects was 3.76 meq/g as they lost weight, a value considerably higher than that found in normal individuals. Burkinshaw and Morgan (1985), using neutron activation, found a lower K/N ratio in ill patients (by about 10%) compared to healthy individuals. In the analyses of rats kept on protein-free diets for 31 days Uezu et al. (1983) found that 71% of the nitrogen lost came from muscle and bone and only 17% from skin and hair. Hence muscle, which has a much higher K/N ratio than skin, contributed a major share of the total nitrogen loss. Another factor is the loss of glycogen from liver and muscle. Each gram of glycogen deposited in liver is accompanied by 0.35 meq K, but no nitrogen (Fenn, 1939), so losses of glycogen would be expected to contribute to this phenomenon.

The graph of one patient shown in the paper by Wynn et al. (1985) shows that negative N balance persisted for at least 150 days during weight loss. Another graph shows that the average N balance for their 13 subjects remained negative for 68 days. Bortz et al. (1967) state most of their obese patients on 800-kcal diets remained in negative N balance for as long as 110 days.

Confirmation of the relationship of the composition of tissue lost during weight reduction to initial body fat content as shown in Figure 7.8 is provided by N balance studies of Durrant et al. (1980); they found an inverse relationship between $\Delta N/\Delta W$ and initial body fat in their subjects.

All of the data referred to thus far support Drenick's (1967) early contention that protein loss "cannot be avoided with this type of diet." It appears reasonable to conclude that significant weight reduction, even in the face of a protein intake that meets the requirement for normal individuals (0.6–0.8 g/kg), is associated with some loss of lean tissue. Protein equilibrium is dependent on an adequate energy intake. The magnitude of the N loss (and the weight loss) will depend on the magnitude of the energy deficit.

Göranzon and Forsum (1985) have done an interesting experiment. Young adult men and women were fed a mixed diet providing 0.57 g protein/kg/day, and after a baseline period an energy deficit was produced by one of two maneuvers: (1) a reduction in energy intake of about 16% while physical activity remained nearly constant; and (2) an increase in energy expenditure (by physical activity) of about 24% while energy intake remained constant. Both maneuvers produced weight loss and a negative nitrogen balance, and what is most significant is the fact that the change in N balance, calculated as mg N/day/kcal change in energy balance, was about the same for both, namely about 2 mgN/kcal. It is the magnitude of the energy deficit that is important, not its cause; and hence experiments looking for the effects of diet on nitrogen balance, and on body composition, must control for energy expenditure as well as energy intake.

Large numbers of obese individuals have been subjected to gastrointestinal surgery—intestinal bypasses of various types, and gastric stapling procedures. The former is designed to decrease the absorptive area of the intestine, and patients soon learn that diarrhea ensues unless food intake is curtailed; the latter effectively reduces stomach capacity. Both procedures result in significant weight loss, a loss that frequently surpasses that achieved by other forms of treatment, such as diet, drugs, or behav-

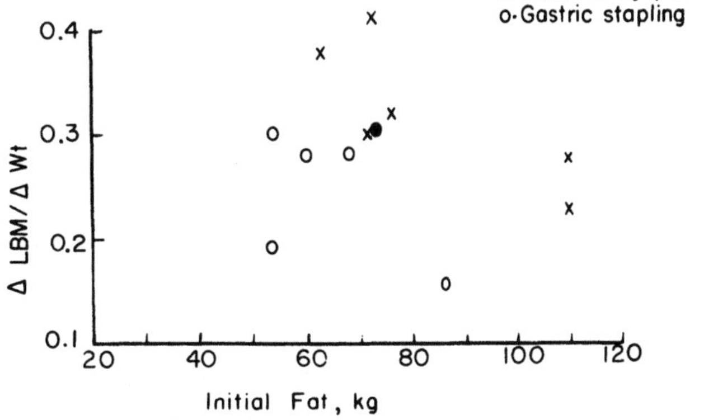

FIGURE 7.9. Fraction of weight loss due to LBM in obese individuals following intestinal surgery, arranged according to body fat content. Data sources as follows: intestinal bypass (x): Scott et al., 1975 (n = 7); Spanier et al., 1976 (n = 8); Kral et al., 1977 (n = 15); Brochner-Mortensen et al., 1980 (n = 8); Cheek et al., 1975 (n = 1). Gastric stapling: (●) Olsson et al., 1984 (n = 15); (○) G. Forbes and P. Schloerb, individual patients, unpublished.

ioral modification. What is the composition of the weight loss in such patients?

Figure 7.9 shows that when the obese are subjected to surgery of the gastrointestinal tract they also lose LBM. The data points represent only those individuals who had body composition assays both prior to and subsequent to surgery, and who had been followed for at least 4 months. The general result shown here is roughly similar to that shown in Figure 7.8, in that the fraction of the weight loss due to LBM, namely $\Delta LBM/\Delta W$, is inversely related to initial body fat content. An additional point is that for a given body fat content those individuals who had an intestinal bypass procedure lost proportionally more LBM than those who had the gastric stapling operation. This difference may relate to the observation that the former patients often have some degree of malabsorption; or it is possible that they ate less. Both procedures result in loss of LBM as well as fat.

Studies on the obese Zucker rat show that they also lose weight after intestinal bypass surgery. Although their body weight 10 months after surgery was only 56% that of the nonoperated controls, these animals still had an excess of body fat, namely 44% of body weight compared to 15% for the lean control animals (Greenwood et al., 1982). About a third of the weight differential consisted of lean tissue.

These data support the general theme of this section: whenever significant amounts of body weight are lost, both LBM and body fat participate in the weight loss process. One might say that these two body components are companions: a change in one is accompanied by a change in the other; they are not completely independent entities. The extant data do not show that only body fat can be lost during significant weight reduction, nor do they show that it is the sole component of weight gain. Finally, the relationship between LBM and fat shown in Figure 7.2, and described by Equations 2 and 3 for stable individuals of varying body compositions serves to predict reasonably well the relative changes in body composition when body weight is gained or lost.

However, it must be admitted that there is one clear exception to this rule—the hibernating bear. In the absence of food and water (!) this remarkable animal has been observed to lose 17 kg (13% of initial weight) during 60 days of hibernation without a change in LBM, and in the face of only a slight drop (1°C) in body temperature (Nelson et al., 1973, 1975). They accomplish this by reabsorbing nitrogen, water, and electrolytes through the bladder. This remarkable adaptation to fasting and thirsting occurs only during winter sleep, for when bears are fasted (but not thirsted) in summer they behave as do other mammals and man in losing LBM as well as fat. Other hibernators such as the marmot and hedgehog do lose LBM as well as fat, despite a significant drop in body temperature and in metabolic rate (Morgulis, 1923; Weinland, 1925).

Body composition has also been measured in birds prior to and after migration. Though they appear to be more efficient at conserving LBM than are humans, they cannot match the hibernating bear in this regard.

Most but not all of their weight loss during migration is fat. The recorded $\Delta LBM/\Delta W$ ratios are 0.10 for hummingbirds (Odum and Connell 1956), 0.23 for palm warblers (D.W. Johnston, 1968), and 0.28 for white-throated sparrows (Odum and Perkinson, 1951). In view of the inverse relationship between body fat and the fraction of weight loss due to LBM in man, it is of more than passing interest to note that the same holds true for these three groups of migrating birds. The premigration body fat content of the hummingbird is 43%, for the palm warbler 21%, and for the sparrow 16%. So once again, the fatter the subject the less is the relative contribution of LBM to the total weight loss, this time in the face of what must be vigorous exercise!

8

Influence of Physical Activity

> Living systems are worn out by inactivity and developed by use.
>
> *A. Szent-Györgyi*

> I am convinced that anyone interested in winning Olympic Gold Medals must select his parents very carefully.
>
> *P.O. Åstrand*

Athletes are in the news today, and so, too, are the astronauts: the former strive for increased muscle mass and lesser fat mass, and the latter must contend with the catabolic effects of weightlessness; both are held in high esteem, and today the adulation of the young who aspire to athletic prowess has been intensified by the large incomes that some athletes now enjoy. More women are engaged in athletics than ever before, there are several journals devoted to exercise physiology and sports medicine, and the sports sections of the daily newspapers have expanded.

Do athletes differ from nonathletes in body composition? Do various athletic types differ in body composition? Certain prerequisites must be met before these questions can be answered. It is known that body composition differs between the sexes, that it is altered by age, both in the child and in the adult, and that LBM is a function of stature at all ages thus far exmained. Hence any comparisons between the athlete and the nonathlete *must be controlled* for all three of these influences. Unfortunately, there are several reports that are deficient in this regard and so cannot be evaluated.

An additional point concerns the need, when recording changes in body composition the result of exercise, training, or for that matter decreased physical activity, also to record concomitant changes in body weight. Then there is the matter of dietary intake. Nutrition can alter body composition, and as was shown in the previous chapter the effects of grossly abnormal energy intakes can exceed any that might accrue to changes in physical activity *per se,* and diets deficient in essential nutrients are known to lead to negative nitrogen balance in the face of an adequate supply of energy. A final consideration is the use of medication, especially anabolic-androgenic steroids. As will be shown later, the use of these compounds will promote nitrogen retention, and thus effect an increase in LBM, even in individuals who do not exercise.

Body Composition in Athletes

An early report by Brozek (1954) showed that 50-year-old men who were physically active differed in body composition from their sedentary age and height peers. They were heavier, on average, by 3 kg; their lean weight was 4 kg greater, and body fat 1 kg less. Cook et al. (1969) did assays of plasma volume, extracellular fluid (ECF) volume, and total body water in 20-year-old male college students. On average, wrestlers had 7 kg more lean body mass (LBM) (as estimated from total body water) and 4 kg less fat than controls of the same height; competitors in track events had 4 kg more LBM and 6 kg less body fat than controls. Although ECF volumes were higher in the athletes, values for plasma volumes were slightly lower. Aloia et al. (1978) compared 40-year-old marathon runners to sedentary men of the same age and height: the former had 4% more total body K, 15% more total body Ca (by neutron activation), and 4% more bone mineral content of the forearm bones, although they weighed 7% less.

Parizkova (1977) assayed five groups of trained athletes by densitometry, together with controls of the same age, height, and weight. Included were adolescent boys engaged in track and field events, adolescent girl gymnasts, young adult male wrestlers and female gymnasts, and 67-year-old men who were physically active. In all five groups average body fat was less in the trained subjects than in those who were relatively inactive, the difference in body fat ranging from 4 to 10%. The difference was most striking in the two groups of young adults: male wrestlers had only 5% body fat and female gymnasts only 10%, compared to 15% and 23%, respectively, for the controls.

Parizkova (1977) did a 5-year longitudinal study of adolescent boys, beginning at age 11 years. Densitometry was performed at yearly intervals. Those boys who were actively engaged in sports were generally taller than those who were not so engaged, had a greater LBM, and had less body fat than the others throughout the entire 5-year period. Since there was a height difference, the ratio of LBM to height is a better index than LBM *per se*. When this is calculated, the physically active boys had from age 12 on a larger LBM/height ratio (by 1–7%) than the others, the largest increment being seen in the oldest group. The active boys also had a higher maximal oxygen consumption and larger heart volumes.

Dalén et al. (1974) measured the limb bones of cross-country runners who had been actively engaged in this sport for at least 25 years. As estimated by x-ray spectrophotometry the runners had 6–19% greater density of the long bones in both the upper and lower extremities.

Huston et al. (1985) have published a detailed review of the athletic heart. Studies have included echocardiography as well as electrical and radiographic techniques. Left ventricular mass is increased in both "isotonic" and "isometric" performers: both have a thicker septum and ventricular wall, but the former has in addition a larger ventricular end-diastolic diameter. The greater ventricular size in the latter is related to their larger LBM.

Maughan et al. (1983) did CAT scans of the mid-thigh region and measured muscle strength in three groups of adult males. The marathon runners had about the same muscle cross-sectional area and the same muscle strength as the nonathletes, whereas the sprint runners had about 10% greater muscle area and 18% greater strength than the other two groups. There was a modest correlation ($r = 0.58$) between muscle area and strength.

All of these studies support the contention that athletes do indeed have a larger LBM, a larger skeletal mass, and less body fat than nonathletes of the same age, sex, and stature.

Stature an Important Parameter

Generally speaking, tall individuals have a larger LBM than short individuals. Lean weight is a function of stature for all ages thus far examined, including infants. The regression slopes for LBM on height are not as steep in infants and children as they are in adults, and they are less steep in females than in males (Forbes, 1974). This well established influence of stature explains in part the observation that Orientals have smaller lean weights than Caucasians. Figure 8.1 shows some selected data for normal adults and for professional athletes. The influence of sex is readily apparent, and so too is the difference between the nonathlete and the professional athlete, the slopes being much greater in the latter groups. The

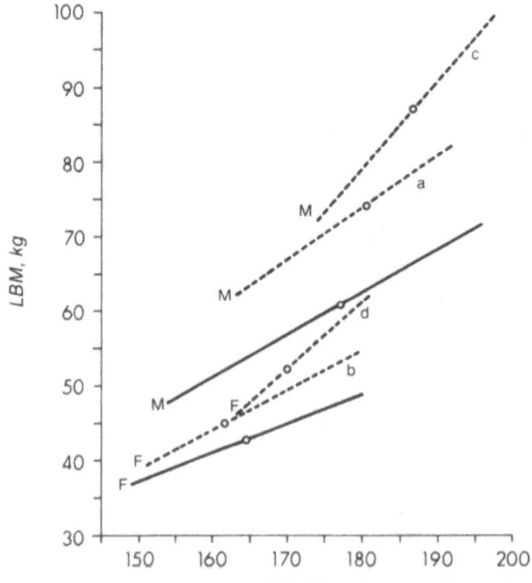

FIGURE 8.1. Plots of LBM against stature for 20–30-year-old men and 18–30-year-old women. Heavy lines: normal men (n = 161, slope 0.57 kg/cm, r = 0.56) and women (n = 135, slope 0.39 kg/cm, r = 0.56), both from author's laboratory. Line a: male athletes (n = 36, slope 0.72 kg/cm, r = 0.52) and b: female athletes (n = 8, slope 0.53 kg/cm, r = 0.71) both from author's laboratory. Line c: male athletes (n = 64, slope 1.15 kg/cm, r = 0.79) and d: female athletes (n = 12, slope 0.85 kg/cm, r = 0.51) both from Behnke and Wilmore (1974). The symbol (○) indicates means for each group. [From Forbes (1985a), with permission.]

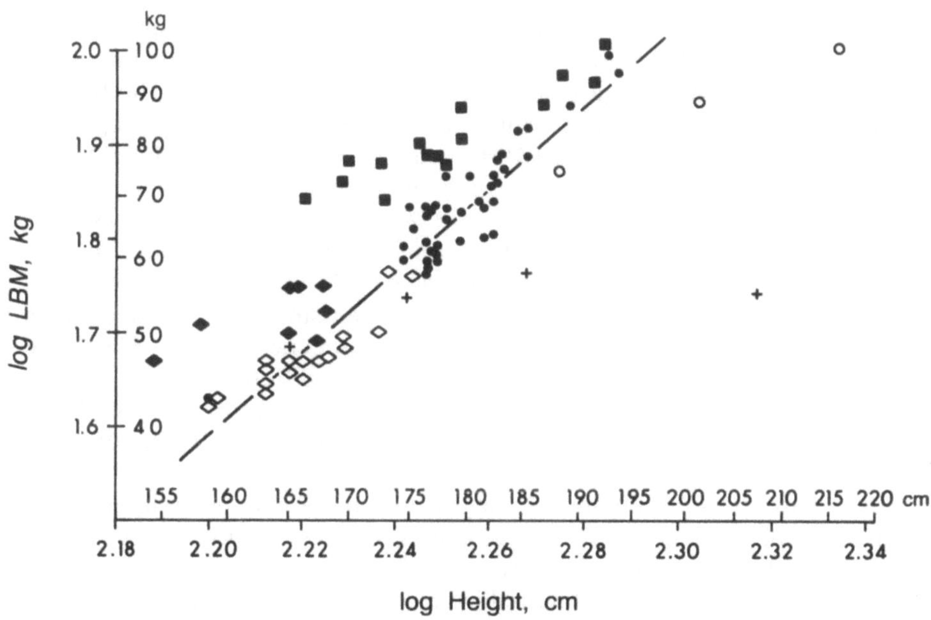

FIGURE 8.2. Plots of LBM (kg) against stature (cm) for various types of athletes, double logarithmic scale. Symbols: weight lifters, body builders, shot putters, discus throwers, Japanese Sumo wrestlers: (■) males, (◆) females. Basketball players: (○) males, (+) females. All others, including canoe racing, football, gymnastics, hockey, jockey, pentathlon, rowing, skiing, swimming, track and field, tennis, volleyball, wrestling, and baseball: (●) males, (◇) females. Regression equation for these is log LBM $= -8.11 + 4.41 \log H$, $r^2 = 0.94$ ($n = 57$). Data sources: Wilmore (1983); Tanaka et al. (1979); Walsh et al. (1984, personal communication)—48 female basketball players, grouped by height; 11 body builders and weight lifters studied by author.

importance of stature, and its related parameter LBM, to the athlete is evident from the analysis made by Khosla (1968) of winners of Olympic contests. Many were taller than the average of all participants for those events, so there is a bias in favor of the tall contestant. The associated increase in LBM thus confers an advantage for many athletic events.

Wilmore (1983) has collected data on stature and body composition of various types of athletes from the literature, and had added his own observations and those of his associates. Using these data together with those from other sources we constructed a plot of LBM against stature (Figure 8.2). Athletes over 50 years of age were excluded. As shown in the figure the trend for the majority of athletes, including both sexes, is nicely described by a double-logarithmic function, with a slope of 4.41. This says that LBM is a function of height to the 4.4 power. In this respect athletes differ from the general population, for a compilation of data from several sources, in which a variety of assay methods were used, showed that adult LBM was related to the third power of height (Forbes, 1974).

However, there are two groups of athletes who do not adhere to this general trend, and these are identified by a different set of symbols on the graph. Starting from a position comparable to the others, both male and female basketball players exhibit an LBM-height regression slope that is less steep. Their maximum LBM occurs at a much greater height; and obviously they have been selected for height rather than muscle mass.

The other deviant group includes the weight lifters, body builders, Japanese Sumo wrestlers, shot putters, and discus throwers. The shorter members of this group have a much larger LBM than the other athletes, but the LBM-height regression slope is not as steep, so the taller ones have an LBM that is not far above that of the general trend. These are the stocky "muscle men." Note that the same difference holds for the female body builders, shot putters, and discus throwers. If one uses LBM as an indicator of strength and possibly endurance, it is to the athlete's advantage to be tall.

The data shown in Figure 8.2 suggest that there may be an upper limit for LBM in males of about 100 kg, and in females of about 60 kg. These limits also hold for massively obese individuals. Bortz's (1969) 304-kg male had an LBM of 80 kg, as estimated from urinary creatinine excretion, and Cheek's (Cheek et al., 1975) 210-kg adolescent boy had one of 100 kg, as estimated from total body water. Values as high as 65 kg LBM have been recorded for very obese women (Chapter 7). Although there may be rare exceptions, these tentative upper limits for LBM are approximately 1.5 times the average values for normal adults of both sexes. However, these limits may be exceeded somewhat by males who take large doses of anabolic steroids. The presumption is that none of the athletes shown in Figure 8.2 were taking such steroids.

None of the athletes included in this compilation were very fat; quite the contrary, most were thinner than the general population. Only one male, a cross-country skier, had a body fat content greater than 20%. The fat content of the Japanese Sumo wrestlers, whose training includes generous amounts of food, was 24 ± 6.7%. Likewise, the female athletes were thinner than average. Figure 8.3 shows the distribution of percent body fat for the male and female athletes shown in Figure 8.2 (Sumo wrestlers excluded). The sex difference in percent body fat, so evident in the nonathletic normal population, persists among these trained athletes. Apparently, a program of vigorous training cannot erase the sex difference in body fat; rather, values for each sex have been shifted downward. Twenty-one of the 32 females have body fat contents of less than 21%, which is typical of normal males.

Are the differences in body composition manifested by trained athletes the result of training and exercise—many do train long and hard—or do they merely represent a different genetic endowment? The latter hypothesis cannot be dismissed out of hand. Identical twins are more concordant for height and weight than like-sexed fraternal twins, even though reared apart. Twin studies have also shown that identical twins are more concordant

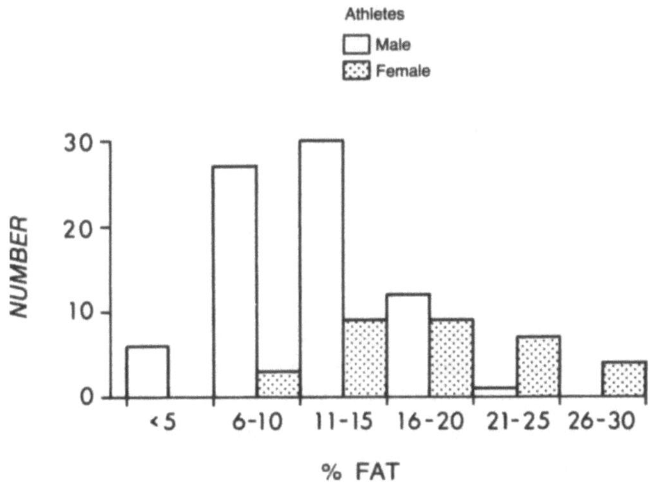

FIGURE 8.3. Frequency distribution diagram of percent body fat for male and female athletes. Data of Wilmore (1983) (n = 97) plus results of author's assays on 11 body builders and weight lifters.

than fraternal twins for bone density (Smith et al., 1973), skinfold thickness (Brook et al., 1975), and fat-free body weight (Bouchard et al., 1985). These last authors also found that the distribution of subcutaneous fat has a significant coefficient of heritability.

Is nutrition a factor? As was pointed out in Chapter 7, an increased intake of food augments the size of the LBM both in man and animals, especially so in young animals. However, it also serves to increase body fat content, whereas many athletes have a lower fat content than their sedentary peers. Nevertheless, adequate nutrition is necessary for LBM growth during childhood.

Athletes and their coaches have great faith in physical training and exercise as a means of improving athletic performance. It is well known that this can improve maximum oxygen consumption and endurance.

The investigations of Hoppeler et al. (1985) on male and female athletes subjected to 6 weeks of a vigorous training regimen showed an increase in maximum oxygen consumption of 19% in females and 10% in males. CAT scans and biopsies were made of the thigh muscles. No change was found in cross-sectional muscle area; however, the percent of total area occupied by fat tissue decreased and that of muscle increased, but only in the female subjects. There was no change in individual muscle fiber area. Alén et al. (1985) also failed to find a change in either total muscle area or individual fiber area in athletes who engaged in "power training" for 24 weeks. More importantly, Hoppeler et al. (1985) did find an increase in mitochondrial density, an increase in intracellular lipid, and an increase in capillary vessels relative to muscle fiber size. These changes were roughly related to the observed increase in muscle power. MacDougall et al. (1977) found a 24% increase in cross-sectional area of the mid-arm region and a 28% increase in extension strength after 5 months of heavy resistance training. They also documented increases in muscle creatine phosphate, ATP, and glycogen.

Changes in Body Composition with Exercise and/or Training

We proceed now to examine those studies that have dealt with body composition *per se*. The emphasis will be on continued exercise and/or training programs; acute changes in plasma volume, such as occur during a marathon run (L. G. Myhre et al., 1985) will not be considered, since the important consideration is what happens in the long term. Neither will we be concerned about the possibility of a differential effect on "fast-" and "slow-" twitch muscle fibers, about which there is conflicting evidence.

Studies in Animals

Young rats who are encouraged to run 2000 m/day for 10 weeks develop limb muscle and bone hypertrophy. Muscle weight was about 6% greater than in control animals of the same body weight, and bone weight was about 5% greater, so both increase to about the same degree (Saville and Whyte, 1969). Vigorous exercise for several weeks served to increase blood volume by about 20% in dogs, in the absence of a change in body weight (Davis and Brewer, 1935). Ninety-day-old male rats trained to run a treadmill daily for 6 weeks had less body fat than sedentary controls (16% vs. 13%), but they also had less LBM, due no doubt to the fact that they weighed less (Stone et al., 1978). In their 3-month study of 7-week-old male rats exercised on a treadmill Crews et al. (1969) also found that the exercised animals weighed less and had less LBM and less fat than the sedentary controls; however, when the exercised animals were compared to sedentary animals of the *same body weight* the former had less body fat but more LBM than the latter. Oscai et al. (1973) also found an increase in LBM and a decrease in body fat when 6-week-old female rats were forced to swim daily for a 21-week period, compared to sedentary controls of the *same body weight*. These authors also assayed individual organ size: of the 20 g difference in lean weight, skin and subcutaneous fat contributed 3.7 g, musculature 6.8 g, viscera 4.4 g, tail 0.7 g, and heart, lungs, liver, kidneys, and blood together 4.7 g. Hence all body organs were involved.

Gollnick et al. (1981) produced muscle hypertrophy in selected muscles in the rat by ablating the synergistic muscles; for example, changes in the plantaris and soleus were studied some weeks after ablation of the gastrocnemius. Muscle weight increased by 29–45% without a change in

TABLE 8.1. Changes in body composition (adult rat).

	Initial	Sedentary (food *ad libitum*)	Swim	Sedentary (paired weight)
Weight, g	676	795 (+119)	496 (−180)	498 (−178)
LBM, g	435	449 (+14)	384 (−51)	365 (−70)
Fat, g	241	346 (+105)	112 (−129)	133 (−108)
Percent fat	35.6	43.5 (+7.9)	20.6 (−15)	26.7 (−8.9)

Values in parentheses are changes from initial status.
Data of Oscai and Holloszy (1969).

muscle fiber number or in muscle water content. Treadmill exercise produced an even greater increment in size, namely 41–88%. The conclusion is that true muscle hypertrophy was produced by the technique employed by these investigators.[1]

Mice who were required to pull weights repeatedly in order to obtain food developed hypertrophy of the biceps brachii muscle over a period of 25 days. Total muscle weight was increased by 10–30%, and individual muscle fiber diameters by 17–22%; the number of myofibrils per muscle fiber was also increased (Goldspink, 1964).The calf muscles of guinea pigs who exercised vigorously for 4 months were 3.2% larger in weight than controls, and in sedentary animals these muscles weighed 4.3% less (Helander, 1961). Since all three groups of animals weighed the same at the end of the experiment, muscle made up a larger fraction of body weight in the exercised animals and a smaller fraction in the sedentary animals compared to the controls. An additional finding was the increase in myofilamental nitrogen in the exercised muscles and a decrease in sarcoplasmic nitrogen.

Most of the animals used in the above cited experiments were young. Oscai and Holloszy (1969) also studied adult rats who were forced to swim daily for 18 weeks, and compared their body composition to sedentary animals fed *ad libitum* and to sedentary animals maintained at the same body weight as the swimmers. The results can be best shown in the form of a table (Table 8.1).

The first thing to notice is the fact that the sedentary animals fed *ad libitum* gained weight, and that about 12% of their gain consisted of LBM. The swimmers lost weight (their food intake was less, and their energy output greater) and about 28% of the loss consisted of LBM. They were much thinner than the sedentary animals, but they also had less LBM.

[1]An excellent example of work-induced muscle hypertrophy is provided by Bernays (1986). Caterpillars raised on a diet of hard grass have mandibular muscles twice the size of those fed artificial diets.

However, the sedentary animals whose body weights matched those of the swimmers lost proportionately more LBM, this body component comprising 38% of the total weight loss. At the same body weight the exercising animals had a larger LBM and less body fat than the sedentary ones. Thus the exercise program succeeded in reducing the decline in LBM occasioned by the reduction in food intake, but it did not spare LBM completely.

The exercise programs cited above involved vigorous exercise over a prolonged period. One wonders if these animals were in a state of chronic fatigue. The decreased food intake recorded for the swimmers studied by Oscai and Holloszy (1969), which was only 89% of the sedentary animals, suggests such a possibility. Unusually severe exercise programs accompanied by significant weight loss (the swimmers in Table 8.1 lost 27% of their body weight) cannot be expected to preserve LBM, much less to cause it to increase.

Studies in Man

It may be instructive to consider first a "natural" experiment. The dominant arm of human beings enjoys a lifetime of greater usage. Both Christiansen and Rödbro (1975) and Exner (1979) found the bones of the dominant arm to have about 4% more bone mineral (by photon absorptiometry) than those on the nondominant side. On the other hand Maughan et al. (1984) found no difference in either muscle, bone, or fat cross-sectional areas between the two arms, as estimated by computed tomography.

Studies of professional tennis players leave no doubt as to the effect of long continued exercise: their dominant arms have thicker bones and larger muscles, though subcutaneous fat thickness is about the same (Buskirk et al., 1956; Gwinup et al., 1971, Jones et al., 1977). It should be remembered, however, that such differences are the result of many years of vigorous exercise.

However, the goal of most exercise and training programs is to augment lean weight and to reduce body fat content. There are a number of statements in the literature to the effect that such programs can accomplish both of these goals, but what is the evidence for such statements? The review to follow will include only those studies in which body composition assays were done both prior to and at the end of the training and/or exercise period, since anthropometric techniques lack sufficient precision. Granted that local muscle hypertrophy can be produced by exercise, the point to be considered here is the effect of exercise on the LBM as a whole, and on body fat content.

The animal data presented above are a reminder of the importance of including information on body weight in the evaluation of changes in LBM that may result from exercise and/or training programs. To this purpose the changes in LBM subsequent to exercise and shown in Figure 8.4 are plotted against the observed changes in body weight. The diagonal dashed

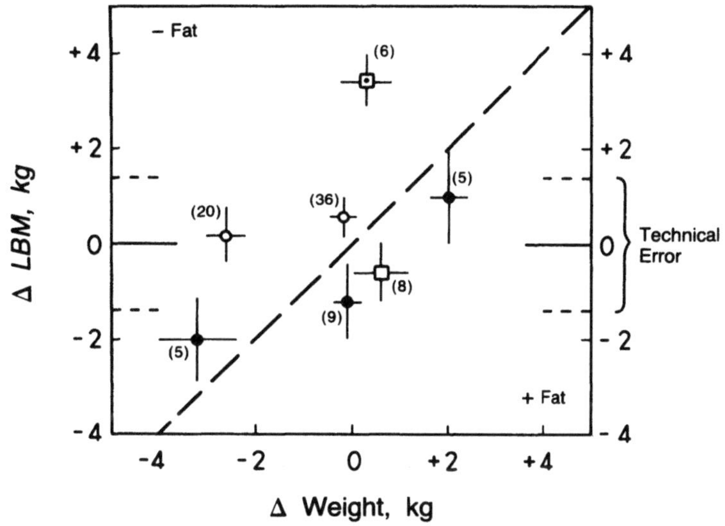

FIGURE 8.4. Effect of physical exercise and/or training on body composition. Plots of change in LBM against change in body weight. Symbols: (○) data compiled from the literature by Pollock et al. (1984) for 39 groups of males and 17 groups of females; (●) author's data on 19 West Point football players after 6 weeks' spring training plus thrice weekly sessions on the Nautilus machine (Forbes, 1982); professional soccer players at mid-season; pre-season heavy training (□), no pre-season training (⊡) (Boddy et al., 1974). Subjects arranged according to change in body weight; horizontal and vertical bars are SEM; numbers of subjects or groups in parentheses.

line is the line of identity: values that fall above this line indicate a fall in body fat content, those below a rise in body fat content. The short dotted lines that project horizontally from each side of the diagram indicate the analytical error—estimated at 2% for a 70-kg person—of the body composition assays.

In this figure the subjects are grouped in accordance with the observed changes in body weight. Of the two groups who lost some weight during their exercise program, one showed a slight loss of LBM, the other no change; both lost some body fat. However, the magnitude of the standard errors indicates that there was considerable individual variability in the response. Of the four groups whose body weights did not change appreciably, only one achieved a significant rise in LBM. These were the professional soccer players who began the playing season without having engaged in pre-season training. By mid-season they had achieved an average increase in LBM of 3.5 kg, or about 6% above the initial value, without a significant change in body weight. However, their teammates who had some pre-season training did not show an increase in LBM at mid-season.

This difference deserves comment. Those athletes, such as the West Point football players shown in Figure 8.4, who are in top physical con-

TABLE 8.2. Change in LBM and weight with exercise.

Subjects, sex	n	Program	Duration	Av. ΔW (kg)	ΔLBM (av.) (kg)	References
Adult F	13	Physical conditioning	12 wk	−2.6	−0.3	Franklin et al. (1979)
Adult M	13	Aerobic training	20 wk	−2.4	+2.0	Després et al. (1985)
Adult F	1	Body builder	3 mo	−1.8	+2.0	Author
Postmenopausal F	14	Aerobics, walking	9 wk	−1.1	−0.2	Cowan and Gregory (1985)
Obese F	23	Physical conditioning	12 wk	−0.4	+0.1	Franklin et al. (1979)
Gymnasts M	7	Training for Olympics	4 mo	-0-	+1.8	Parizkova (1977)
Gymnasts F	8	Training for Olympics	4 mo	-0-	+1.2	Parizkova (1977)
Adult F	16	Aerobics, walking	9 wk	-0-	+0.7	Cowan and Gregory (1985)
Elderly M, F	21	Endurance training	1 yr	-0-	+1.4	Sidney et al. (1977)
Adult F	14	Run 50 miles/wk	13 mo	-0-	+1.7	Boyden et al. (1982)
Adult M	10	Weight training	4 mo	+1.0	-0-	Tanner (1952)
Adult F	1	Run 25 miles/wk	7 mo	+1.9	+2.6	Author
Adult M	1	Cross-country bicycle trip	61 day	+3.9	+3.4	Nelson and Craig (1978)

dition prior to entering an exercise program, could not be expected to achieve much, if any, further increase in LBM.

Table 8.2 lists data from the literature and from the author's experience which were not included in the review by Pollock et al. (1984) nor shown in Figure 8.4. Three groups of subjects lost more than 1 kg in weight during their exercise programs; while they lost fat, LBM did not change appreciably. The one body builder did gain LBM in the face of a 1.8 kg loss of weight, but it is very likely that her program of exercise and training was more intense than the others. Of the next six groups, who did not have a significant change in body weight, four had a slight increase in LBM; the weighted average of all six groups was +1.0 kg LBM, and so fat content decreased by a like amount. There are too few subjects who gained weight with exercise and/or training to warrant a firm conclusion. The two subjects studied by myself did gain LBM in response to a vigorous exercise experience. Tanner (1952) estimated body composition from urinary creatinine excretion; all of the others were assayed by densitometry or ^{40}K counting. Of some interest is the observation that one of the two groups of elderly subjects achieved as much of an increase in LBM as the others.

It is unfortunate that most of the data shown in Figure 8.4 and Table 8.2 are reported as averages of weight and LBM before and after the exercise period, so there is limited opportunity to judge the variability in the change in weight and change in LBM. It is likely that there is a certain amount of individual variability in the response to exercise, although some of this may reflect technical error rather than true biological variability. Another factor is dietary intake. The only individual for whom this is known is the young man who rode his bicycle across the United States (Table 8.2): he kept a complete dietary record, which showed that he took 3200 kcal/day in excess of his usual diet during this long and arduous trip. Despite this almost all of his weight gain consisted of LBM. If he had eaten the same amount of food without exercising, in all likelihood he would have gained a lot more weight, and a majority of the gain would have been fat.[2] This subject also had a gain in thigh and leg circumference of about 6%, an increase of about 22% in leg muscle strength, and a gain of 16% in maximal oxygen consumption at the end of his bicycle trip (Nelson and Craig, 1978).

A question of current interest is whether a vigorous exercise program can serve to conserve LBM in the face of weight loss. Some of the data shown in Figure 8.4 and Table 8.2 suggest that it can, if the degree of weight loss is not very great. However, the group of five West Point foot-

[2]3200 kcal/day × 61 days, less 5% for fecal wastage, equals a total of 185,440 excess kcal; divide by the energy cost of weight gain (8050 kcal/kg) = weight gain of 23 kg, of which 62% (14 kg) would be fat, and 38% (9 kg) would be LBM, according to the calculations shown in Chapter 7, p. 226.

ball players who lost an average of 3.3 kg in weight also lost LBM (Figure 8.4). Harvey et al. (1979) did body composition assays (total body K and N) on a group of 14 men before and after a 23-day Himalayan mountain climb in which they ascended from 1000 to 5400 m. Despite an adequate food intake they lost an average of 3.3 kg, and about two-thirds of the loss consisted of LBM. The subjects studied by Keys et al. (1950) were encouraged to exercise vigorously during a 24-week period while consuming a 1600-kcal diet: they lost an average of 16 kg, and 57% of this loss consisted of LBM.

A recent study of overweight men and women showed that a program of exercise did not prevent loss of LBM as they lost weight on 1200-kcal diets (Hagan et al., 1986). In point of fact those who exercised lost somewhat more weight—and more LBM—than those who took the same diet but did not exercise. Since exercise serves to increase the energy deficit in individuals given low-energy diets such a finding could have been anticipated.

However, the data of Rigotti et al. (1984) on female patients with anorexia nervosa suggest that physical activity can act to conserve skeletal mass during weight loss. Photon density measurements of the radius were comparable to those of normal women in those patients who were physically active, while being reduced by about 14% in those who were not.

Krølner et al. (1983) made measurements of lumbar spine density (^{153}Gd photons) in a group of adult women who had sustained a fracture of the wrist. Those who engaged in an exercise program had a 3.5% increase in vertebral density over a period of 8 months, while randomized nonexercising controls had a decrease of 2.7% during this same period. However, there was no effect of exercise on the mineral content of the forearm bones.

Physical activity requires energy, and so will increase the energy deficit in those who were formerly in energy balance, and will serve to decrease energy storage in those who were in positive energy balance. Earlier mention was made of the experiments of Göranzon and Forsum (1985), in which the superimposition of a program of exercise resulted in negative nitrogen balance in adult men who were being maintained on a constant diet (see Chapter 7). It is reasonable, then, to assume that the energy deficit pursuant to exercise should not be too great if LBM is to be preserved, and that the energy excess should not be too large if exercise programs are to yield an increase in LBM without a concomitant gain in body fat. Although there is some support from animal experiments for these conjectures, the extant data on human subjects are far too few to draw firm conclusions; indeed more work will have to be done to determine whether one of the coveted goals of exercise programs, namely a significant weight loss—and by this I mean more than 2 or 3 kg—can be achieved without a loss of LBM.

The other goal of exercise programs—that is, to augment LBM by a significant amount—has thus far proved to be elusive. The best results shown in Figure 8.4 were obtained by the professional soccer players, namely a 3.5 kg increase in LBM. One of the groups of athletes listed in the review by Pollock et al. (1984) gained an average of 3.1 kg. Only two of the subjects listed in Table 8.2 gained more than 2 kg. Such increases amount to only 5–6%, a significant though hardly impressive increase in LBM. It should also be noted that many of the subjects shown in Figure 8.4 and Table 8.2 failed to achieve an appreciable change in LBM, although some body fat was lost. The explanation probably lies in the rather modest exercise programs that were employed.

In athletes who were engaged in vigorous training programs neither Hoppeler et al. (1985) nor Alén et al. (1985) could detect a change in cross-sectional area of the thigh muscles by CAT scan; nor did they find a change in muscle fiber area in biopsy samples.

The interpretation of these results must be made in the context of the prior training of the subjects. In the two studies cited above the subjects were athletes, and so may already have attained their maximum muscle mass prior to the experiment. It was previously noted (see Figure 8.4) that professional soccer players achieved a larger LBM by mid-season only if they had begun the season without prior training.

In this connection, the results of the previously mentioned animal experiments are of interest. Compared to the sedentary paired-weight animals shown in Table 8.1, those who swam had only a 5% increase in LBM. Saville and Whyte (1969) recorded a 6% increase in muscle weight in rats who ran rather long distances for many days; an increase in LBM of 10% was recorded by Crews et al. (1969) in rats who exercised on a treadmill, and one of 10% by Oscai et al. (1973) for rats who swam daily for many days. The somewhat better performance of some of the animals in this regard is probably due to the intensity and duration of their exercise regimens.

More striking results have been obtained when the training programs have been aimed at specific muscle groups. MacDougall et al. (1977) did find a significant increase in cross-sectional area of the mid-arm region (by tape measure) after 5 months of heavy resistance training but they did not measure muscle area *per se*. The experiments on rats referred to earlier showed that significant hypertrophy of the plantaris and soleus muscles developed after ablation of the synergistic muscles. Biceps hypertrophy in mice results from repeated application of heavy loads, and a clear example in man is the greater muscle mass of the dominant arm of professional tennis players.

Hormonal Responses

A bout of vigorous exercise is known to result in acute elevations of the serum levels of a number of hormones: testosterone, androstenedione, dehydroepiandosterone, progesterone, estradiol, thyroid stimulating hormone, catecholamines, cortisol, and growth hormone. Indeed, exercise is used as a provocative test for growth hormone deficiency. Levels return to normal rather quickly afterwards. Free thyroxine levels increase during exercise, and also the rate of disappearance of labelled T_4 from plasma is faster (Shephard, 1982); the latter feature suggests that there may be a more rapid conversion to triiodothyronine (T_3), which is the active form of the hormone at the tissue level.

Of more interest are the long-term changes, an interest that has been heightened by the observation that some female athletes have menstrual irregularities. This occasionally takes the form of abnormal bleeding, but the common findings are oligomenorrhea or amenorrhea. The incidence of these latter two conditions ranges from 1 to 44% (Loucks and Horvath, 1985). What is the cause?

The proponents of the Frisch hypothesis (see Chapter 4, p. 162) point out that many female athletes are thinner than normal women, and this is nicely shown in Figure 8.3. Though far from proven, this hypothesis states that a minimal level of body fatness is necessary for continued menstruation. There are two recent reports that address this question. Carlberg et al. (1983) assayed two groups of female athletes by densitometry: one group of 28 women had regular menses, the other ($n = 14$) had fewer than five menstrual periods per year; both were engaged in the same exercise program. The former group had an average of 16% body fat, the latter 13% body fat. Drinkwater et al. (1984) report an average of 17% body fat in eumenorrheic athletes and 16% in those with amenorrhea. They also found a lower density of the lumbar vertebrae in the latter group; however, there was no difference in density of the radius. Boyden et al. (1982) encouraged a group of female runners who had been running less than 25 miles per week, and who were menstruating normally, to increase their running distance to 50 miles per week over the ensuing year. The result was a modest increase in LBM (see Table 8.2) and a drop in body fat from 25.5% to 22.4%, without a change in body weight. Thirteen of the 14 subjects developed oligomenorrhea; in none did menses cease entirely. Bullen et al. (1985) induced untrained women (body fat content 26%) to engage in rather strenuous excercise for 6 weeks, as a result of which they lost some weight. Twenty-six of the 30 subjects developed abnormal menses, and 18 had a diminution in serum levels of luteinizing hormone; and this occurred more frequently in those who had lost the most weight. Six months later, all of the subjects reported normal menses.

Drinkwater et al. (1984) found lower serum levels of estradiol, progesterone, and prolactin in their amenorrheic athletes, and slightly higher levels of testosterone, in comparison to athletes with normal menses. However, in their review of the literature Loucks and Horvath (1985) could find no indication of a consistent increase in testosterone, androstenedione, prolactin, or dehydroepiandosterone sulfate in oligo-amenorrheic athletes; however, estradiol levels tended to be lower. They did find a tendency for a higher incidence of amenorrhea in runners and ballet dancers than in swimmers and bicyclists. They also uncovered a feature that had not been commented on previously, namely that a higher proportion of amenorrheic athletes had a history of irregular menses prior to training and more were nulliparous than was true for the eumenorrheic athletes. Apparently, vigorous exercise can serve to accentuate a preexisting situation.

The fact that there are many athletes who are eumenorrheic despite a lower than normal body fat content—indeed much lower than the Frisch hypothesis would suggest—speaks against a primary role for body fat *per se*. It is safe to say that women in general—even today—are not as inclined to engage in athletic endeavors as men, and this is particularly true of programs that demand a high level of physical activity over a protracted period of time. This represents, therefore, a distinctly abnormal "lifestyle" for many women. Moreover, some programs are highly competitive. The fact that amenorrheic ballet dancers resume menses during vacation from ballet school (Abraham et al., 1982) suggests that the strenuous, disciplined, and highly competitive nature of their program has altered the function of those central nervous system centers responsible for the cyclic production and/or secretion of pituitary gonadotropins.

Turning now to male athletes, Remes et al. (1979) studied 39 Army recruits who were subjected to a rather vigorous training program. All blood samples were taken at 7 a.m. At the end of 1 month serum testosterone levels had increased by 21%, androstenedione levels by 25%, and luteinizing hormone levels by 25%; there was no change in hormone binding globulin. Maximum oxygen consumption rose by 16%. There were no further changes in any of these parameters during the succeeding 5 months of training. One might speculate that the observed increase in anabolic hormone levels facilitates the increase in LBM that accompanies exercise and/or training programs, and that the increases in catecholamine and growth hormone levels (both of which have lipolytic activity) facilitate the loss of body fat.

Immobilization, Bed Rest, Space Travel

The decrease in muscle mass and bone density in immobilized or paralyzed limbs is a well-known phenomenon. Anyone who has ever had a plaster cast applied to a broken arm or leg can testify that the cast seems to be too large after a few weeks, and the shrunken limbs of children with poliomyelitis was once a vivid event. The space programs have engendered fears that prolonged weightlessness could lead to wastage of body nitrogen and calcium.

Denervation leads to change in muscle composition that are similar to those seen in malnutrition: the ratio of ECF volume to total water content increases, and potassium content falls (Hines and Knowlton, 1937). Greenway et al. (1970) did longitudinal studies of patients with paraplegia and quadriplegia with respect to total body water, sodium "space," and plasma volume. Over a period of 15–51 months there was a progressive decline in calculated LBM, and an increase in the ratio of Na space to total body water. Unfortunately, the initial observations were not done until some time after the paralysis occurred, so there is no way to compare these values to the normal status of these patients.

Ryan et al. (1957) studied three male patients with poliomyelitis. Total body water was reduced by 20% and urinary creatinine excretion by 55%, and calculated body fat was increased by 19% compared to healthy men of roughly the same stature and age. The ECF/total water ratio was higher (0.37) than that of the healthy subjects (0.32), as determined by radiosulfate and antipyrine dilution.

In a landmark experiment Deitrick et al. (1948) placed four healthy young men in plaster spicas from the waist down, and kept them there for 6–7 weeks. During this period they lost an average of 0.2 kg on diets providing 2500–2800 kcal/day. Basal metabolic rate dropped about 7%, blood volume fell by about 5%, and thigh girth declined. Metabolic balances were done. These revealed a total nitrogen loss of 54 g, a calcium loss of 14 g, (160–380 mg/day), and a phosphorus loss of 8.6 g. Assuming an initial body content of 1800 g N, 1100 g Ca, and 550 g P, and taking into account unmeasured N losses, these subjects lost roughly 4% of their body N, 1.3% of body Ca, and 1.6% of body P. Metabolic balance determinations were continued during a 3–6-week recovery period, during which the subjects retained 36 g N, and so had not yet made good this loss.

Greenleaf et al. (1977) assayed body density in male subjects before and after a 14-day period of absolute bed rest. The diet provided 3100 kcal, 3.8 g Na, and 150 g protein. These subjects lost an average of 0.4 kg in body weight, and 0.8 kg LBM; urine urea, K, Ca, and P excretion all increased. The interpretation of the body composition changes is made difficult by the fact that 2.1–2.4 kg of weight were lost during the 3-day control period immediately preceding the period of bed rest. Other subjects kept in bed had daily isometric or isotonic exercises; this did not prevent the losses of body constituents.

In an earlier experiment by Lynch et al. (1967), men were confined to bed for 28 days. Body weight did not change, but the subjects lost a total of 40 g of nitrogen, 4.2 g of phosphorus, and 5 g of calcium during this period, together with some Na, Cl, and K. The average daily increase in urine calcium was 180 mg, which is somewhat less than the values found by Deitrick et al. (1948) in immobilized men.

Krølner and Toft (1983) made photon absorptiometric measurements of the lumbar spine in 17 adult men and 17 adult women who were kept in bed for 11–61 days because of back pain. They were allowed up only for toileting, and exercises were done in the recumbent position. Changes in vertebral density varied considerably (some actually gained density); the average loss for the 34 subjects was 3.6%. Following several weeks of reambulation vertebral density increased but did not quite reach the values recorded prior to bed rest. There was no change in urinary calcium excretion.

A clear exception to this tendency to lose nitrogen and minerals when individuals are put at rest is the hibernating bear. This remarkable creature can remain at rest for as long as 60 days without losing LBM during winter sleep.

A number of studies have been done on astronauts. The loss of LBM during the 84-day space flight varied somewhat according to the method used: 3.4 kg by ^{42}K dilution, 1.2 kg by densitometry, 2.6 kg by nitrogen balance, and 1.1 kg by total body water. The average is 2.1 kg in the face of a 2.8 kg loss in body weight (Leonard et al., 1983). Blood volume declined by 10%. The interpretation of these results is difficult because of the occurrence of mild hyponatremia, suggesting a change in body fluid osmolality, the existence of a strong positive N balance prior to the flight, and the fact that body weight was rapidly regained on returning to earth.

Mineral balances have also been done on the astronauts—a remarkable feat in view of the difficulties in collecting urine and feces in a weightless environment! The average results for a 60-day flight were a negative calcium balance of 248 mg/day, phosphorus 222 mg/day, and nitrogen 4 g/day (Whedon et al., 1976). The losses of calcium were similar to those recorded for immobilized men by Deitrick et al. (1948).

Pitts and co-workers (1983) found an 8% drop in the fat-free weight, a 20% drop in total body Ca, a 6% drop in total body N, and an increase in body fat in rats who had spent 18 days in space. Smaller decrements in total body K, P, Na, and Mg were also detected by carcass analysis. These male animals were 113 days old at launch. One problem in analyzing these results is the fact that the control animals were 2–5 days older than the flight animals at the time of sacrifice, and had been fed a somewhat different diet.

Morey and Baylink (1978) did tetracycline labelling of 63-day-old rats who were sent on a 20-day space flight. This antibiotic stains the bone at the time it is given, and so the distance between two labels, given at different times, can be determined on microscopic sections. The calculated rate of periosteal bone formation in the space flight animals was only about half that of the controls. There can be no question as to the effect of weightlessness on body composition. One wonders how well the human body would fare were space flights to become very long in duration, and how long human beings could tolerate life in "space stations" were such to become a reality.

A question of more than theoretical interest is: why does the fetus *in utero* do so well? Its muscles are well innervated, but there is very little opportunity for movement within its confined and cramped environment. Moreover, it is in a state that approaches true weightlessness for its effective weight in amniotic fluid is only one-thirtieth of what its weight would be in air. Yet in this environment, which allows only a modicum of physical activity and greatly diminishes the effect of gravity, the fetus grows and enjoys a strongly positive balance of nitrogen and minerals. One could extend this question to include the very young infant, who spends most of its days at bed rest and yet manages to retain nitrogen and minerals. The young organism's propensity for growth must outweigh those forces which act in such a deleterious manner in the adult.

Muscle paralysis produces effects that differ somewhat from those seen in the astronauts. In comparing the urinary constituents of quadriplegic patients with those of astronauts in space Claus-Walker et al. (1977) found that although both had hypercalcuria, the former excreted about twice as much hydroxyproline. Furthermore, the observed ratio of glucosyl-galactosyl to galactosyl hydroxylysine residues in the urine suggested that both skin and bone collagen were being degraded in the quadriplegic patients, whereas only bone collagen was involved in the astronauts. Although it is tempting to invoke a different mechanism to explain this difference, it should be remembered that the patients had extensive paralysis, the lesion being at the level of 4th to 6th cervical vertebrae, whereas the only problem for the astronauts was the absence of gravity. Incidentally, the superimposition of an exercise program in space did not prevent the loss of calcium and nitrogen by the astronauts.

Finally, note should be taken of the recent report by Horber et al. (1985) on the use of resistance training for patients with prednisone-induced myopathy. Measurements were made of cross-sectional area of the mid-thigh region (by CAT scan) and of leg strength. Patients who had had renal transplants and had been treated with prednisone (and other drugs) for at least 6 months were compared to normal controls. Muscle area was reduced by 16%, fat area was 11% greater, and muscle strength was 20% less than the controls, and in some of the patients the thigh muscles had a mottled appearance on CAT scan, similar to that reported by Heymsfield et al. (1979) in malnourished subjects. When one leg of these patients was exercised thrice weekly for 7 weeks on a special isokinetic dynamometer device, muscle area increased by 8%, fat area decreased by 7%, and muscle strength increased by 29%. However, the untrained leg was also affected, though to a slightly less degree, suggesting that the patients may have increased their general level of physical activity as well.

9

Influence of Hormones

Greek *hormaein:* to excite, urge on, arouse, stimulate

Hormones are the facilitators and the regulators of metabolic processes. Through their action on target organs, where they combine with specific receptors, they stimulate the formation of cyclic AMP or messenger RNA so as to affect cell function. Included are the regulation of body fluid volume and electrolyte metabolism; of gastrointestinal absorption and renal excretion; and of carbohydrate, protein, and fat metabolism. They are concerned with tissue growth and differentiation, and with the resting metabolic rate of individual tissues and of the body as a whole. A deficiency or an excess of certain hormones can affect body composition.

In assessing the effects of individual hormones on body composition it should be remembered that a number of interrelationships have been identified. For example, testosterone is required for the formation of somatomedin (a growth hormone-induced compound), thyroid hormone is required for the release of growth hormone by the pituitary and estrogens also stimulate its release, excess cortisone inhibits the formation of somatomedin by the liver, progesterone blocks the action of aldosterone, and individuals with hypothyroidism fail to respond appropriately to the administration of growth hormone. Although the action of a hormone at the level of a specific target tissue can be assessed, the effect on the body as a whole is influenced by—and contingent upon—these various interrelationships. Adequate nutrition is obviously a prerequisite: growth hormone cannot stimulate growth in the face of nutritional deprivation, nor can insulin or thyroid hormone or androgens.

Testosterone

Testosterone is the principal androgenic-anabolic hormone. It stimulates amino acid incorporation in muscle and increases muscle size in both normal and castrated animals; it acts to increase renal cell size and to stimulate erythropoietin secretion by the kidney; and in the intact animal these actions are reflected in the achievement of a positive nitrogen balance. The masculinization of the male fetus is due to the production of this hormone by the fetal testis, for castrated fetuses develop female external genitalia.

The sex difference in lean body mass (LBM) and in muscle strength that develops during adolescence is due in large part to the greater production of testosterone by the male. Kochakian's (1976) book provides a complete review of testosterone and its analogs.

Of interest to the process of adolescent growth is the demonstration that in addition to its anabolic activity testosterone promotes the production of somatomedin, a hormone that stimulates the growth and maturation of epiphyseal cartilage (Parker et al., 1984). This effect is probably mediated via the pituitary, since testosterone administration does not alter somatomedin levels in growth hormone-deficient children. When testosterone is given to children with hypogonadism there is a fairly prompt acceleration in stature and in bone maturation, as well as a retention of nitrogen.

Castrated male mice weigh more, and have 2–3 times as much body fat and 13% less body protein than controls (Hausberger and Hausberger, 1966). Castrated female mice are also fatter than controls, but total body protein content is about the same. Human males with a 47,XXY chromosome complement (Klinefelter's syndrome) have on average 9.4 kg less LBM than normal males, and are lighter in weight by 7.3 kg (East et al., 1976). These patients have serum testosterone levels that are roughly midway between those for normal males and females.

The administration of testosterone or one of its analogs produces an anabolic effect in a variety of situations: castrated animals, women, both normal and hypogonadal men, children with renal failure (Jones et al., 1980), patients with poliomyelitis (Whedon, 1956), and patients with myotonic dystrophy (Griggs et al., 1985). This last group (nine males) had ^{40}K counts and measurements of urinary creatinine excretion before and after 10–13 weeks of intramuscular testosterone injections. Total body K increased by 16% and urinary creatinine excretion by 28%; however, neither had reached normal values for age and stature by the end of the treatment period. The children with renal failure, who were being dialyzed at regular intervals, gained stature and total body water; the ratio of extracellular fluid (ECF) volume to total body water did not change, so the increase in total body water can be taken as indicating a gain of LBM.

Animal experiments show that the effect of anabolic steroids is dose-related, some studies indicating that the magnitude of the response is a function of the logarithm of the dose (Kochakian, 1976). It was of some interest, therefore, to search for such a relationship in man. The literature provides a number of experimental studies in which nitrogen balances were done, and several in which body composition assays were carried out; together these span a wide range of steroid doses. When the observed changes in LBM are plotted against the *total dose* of steroid given on linear-log coordinates, the result is as shown in Figure 9.1. At relatively low doses, i.e., less than 500 mg total, the effect on LBM is rather small—a gain of 1–2 kg on average. However, larger doses—such as those taken by some athletes—result in a significant augmentation of the LBM, and

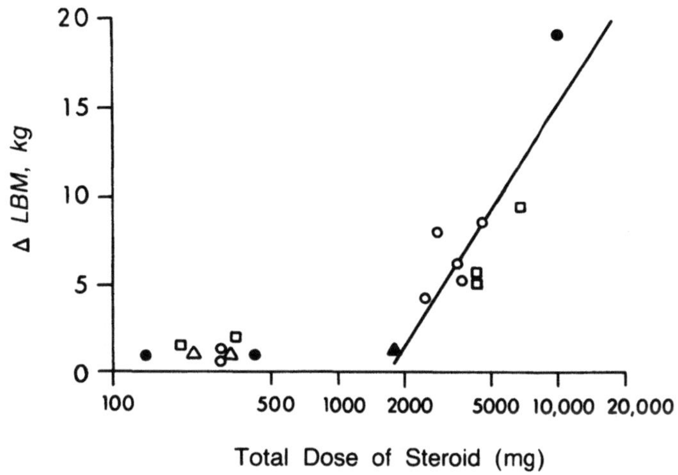

FIGURE 9.1. Change in LBM induced by androgens plotted against logarithm of total dose administered. Symbols: ○, testosterone, ●, oxandrolone, □, dianabol, △, androstalone, ▲, nandrolone. Data sources are listed in Table 9.1. Equation for regression line is: $y = 20.0 \log x - 65$ ($r = 0.86$).

as is shown in the figure, this appears to be a function of the logarithm of the total dose administered.[1] Although there is some indication that the various anabolic steroids vary somewhat in potency when tested in animals (Kochakian, 1976), the plot shown in Figure 9.1 fails to show a significant difference with regard to their effect on LBM in man.

Data sources for Figure 9.1 are listed in Table 9.1. Although anabolic steroids have been used in a number of clinical situations—burned patients, those undergoing extensive surgery, even prematurely born infants—the list was restricted to those circumstances where either body composition assays or complete nitrogen balances were done, and where the total dose of anabolic steroid was known. Included are normal men, patients with hemiplegia and with pulmonary tuberculosis, and athletes preparing for contests. Some exercised vigorously, whereas others maintained their usual—and rather modest—exercise level. The majority were normal men.

Although in none of the sources quoted was there a clear statement as to food intake during the experimental period, all of the subjects must have been in positive energy balance for they all gained weight. But there is an important difference between weight gain due to overeating *per se*, and that which results from the administration of anabolic steroids. Both circumstances demand a positive energy balance, but in the former situation the weight gain comprises both lean and fat, whereas in the latter *lean is gained while body fat is lost*. The listing in Table 9.1 shows that

[1]The derived equation ($y = 20.0 \log x - 65$), based on the 10 points shown in Figure 9.1, is slightly different from the one published by Forbes (1985b) which was based on only 6 points, namely $y = 29 \log x - 99$.

TABLE 9.1. Effect of anabolic steroids on weight and LBM.

Subjects (no./sex)	Dose rate	Total dose (mg)	Weight (kg)	LBM (kg)	Method	Remarks	Reference
4 M	25 mg/day I.M.[1]	225–350	+2	+1.3	N balance	Eunuchs	Kenyon (1938)
2 F, 2 M	25 mg/day I.M.[1]	225–375		+0.8	N balance	Normals	Kenyon (1940)
1 M	25 mg/day I.M.[1]	200	+1	+1.4	N balance	Eunuch	Knowlton (1942)
3 F, 6 M	10–30 mg/day p.o.[2]	140–420		+1.0	N balance	Hemiplegics	Albanese (1962)
8 M	10 mg/day p.o.[3]	350	+0.3	+2.0	Density	Normals, exercise	Ward (1973)
14 M	75 mg/wk I.M.[4]	225	+1.6	+1.1	Density	Normals, exercise	Fahey (1973)
9 M	100 mg/wk I.M.[1,4]	300	+0.6 (±0.4)[a]	+1.0 (±0.4)[a]	Density	Normals, exercise	Crist 1983[b]
15 F	100 mg/day p.o.[5]	1800	+0.7	+1.2	N balance	Pulmonary tuberculosis	Harris (1961)
1 M	190 mg/wk I.M.[1]	2470	+2.2	+4.2	^{40}K counting	Normal	Author
1 M	225 mg/wk I.M.[1]	2940	+3.2	+8.0	^{40}K counting	Normal	Author

11 M	100 mg/ day p.o.[3]	4200	+3.5	+5.1	Density, ^{40}K	Normals, exercise	Hervey (1976)
7 M	100 mg/ day p.o.[3]	4200	+3.5	+5.2	Density, ^{40}K, body N	Normals, exercise	Hervey (1981)
1 M	260 mg/ wk I.M.[1]	3660	+4.0	+5.2	^{40}K counting	Normal	Author
1 M	300 mg/ wk I.M.[1]	3500	+3.0	+6.2	^{40}K counting	Normal	Author
1 M	350 mg/ wk I.M.	4500	+1.1	+8.5	^{40}K counting	Normal	Author
1 M	Variable, 125 day p.o. I.M.[1,3]	6720	+5.1	+9.1	^{40}K counting	Weight lifter, vigorous exercise	Author
1 M	Variable, 140 day p.o.[2]	9760	+9.7	+19.2	^{40}K counting	Body builder, vigorous exercise	Author

[1]Testosterone
[2]Oxandrolone
[3]Dianabol
[4]Nandrolone
[5]Androstalone
[a]SEM
[b]Also personal communication

such is the case in almost every instance. This is of course precisely what many athletes desire.

Figure 9.2 shows the relationship between change in body weight and change in LBM for the subjects given androgens (see Table 9.1). Excluding one outlier, the correlation between these two variables is very high (r = 0.96). On average, the increase in LBM in response to androgen administration is almost twice as great as the increase in body weight. Looked at in another way, for each kilogram increase in LBM, body fat decreases by 0.52 kg.

Athletes also desire to increase their strength and endurance. The recorded increases in LBM referred to above suggest, but do not prove, that such is the case. Two of the subjects studied by myself (Forbes, 1985b) experienced a gain in hand grip strength; other investigators have obtained equivocal results. A recent report by Alén et al. (1984) is of interest in this regard. These authors studied professional athletes who engaged in a program of "power training" for a period of 24 weeks. Five athletes were given a total dose of 4300–5200 mg of anabolic steroids (methandienone, stanozolol, nandrolone, or testosterone) during this time; six others served as exercise controls. The treated subjects increased their isometric muscle force by 15%, individual muscle fiber area by 16%, and their body weight by about 6%, while skinfold thickness diminished. The exercise controls increased their muscle strength by only 6%, and there was no change in muscle fiber area or in body weight. In this connection it should be noted that Mahon et al. (1984) have reported a considerable difference in fiber size, and in proportion of "fast-twitch" and "slow-twitch" fiber types among normal individuals and even between different biopsy sites in the same individual. The proportion of "fast-twitch" fibers varied from 32 to 58% and that of "slow-twitch" fibers from 38 to 66%. The coefficient of variation for fiber size within individuals was as high

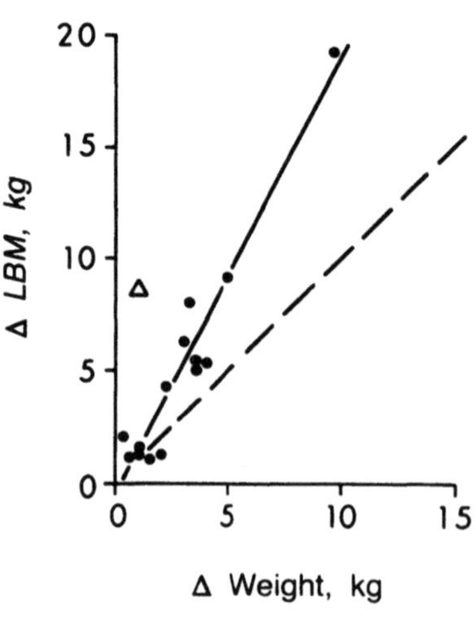

FIGURE 9.2. Plot of change in LBM (kg) against change in weight (kg) for subjects given anabolic steroids. Data sources listed in Table 9.1. Solid line is calculated regression line: $y = -0.56 + 1.93x$ ($r = 0.960$); triangle point not included in the regression calculation. Dashed line is line of identify.

as 20%. Inokuchi et al. (1975) also found a twofold variation in muscle fiber diameter. Such data make for caution in interpreting presumed changes in muscle fiber area.

In both sexes serum testosterone levels rise during adolescence, but the rise is far greater in boys, to rather quickly achieve values that are about 10 times greater than girls'. This accounts for the virilization of boys, and it undoubtedly is the principal reason why boys eventually achieve a 40–50% larger LBM than girls.

Taking a clue from the relationship shown in Figure 9.1, I have attempted to relate the development of the sex difference in LBM to the sex difference in endogenous testosterone production. Measurements of production rates in adults show that the males produce about 6 mg more testosterone per day than females (Horton and Tait, 1966). Under the assumption that this difference is established by age 13 years—at which time the sex difference in serum levels of testosterone is 5–10-fold—one can calculate the sex difference in total production of the hormone during the adolescent years. By age 20 years, this amounts to 6 mg/day × 365 day/yr × 7 yr = 15,330 mg. During this period the sex difference in LBM will have increased, on average, from 1–2 kg to about 19 kg.

Figure 9.3 is a plot of the sex difference in LBM against the logarithm of the sex difference in total testosterone production from age 13.5 to 20 years. The shape of this plot is quite similar to that shown for adults given large amounts of testosterone (Figure 9.1), as is the slope of the calculated regression line and its intercept. Before dismissing this good correspondence as fortuitous it should be pointed out that the assumption of another value for the sex difference in testosterone production rate—say 4 mg/day instead of 6 mg/day—would not change the slope of the regression line, only shift the line to the left.

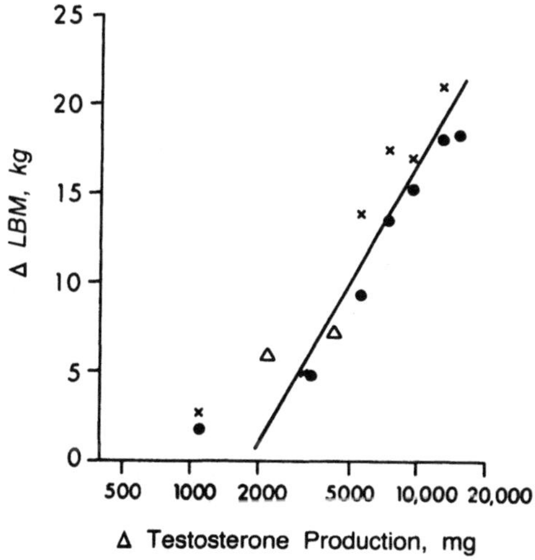

FIGURE 9.3. Male-female difference in LBM for adolescents plotted against logarithm of sex difference in total testosterone production, assumed to be 6 mg/day beginning at age 13 years. Symbols: (●) Burmeister and Bingert (1967) (n = 3522); (△) Novak et al. (1973) (n = 211); (x) author (n = 210). Equation for regression line is $y = 22.4 \log x - 73$ (r = 0.94).

Krabbe et al. (1984a) did a longitudinal study of adolescent boys, making measurements of serum androgens and alkaline phosphatase, and bone mineral content (BMC) of forearm bones at 3-month intervals for several years. They found a good correlation between the time of maximal increase in serum testosterone and the time of maximal increase in BMC, the former preceding the latter by about 5 months. The time of maximal increase in serum alkaline phosphatase lagged behind that of serum testosterone by only 1 month. Their data reveal a close relationship between serum testosterone, osteoblastic activity, and mineralization in normal male puberty. Assays were also done for serum dehydroepiandrosterone (DHEA) and androstenedione (both of adrenal origin); neither could be correlated with serum alkaline phosphatase or with BMC. The authors conclude that testosterone is the dominating androgen at the time of the adolescent spurt in alkaline phosphatase, BMC, and growth velocity. In another study Krabbe et al. (1984b) found a correlation between plasma somatomedin-A and testosterone levels, but not with DHEA levels, during the course of male puberty. Butenandt et al. (1976) found a relationship between height velocity and 24-h integrated levels of testosterone in prepubescent and pubescent boys, but not with growth hormone levels.

Testosterone thus plays a major role in the events of male puberty, and, as noted above, when given in sufficient amounts to adults it augments LBM to a significant extent; indeed the effect in this respect exceeds those that have been recorded thus far from deliberate overfeeding or from exercise programs. It is easy to understand why athletes take androgens. Their striving for increased muscle mass and strength outweighs the specter of the toxic effects of these drugs, namely decreased gonadotropin production, impaired spermatogenesis, and (rarely) hepatic damage.

Estrogens, Progesterone

The temporal relationship between the rise in serum estradiol levels at age 11–12 years and the increase in body fat during female puberty suggests that this hormone, the most potent estrogen produced by the ovary, is responsible. Girls with a 45-XO chromosome complement, and who therefore lack functional ovaries, do not give the appearance of putting on subcutaneous fat at the age when this normally occurs. However, when they are given estrogens breast development occurs and subcutaneous fat depots become prominent.

In acute experiments on castrated female rats Cole (1952) found that estradiol produced an increase in ECF volume and a decrease in intracellular fluid (ICF) volume of both muscle and liver. Preedy and Aitken (1956) gave estradiol to human subjects. After seven daily injections of 10 mg each the subjects had gained an average of 0.6 kg in weight, and ECF volume had increased by 0.8 liters; serum protein and hematocrit both fell, and there was a retention of sodium and chloride. Estradiol administration increases the aldosterone secretion rate in man (Kata and

Kappas 1967), and this may be the principal reason for the sodium retention that occurs in pregnancy.

The administration of diethylstilbestrol (a synthetic estrogen) accelerates growth in beef cattle (Trenkle and Willham, 1977), and when compared to control animals, this difference in growth is due to a greater increase in LBM. In fact, the treated animals gained less body fat than the controls. For example, in one series of animals the treated animals gained 33 kg more than the controls, but 5 kg less fat. Such a result is one that would be expected from androgens rather than estrogens. In a review of this topic Preston (1975) suggests that this could be due to the increased amounts of growth hormone and insulin that are found in the treated animals. It may be noted here that women have slightly higher fasting levels of growth hormone than men, and that prior treatment with stilbestrol increases the plasma growth hormone response to insulin and arginine in boys. Preston (1975) goes on to say that stilbestrol produces the above noted effect in cattle and lambs, but not in other species. Estrogens given to bulls and cockerels augment body fat.

Estrogens also play a role in skeletal integrity, for the age-associated loss in bone mineral content is accelerated after the menopause.

Divergent results have been reported for progesterone administration. The work of Landau (1973) and others shows that this hormone increases the urinary excretion of nitrogen and sodium in humans, due to the blocking of aldosterone. On the other hand, when given to female rats, it acts to increase body weight, and of this increase 31% is fat, 43% is lean tissue, and 26% is excess water (Hervey and Hervey, 1967). These gains were all proportional to the (log) dose of progesterone, which ranged from 0.2 to 10 mg/day. However, these amounts were far greater on a body weight basis than the 50–100 mg doses used by Landau (1973) in humans; since progesterone can be converted to testosterone and to cortisol, it is possible that large doses exert their effect via these conversions. Hervey and Hervey (1967) also recorded increases in size of liver, heart, kidney, and gastrointestinal tract; uterine size diminished. However, when similar doses were given to male rats there were no changes in body fat or LBM, except for a slight increase in the water content of the LBM.

Blyth et al. (1985) gave stilbestrol to men with prostatic carcinoma, and recorded an average increase of 540 ml in plasma volume but only a 0.48 kg increase in body weight during a 3-month period. Levels of renin, aldosterone, and angiotensin II also increased. The disparity between the increase in plasma volume and the increase in body weight suggests that fluid retention occurred. However, the authors could find no evidence for sodium retention by the metabolic balance technique.

McAreavey et al. (1983) did assays of total body sodium and potassium in women with essential hypertension who were taking oral contraceptives (estrogen-progesterone combination). Some had sustained an increase in blood pressure, others did not. Total body Na and K were normal in all.

Andersch et al. (1978) studied a group of women who had premenstrual tension, and an equal number who did not. Body weight and total body water were measured in the early follicular phase and again in the late luteal phase of the menstrual cycle. These authors failed to find a significant change in body weight or total body water in either group, and they make the bold statement that there is no evidence "to support the widely held belief that normal women or women with premenstrual tension gain weight premenstrually. In neither group could we find a substantial increase in body water during the luteal phase." Serum progesterone levels are higher in the luteal phase of the cycle than in the follicular phase.

However, "widely held beliefs" cannot be lightly dismissed. In the study quoted above some women weighed as much as 4 kg more during the luteal phase, others as much as 5 kg less than in the follicular phase of the cycle. Differences in total body water ranged from +3.4 liters to −4.5 liters. Diet was not controlled, and without a knowledge of energy intake between the first and second assays one cannot draw firm conclusions. Such wide variations in body weight suggest that energy intake did indeed vary.

However, if fluid retention does occur during the premenstrual phase, the magnitude of the change in body water content may not be sufficient to be detectable by the isotopic dilution technique.

Growth Hormone

Growth hormone acts to increase amino acid uptake in muscle, and through stimulation of somatomedin production by the liver, to enhance the uptake of sulfur and amino acids in epiphyseal cartilage. Administration of the hormone to children with hypopituitarism, to normal individuals, and to patients with severe burns results in positive balance of N, Na, and K. Indeed in this last group a good correlation was observed between the amount of nitrogen retained and the dose of growth hormone (Wilmore et al., 1974). In castrated rats growth hormone and testosterone act synergistically in promoting N retention (Kochakian, 1960).

Several investigators have done body composition assays in children with growth hormone deficiency. Prior to treatment total body water and ECF volume are reduced for age but normal for height (Parra et al., 1979). With treatment there is an increase in LBM, ECF volume, and urinary creatinine excretion, and in some a decrease in body fat and of course an increase in height (Tanner and Whitehouse, 1967; Novak et al., 1972; Collipp et al., 1973; Parra et al., 1979).[2]

[2]The effect of removal of the pituitary gland, a procedure that deprives the animal of all of the hormones produced by this gland, has been studied by Plocher and Powley (1976) in both normal and genetically obese *(A^y/a)* mice. Prior to surgery the obese mice weighed 29–58% more than the nonobese animals, and had a slightly greater LBM. Hypophysectomized animals in both groups weighed less, and LBM accounted for 51–76% of the total weight difference in the obese animals, and 65–71% of the loss in the nonobese.

Growth hormone excess, as seen in patients with acromegaly, leads to an increase in total body N, Ca, P, Na, Cl, and K (Aloia et al., 1972) as determined by neutron activation. In this series of 10 patients total body calcium was about 20% higher than these authors have recorded for normal subjects of comparable height; there was no indication of osteoporosis as has been noted by others in some patients with acromegaly. Ikkos et al. (1954) found an increase in total body water as a percent of body weight, and an increase in ECF volume and in Na_e in relation to total body water. A number of their subjects as well as those of Aloia et al. (1972) were tall (>170 cm in women and >180 cm in men), suggesting that the disease had begun prior to the time when growth in stature normally ceases. In a subsequent paper Ikkos et al. (1956) reported assays for Na_e, K_e, and Cl_e in patients with acromegaly, and kindly provided comparative data on normal subjects. The ratio of ECF volume to total exchangeable potassium was increased, whereas the ratio of ICF volume, whether estimated by inulin dilution, ^{82}Br dilution, or ^{24}Na dilution, to K_e was normal. A plot of K_e against height revealed the usual relationship in their normal subjects; when values for K_e and stature for the acromegalic patients were plotted on the graph for normal subjects, five of the nine patients had values for K_e that were in excess of normal for height, one was normal, and three were below. Hence, it appears that the distribution of body cell mass values in these nine patients was not abnormal, a finding that is at variance with those of Aloia et al. (1972).

Thyroid Hormone

Thyroxine (T_4) and triiodothyronine (T_3) are produced by the thyroid gland. In peripheral tissues T_4 is converted to T_3, and it is this compound that stimulates mitochondrial function, protein turnover, cell growth, and metabolic rate. Edmunds and Smith (1981) have done ^{40}K assays on a number of adults with both hyper- and hypofunction of the thyroid gland, the latter being under treatment. Total body K was reduced in those with hyperthyroidism, and there was a rough relationship between the reduction in body K and the degree of muscle weakness. With treatment these patients gained weight and body K, the ratio being an average of 69 meq K/kg of weight gain. Since the K content of the LBM is 64–68 meq/kg (see Chapter 2), this suggests that all of the gain in weight represented lean tissue. However, if intracellular K is reduced in this condition, this conclusion is not valid.

These authors then proceeded to relate the degree of body K deficiency to the serum level of thyroid hormone at the time the ^{40}K count was made. Serum T_4 levels ranged from 40 to 250 nmol/liter (3.2–20 μg/dl) and serum T_3 from 1.5 to 7.4 nmol/liter (120–580 ng/dl), i.e., from low normal to clearly excessive values. In this series of 42 patients the authors found an inverse correlation between total body K and the serum T_3 level ($r = -0.48$). Patients with normal T_3 levels had a normal body content of potassium, and hence a normal LBM, and with increasing degrees of hyperthyroidism body K progressively fell to about 80% of the expected value. Moreover, the degree of muscle weakness was greatest in those patients with the lowest body K content. There was no correlation with serum T_4 levels. These observations support the contention that T_3 is the active form of the hormone—T_4 being considered a prohormone—in keeping with other data, which show that effects such as synthesis of growth hormone is better correlated with T_3 than with T_4 levels. Data obtained by body composition assays thus confirm the findings of others.

Patients with hypothyroidism show unexpected findings. Children have increased body K, ECF volume, total body water, and urinary creatinine excretion (Cheek, 1968), and adults have normal values for urinary creatinine (Kuhlback, 1957). With treatment, urinary creatinine diminished to the normal range for stature in the children.

The addition of thyroid hormone to a weight reducing diet can increase the rate of weight loss and the data of Lamki et al. (1973) suggest that LBM loss is also increased. These investigators studied four obese men, whose body fat content calculated from total body water was 60–80 kg. They were given either a 600-kcal diet plus 0.3 mg L-thyroxine daily or a 1200-kcal diet with 0.9 mg L-thyroxine for 3–4 weeks; then the diets and thyroid medication were switched for an additional period; and then returned to the original format. The total duration was 12 weeks. During the periods on 600 kcal and the lower dose of thyroid hormone the average weight loss was 497 g/day, nitrogen balance 4.09 g/day, and loss of body water 165 ml/day (tritium dilution). When the 1200-kcal diet and higher thyroid dose was given these were, respectively, 402 g weight per day, 151 ml water per day, and 3.45 g N per day. Except for a couple of days when the energy (and protein) intake was increased all of the subjects were in negative nitrogen balance; however, since the first 7–10 days of the balance were not entered into the calculations it is not possible to relate nitrogen loss to weight loss in a precise manner. When one calculates the ratio of LBM loss (from body water) to weight loss it turns out that about 40% of the weight loss consisted of LBM. This percentage is somewhat higher than those found by others in massively obese subjects maintained on diets of similar energy contents (see Chapter 7, Figure 7.8). So despite the title of Lamki's article, it is evident that the administration of thyroid hormone actually did augment the loss of lean tissue.

Adrenal Hormones

The adrenal cortex manufactures several types of hormones: mineralocorticoids have to do with salt and water balance, glucocorticoids with protein and carbohydrate metabolism, and androgens have anabolic and androgenic activity. Abnormalities of the adrenal gland, whether produced by disease, enzymatic errors, or alterations in the production of corticotropin (ACTH) by the pituitary, can affect any one, or all three, of these adrenal hormones. The adrenal medulla secretes catecholamines that have a variety of functions having to do with vascular tone, cardiac function, oxygen consumption, and carbohydrate and fat metabolism. This small organ, weighing but a few grams, performs via its several elaborated hormones a wide variety of functions.

Glucocorticoid excess, such as occurs in Cushing's syndrome, is accompanied by an increase in body fat and a decrease in LBM (Ross et al., 1966; Ernest, 1967; Aloia et al., 1974; Ellis and Cohn, 1975a). The patients studied by Aloia et al. (1974) had a reduced total body K (88% of the expected value) and a reduced total body Ca (87% of expected). The total body Na/K ratio was increased, whereas the ratios of total body Na/Cl and total body P/Ca were within the normal range. Decreased muscle mass, osteoporosis, and increased body fat are commonly observed features of this syndrome. With treatment there was a rise in body K and P, and a fall in body Na and Cl, but only two of the four patients showed a rise in total body Ca, and in these two the rise in total body Ca was less than that of total body P. It is possible that the observed increase in total body Na/K ratio in the untreated patients was due to excess production of mineralocorticoids. The decrease in bone density is most pronounced in the spine (Chapter 5, Figure 5.10).

Mayer et al. (1956) analyzed the carcass of mice that had been implanted with adrenotropic tumors 10 weeks earlier. The tumor-bearing mice had more body fat (4.2 g vs. 2.2 g) and less LBM than the controls, whereas total body weight was about the same. These changes were not seen in tumor-bearing mice whose adrenal glands had been removed; thus the tumor produced its effect on body composition via adrenal hormones.

Treatment with large doses of cortisone or prednisone can produce similar effects in patients. Horber et al. (1985) found a reduced muscle area and an increased fat area (by CAT scans of the thigh) in renal transplant patients given prednisone. However, Lindholm (1957) found a slight increase in body cell mass (by ^{42}K dilution) in asthmatic patients given cortisone, although there was evidence of osteoporosis. The problem here is that compliance with medication—known to be poor in many situations—is difficult to judge in patients on long-term treatment, so it is impossible to know the doses actually taken by Lindholm's patients. Adinoff and Hollister (1983) studied asthmatic patients by means of photon absorption of the radius, both at the distal end (principally trabecular bone) and in the mid-shift region (cortical bone). When compared to normals, those patients who had been on continuous prednisone treatment had about a 15% reduction in trabecular bone density, but no change in cortical bone density. Those who had been on intermittent treatment had normal values for both. Of clinical interest is the finding that 11% of the 128 patients on continuous treatment had had one or more fractures, whereas none of the 54 patients on intermittent treatment had had a fracture.

Mineralocorticoid excess leads to sodium retention and potassium loss.

Rats given desoxycorticosterone acetate have an increase in total body Na, a decrease in total K, and hypertension; intracellular Na is increased in muscle and aorta, but not in brain; muscle K is reduced. However, these effects are dependent on an adequate intake of salt (Tobian and Binion, 1954; Knowlton and Loeb, 1957). Patients with primary aldosteronism have elevated values for Na_e and ECF volume and reduced K_e (Chobanian et al., 1961).

The administration of desoxycorticosterone acetate (DOCA) or fludrocortisone, in doses of 20 mg/day and 1 mg/day, respectively, to normal subjects results in an increase in ECF volume and Na_e, and a decrease in K_e. During a 15–21-day period Na_e increased an average of 320 meq, and K_e fell by 110 meq. Both of these compounds possess strong mineralocorticoid activity (Chobanian et al., 1961).

Adrenal insufficiency is accompanied by a loss of plasma volume, and of ECF volume (Flanagan et al., 1950; Skrabal et al., 1972). The latter authors did not find a change in total exchangeable Na or K, however. Muscle and serum Na fall, muscle K increases, and intracellular Na drops to zero (Cole, 1953). Other tissues are also affected: tendon, skin, and cartilage show a decrease in Na concentration and an increase in K concentration, as well as muscle (Kriegel and Discherl, 1964). Serum Na falls and serum K rises, sometimes to life-threatening levels.

Insulin

Insulin acts to favor the entrance of glucose into cells, save for brain, and has anabolic and lipogenic activity. Many years ago Atchley et al. (1933) did complete balance studies on two diabetic patients and nicely demonstrated the loss of body nitrogen and electrolytes when insulin was discontinued, and the positive balances when insulin treatment was resumed. The patient who became acidotic had much larger losses of these materials and a more rapid weight loss than the one whose blood pH remained normal. Walsh et al. (1976) measured total body K in newly diagnosed diabetic patients and again after 6–8 weeks of treatment. Those who required insulin gained on average 560 meq K; and those who responded to oral hypoglycemic agents and diet gained 115 meq K; based on the final values body K deficits were 17% and 3%, respectively, at the time of diagnosis.

Studies of adults with established diabetes mellitus have shown that many have slightly reduced values for total body K (Delwaide, 1973) and some reduction in bone density (Levin et al., 1976) and in total body Ca (Ellis and Cohn, 1975a). However, it is difficult to ascertain how many of these subjects were in poor metabolic control, for improper control of the disease can be associated with acidosis and tissue wasting. Diabetes mellitus involves more than insulin deficiency, for in time many patients develop widespread vascular lesions that serve to compromise the function of a number of organs.

Antidiuretic Hormone

Antidiuretic hormone acts on the renal tubule to facilitate the reabsorption of water. Individuals who lack this hormone, and others whose kidneys lack the ADH receptor, experience a profound reduction in body water unless provided with an abundant intake (several liters!) of water each day. Administration of the hormone to the former group reduces urine volume to near-normal quantities.

10

Influence of Trauma and Disease

Diseases are the tax on pleasures.
English proverb

Changes in body composition the result of malnutrition, obesity, and certain endocrine diseases have been discussed in previous chapters. We now proceed to consider the changes that have been observed in other disease states and following physical trauma. In evaluating these data it must be remembered that some of the situations to be discussed involve bed rest, which is known to lead to negative nitrogen balance, and a decreased intake of food, which also is associated with a loss of LBM. It is obviously difficult, and in some situations impossible, to divorce the effects of these two influences from that of the disease process *per se*.

Trauma

Many years ago Cuthbertson (1932) reported that patients with long bone fractures promptly displayed an increase in urinary nitrogen excretion to values in excess of intake and that the nitrogen losses were proportional to the extent of the accompanying hypermetabolism. He also measured urine sulfur and phosphorus, and based on the ratio of the S and P loss to N loss he suggested that muscle was the source of the N loss. In one patient the total urinary N loss amounted to 137 g, or about 7% of estimated total body nitrogen. Later he and his associates (Cuthbertson et al., 1972) reported that losses of Zn and K were proportional to nitrogen loss, and that the extent of these losses was greatest in the most severely injured patients. Whipple (1938) demonstrated the effect of infection and nonbacterial injury on nitrogen catabolism, and distinguished between labile protein and so-called fixed protein reserves.

Surgical operations are a form of trauma. Francis Moore and his associates (Moore et al., 1963) made extensive observations on patients who underwent major operative procedures, and spoke of the catabolic response to surgery: the tendency for ECF volume to increase and for body cell mass (estimated by ^{42}K dilution) to diminish.

Significant trauma, whether surgical or accidental, provokes a chain of endocrine events, the central one of which is a stimulation of adrenocorticotropin (ACTH) production by the pituitary gland, which leads to an increase in cortisol production by the adrenal gland. Indeed, studies of patients who have been burned have demonstrated a significant relationship between the extent of the burn and the serum level of 17-hydroxycorticoids. The stimulus for ACTH production is largely neurogenic, for denervation of the injured area abolishes the response. Other hormones are also involved in the response: renin, catecholamines, antidiuretic hormone, growth hormone, and glucagon. The situation is complicated by blood loss and ECF fluid losses (as into traumatized tissue) and by an increase in metabolic rate. All of these factors play a role in producing the catabolic response to trauma—nitrogen and potassium losses, and sodium retention—which in uncomplicated situations lasts for 3–10 days, to be followed if all goes well by an anabolic restorative phase which may take several weeks for completion.

However the changes which occur at the time of surgery must be distinguished from those which take place during the postoperative period. Shires et al. (1961) measured extracellular fluid (ECF) volume, plasma volume, and total red cell mass both prior to surgery and again two hours later in patients who had general anesthesia but no intravenous fluids. As estimated by radiosulfate dilution, ECF volume fell by an average of 1.4% in those patients having minor surgical procedures and by 13% in those who had major surgery. Since they failed to find a correlation between blood loss and ECF loss the authors concluded that there had been an internal redistribution of body fluids. In this connection the data provided by Bergström (1962) on muscle electrolytes are pertinent. Biopsy samples from human subjects obtained under general anesthesia contained 9% less K and 45% more Na than those from patients who had local or spinal anesthesia; there was no difference in water content.

In a recent study Bessey et al. (1984) described the effects of administering epinephrine, glucagon, and cortisone, singly and in combination, to normal volunteers for periods of 72 h. The amounts given (30 ng epinephrine/kg/min, 3 ng glucagon/kg/min, 2.3 μg hydrocortisol/kg/min) produced blood levels of these hormones comparable to those found in patients with injuries of moderate severity. The subjects were given a normal diet of known composition; intakes of energy, protein, and minerals were held constant. The subjects who received the hormone infusions had an increase in metabolic rate; serum K fell; urine K excretion and N excretion both rose (by 31% and 25%, respectively) whereas urine Na excretion fell by 43%. Nitrogen balance became negative. Heart rate and blood pressure increased, as did forearm blood flow and urinary C-peptide excretion (an index of insulin secretion). The subjects given saline exhibited none of these changes. Hence it would appear that many of the metabolic changes seen in traumatized patients were duplicated by the infusion of these three hormones, though perhaps not to the same degree. Trauma patients of course have the added features of blood loss and tissue damage.

Metabolic balances in patients undergoing major surgery and receiving the usual low energy intravenous feedings without added nitrogen (or relatively little nitrogen) show cumulative losses ranging from 26 to 141 g (Mulholland et al., 1943), Kinney et al., 1968), the average being 15–19 g N/kg of weight loss. However, when body nitrogen is assayed by neutron activation techniques, the losses are considerably greater: Hill et al. (1978) recorded an average value of 39 g N loss/kg weight loss, and Bogle et al. (1985) one of 42 g N/kg in their patients. Total nitrogen losses averaged 160–174 g, values considerably in excess of those found by the balance technique. This serves once again to emphasize the fact that the metabolic balance technique may underestimate losses of nitrogen and minerals in patients who are in negative balance. The multitude of day-to-day problems encountered in the care of surgical patients make for great difficulty in carrying out a proper metabolic balance study. Hence body composition techniques offer a real advantage.

When Mulholland et al. (1943) and Hill et al. (1978) provided their patients with high-calorie–high-nitrogen fluids intravenously, they did not lose weight or body nitrogen during the 2-week period of observation. Thus the effect of the catabolic process can be minimized, even overcome, by providing proper nutrition, and it is now common practice to provide nourishment by the intravenous route for those patients who have extensive surgery and who are unable to take adequate nourishment by mouth.

Bogle et al. (1985) also measured total body K in their patients, both prior to surgery and 2 weeks postoperatively. Body K losses averaged 300 mmol, or 11.4% of initial body K content, and body weight loss was 7.1%. Nitrogen losses averaged 10.8% of initial body content, and the average $\Delta K/\Delta N$ ratio was 2.22 ± 0.42 (SEM). This ratio is close to that determined by cadaver analysis of adults, namely 2.0; so in those patients K and N were lost in amounts proportional to their body content.

Muscular Dystrophy

The Duchenne type is a progressive, sex-linked disorder in which symptoms of muscle weakness first appear in early childhood and that culminates in death in the second or third decade of life. The degree of apparent muscle wasting is not commensurate with the amount of weakness since there is extensive replacement of muscle by connective tissue. Biopsy specimens show that connective tissue represents 26–66% of the muscle cross-sectional area (Edmonds et al., 1985), in contrast to 4% in normal muscle. Since the disease is progressive one would expect that total body K would progressively deviate from normal as these patients age, for as mentioned in Chapter 2, connective tissue has a lower potassium content than muscle *per se*. Figures 10.1 and 10.2 show that such is indeed the case. The ratio of LBM/H shows little tendency to change from age 5 years to age 15 years in these patients, whereas there is a progressive rise

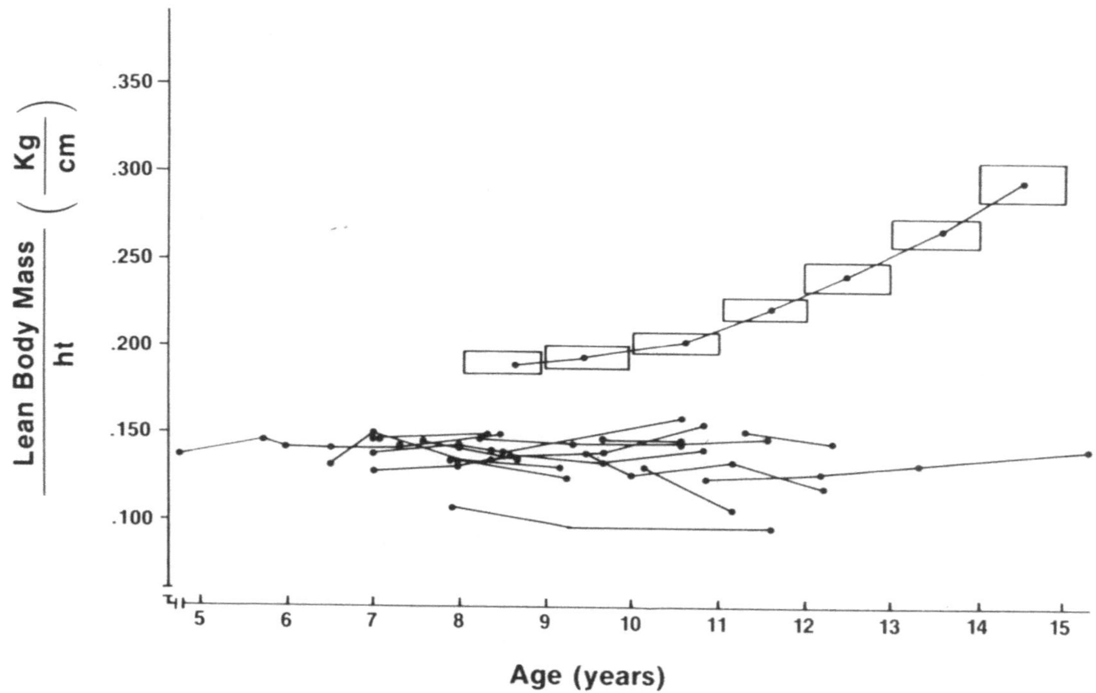

FIGURE 10.1. Sequential assays of "LBM" (by ^{40}K counting) to height ratio in boys with Duchenne muscular dystrophy. Data for normal boys shown in rectangles (±SD). "LBM" is in quotation marks because it is likely that the K content of the LBM is reduced in this disease. Multiply ordinate values by 2.66 to obtain g K/cm. [From Griggs et al. (1983), with permission.]

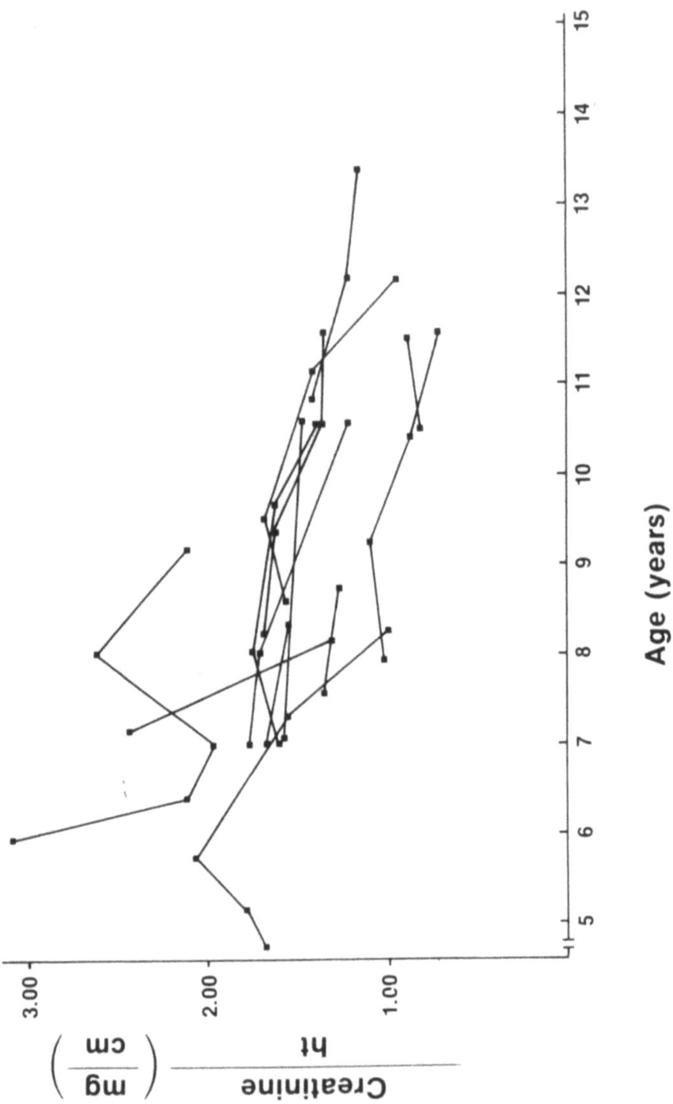

FIGURE 10.2. Sequential determinations of 24-h urinary creatinine excretion: height ratio in boys with Duchenne muscular dystrophy. [From Griggs et al. (1983), with permission.]

in the normal boys. Creatinine excretion per unit height actually falls with increasing age, indicating that muscle mass suffers more than total body K as the disease progresses. As shown in Chapter 4, Figure 4.9, creatinine excretion increases with age in normal children.

Mays et al. (1968) determined organ size in three boys who died in the second decade of life, and analyzed various tissues for potassium. Muscle contained only 10–20% of total body K, a value far lower than the one of about 50% determined for nondystrophic individuals. Except for low values for skeleton and spleen, all of the other organs had a normal concentration of potassium.

Borgstedt et al. (1970) did ^{40}K assays of 16 relatives (mothers, sisters) of boys with muscular dystrophy. On the basis of chance half of these individuals should be carriers of the disease. When compared to normal females of equivalent age and stature they had normal values for total body K. Hence there is no evidence on this basis of heterozygote expression.

Edmonds et al. (1985) have done a more complete study of boys with muscular dystrophy, including total body water and total exchangeable sodium as well as total body potassium and urinary creatinine excretion. They confirmed the progressive deviation from normal of total body K with age shown in Figure 10.1. However, total exchangeable Na (NA_e) continued to rise with age, though remaining below normal for height. The ratio of Na_e to body K was 1.89 (molar basis) compared to one of 0.95 for normal adults, and the ratio of total body K to total body water was 56 meq/liter, considerably below their normal value of 86 meq/liter. Of interest is the fact that Edelman's equation[1] (Edelman et al., 1958) relating $Na_e + K_e$ to total body water and serum sodium holds true for these grossly abnormal subjects. Hence all of the exchangeable Na and K are osmotically active, as is the case for normal individuals. These authors also estimated total body fat from total body water assays. Patients who had progressed to the point of being unable to walk had an average of 39% body fat, whereas those who were still walking had 20%, a value close to the one of 17% for their normal male subjects.

Adults with myotonic dystrophy have total body K values that are only 58–90% as great as normal subjects of comparable age and stature (Griggs et al., 1983); urinary creatinine excretion is 48–92% of the expected value.

Infection

Severe infections are accompanied by fever, hypermetabolism, and often by anorexia. Those involving the gastrointestinal tract also provoke losses of nitrogen and minerals via stool and/or emesis. Consequently, nutritional requirements increase and nitrogen and mineral losses can occur. Many

[1] Serum Na $= 1.11\,(Na_e + K_e)/H_2O - 25.6$ (Chapter 2).

years ago Shaffer and Coleman (1909) documented the occurrence of negative nitrogen balance in patients with typhoid fever, and showed that nitrogen losses could be reduced by increasing the intake of energy and protein.

Beisel and his co-workers (1967) conducted an extensive balance study of subjects with bacterial and viral infections. They documented losses of nitrogen, potassium, magnesium, sodium, and phosphorus; calcium balances were variable. Loss of body weight was 3–6 kg. In tularemia the cumulative balances were −52 g N, −250 meq Na, −200 meq K, −2 g P, and −600 mg Mg. Losses of N, Na, and K were less in subjects who had sandfly and Q fever, but strangely, those of Ca, P, and Mg were greater than with tularemia, which is a more severe infection.

The problem of assessing changes due to infection *per se* is that most patients reduce their food intake. Beisel et al. (1967) also studied normal (noninfected) subjects who were fed amounts of food comparable to that

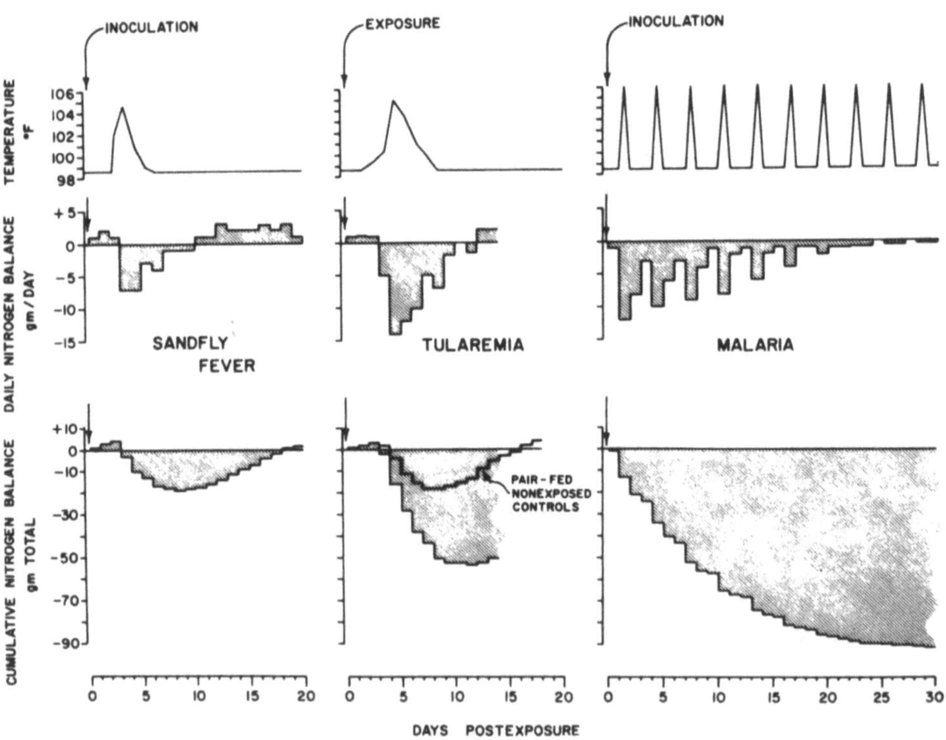

FIGURE 10.3. Time course of fever and nitrogen balance in viral, bacterial and parasitic infections in man: average values for three groups of infected subjects. Values for pair-fed normal control subjects are included for comparison. [From Beisel (1976), with permission.]

consumed by the patients with tularemia. Cumulative loss of nitrogen was 18 g, potassium 20 meq, phosphorus 6 g, and magnesium 300 mg. Sodium was lost quickly (~200 meq) but this loss was quickly made good so that the balance became zero in about 10 days.

Figure 10.3 shows the time course of fever and nitrogen balance for subjects with three types of infection—viral, bacterial, parasitic—and also shows the cumulative nitrogen balance in normal pair-fed subjects. It is apparent that tularemia provokes a degree of nitrogen loss that is in excess of that incident to decreased food intake. Malaria has the most profound effect, sandfly fever the least. The data on malaria were obtained from a report by Howard et al. (1946).

While many aspects of the metabolic response to infection resemble the response to surgery and trauma, there are some differences. The latter is characterized by a retention of sodium, the former by a loss. Furthermore, in some of the subjects studied by Beisel et al. (1967) the negative balance of nitrogen and minerals was established a day or two prior to the onset of clinical symptoms.

Hypertension

Studies of body composition in experimental hypertension in animals and in the disease in humans have produced variable results. Rats made hypertensive by partial nephrectomy show a modest increase in Na content of many organs, an increase in water and chloride content, and a decrease in potassium (Laramore and Grollman, 1950). The increase in muscle Na was 8%, compared to controls; brain Na did not change. Taverner et al. (1985) found a 36% increase in blood volume and a 6% increase in total exchangeable sodium. On the other hand, Gresson et al. (1973) found slightly lower values for total red cell mass, plasma volume, ECF volume, and total exchangeable Na in a strain of rats with hereditary hypertension compared to normotensive rats of comparable body weight.

Turning now to human subjects with essential hypertension, Hollander et al. (1961) found normal values for Na_e and K_e compared to controls matched for age, sex, and body weight, as did Rajagopalan et al. (1980), including plasma volume. In the series of patients studied by Beretta-Piccoli et al. (1982) Na_e exceeded control values by an average of 6%, and there was a modest correlation between the Na_e/K_e ratio and the height of both the systolic and diastolic blood pressure. Bauer and Brooks (1982) found a slightly higher ratio of ECF volume to total body water in their hypertensive subjects; the ratio of plasma volume to total body water was also a bit lower, whereas the ratio of plasma volume to ECF volume was comparable to that of the controls. Their data on absolute fluid volumes cannot be easily interpreted because of the wide age span of both the controls and hypertensive subjects, and because there is no mention of stature.

Williams et al. (1982) did extensive body composition assays on subjects with essential hypertension. They found a small but significant correlation ($r = 0.4$) between Na_e and the height of the blood pressure, but there was no correlation with total body Cl, N, K, Ca, or P.

Patients with malignant hypertension and those who have complicating mild cardiac failure have a 22–24% increase in Na_e and a 9% decrease in K_e in comparison to normal individuals (Hollander et al., 1961), and those with primary aldosteronism have a 14% increase in Na_e and a 31% decrease in K_e (Chobanian et al., 1961); ECF volume (radiosulfate space) is also increased. When the adrenal tumor was removed, Na_e and ECF volume both dropped and K_e rose significantly. These investigators carefully provided age-, sex-, and weight-matched controls.

It is of interest that the excessive amounts of aldosterone produced by the tumor had a greater effect on K_e than on Na_e; indeed these patients had a subnormal level of serum potassium. As mentioned in Chapter 9, the administration of mineralocorticoids to animals and to humans serves to increase Na_e and to decrease K_e.

A number of body composition studies have been done on hypertensive patients undergoing treatment with drugs. Diuretics such as hydrochlorthiazide had no effect on total body K in one study (Graybird and Sode, 1971) whereas other investigators (Wilkinson et al., 1975) found a slight fall. In the study by Sederberg-Olsen and Ibsen (1972), propanol treatment resulted in 5% increase in ECF volume (^{82}Br dilution) but no change in plasma volume or in body weight.

In another study of the effect of diuretic drugs (this time for urinary lithiasis) Williams et al. (1980) found a six percent decrease in total body K, a 7% decrease in total body Cl, but no change in total body Na, Ca, or N, and no change in body weight over a 5–8-month period of treatment with bendrofluazide plus potassium. Total body phosphorus decreased slightly.

Cardiac Failure

Talso and his associates (1953) studied muscle composition in patients with cardiac edema and in normal controls. As would have been expected, the patients had higher values for water, Cl, and Na, and subnormal values for K and N; with treatment all of these constituents reverted to approximately normal values. Of interest was the finding that the K/N ratio in muscle did not vary greatly, the values being 3.19 meq/g in the normal subjects, 3.08 in those with edema, and 3.14 after treatment.

Body composition assays have been done in patients with heart disease. O'Meara et al. (1957) found a great increase in total body water and Na_e and very low values for K_e in their edematous patients. The Na_e/K_e ratio was about twice normal and the ratio of K_e/total body water was about half the normal value. Comparable ratios were recorded for patients with cirrhosis of the liver with edema. Edema fluid resembles normal extracellular fluid in composition, so one would expect low values for K_e. Thomas et al. (1979) found subnormal values for total body K and total body N in patients with chronic heart disease in the absence of overt edema. When the deviations in body K from the mean were plotted against the deviations in body N there was a straight line relationship for both the normal individuals and the patients, the curve for the latter being displaced downward to a slight extent. Hence there was some reduction in the total body K/N ratio.

Cancer

Shizgal (1985) compared changes in body composition in cancer patients with those seen in malnourished patients. Both weighed less than controls, both had lost LBM and fat, both had an abnormally high Na_e/K_e ratio, and both had an excess of ECF volume. Heymsfield and McManus (1985) did a more detailed study, which included CAT scans and ultrasound scans of various organs, and compared the cancer patients with those who had anorexia nervosa. The anorexic patients had a reduction in heart, liver, spleen, and kidney size as well as muscle, all of which were reduced in proportion to total weight loss. Visceral size was better preserved in the cancer patients and hence muscle accounted for a large proportion of their weight loss. The difference may lie in total metabolic rate, which is often elevated in patients with tumors, and usually reduced in those with anorexia nervosa.

Toal et al. (1961) have made an interesting observation on rats bearing the Walker carcinosarcoma, a tumor that has a high sodium content, the average value being 96 meq/kg fat-free weight, compared to a total body Na of about 55 meq/kg in this species. They noticed that the tumor-bearing animals excreted subnormal amounts of sodium in their urine, that their adrenal glands were enlarged, and that urinary Na excretion returned to normal when the tumor was removed. They proceeded to do total body analyses on these animals, and found that total carcass sodium increased with increasing tumor size, from 55 meq/kg when the tumor constituted only 5% of the total weight to 65–70 meq/kg when the tumor mass made up 40% of the total weight. Carcass chloride was higher and K less in the tumor-bearing animals, but an unexpected finding was an increase in body nitrogen.

Body composition studies have been made in other disease states. Those with cirrhosis of the liver, Crohn's disease, and cystic fibrosis of the pancreas have findings that one would expect in malnutrition. Indeed, in appraising the effect of disease on body composition, it is often difficult to separate the effect of the disease *per se* from the accompanying weight loss and undernutrition. As a final note, it may be mentioned that assays have recently been reported by Kotler et al. (1985) on patients with the acquired immunodeficiency syndrome (AIDS). These patients have reduced total body K levels and, surprisingly, a reduced ECF volume when compared to controls of the same stature. Healthy homosexual men had normal values.

Osteoporosis

As mentioned in Chapter 5, Figure 5.10, not all bones are equally involved in this all too common disorder. The spine usually suffers more than the long bones. Total body calcium has been assayed in these patients by neutron activation. Nelp et al. (1972) found a reduction of 25–35% in total body Ca. Ellis and Cohn (1975a) found a reduction in both total body K and total body Ca in their patients. In males the ratio of the decrement in body K to that of body Ca was 0.12 g K/g Ca, which is the same as the body K/Ca ratio in normal males. The female patients had lost somewhat less total body K relative to the loss of body Ca, the ratio being 0.077 g K/g Ca, compared to one of 0.10 in normal females. Or one could say that the reduction in body Ca was relatively greater than that of K.

Attempts to relate measurements at specific bone sites to total body Ca have met with only partial success. Manzke et al. (1975) report a correlation coefficient of 0.94 between photon absorptiometric assays of the radius and total body Ca, yet the standard error of estimate is so high (160 g Ca) as to suggest a considerable imprecision. Aloia et al. (1977) looked at the relationship between total body Ca and various indices of regional bone mass in women with spinal osteoporosis who had had one or more compression fractures. Total body Ca was 81% of the predicted value for age. Three of the 45 subjects had less than 400 g body Ca. The indices used were the metacarpal index, the vertebral biconcavity index, and the femoral trabecular pattern. The correlations with total body Ca ranged from 0.51 to 0.70. It would appear that regional bone mass indices are not as good predictors of total body calcium in patients with osteoporosis as is the case for normal individuals. Viewed in another way, the skeleton is not uniformly involved, the spine and other areas of trabecular bone being more affected than the cortical bone of the axial skeleton (see Chapter 5, Figure 5.10). An example is provided by the report of Drinkwater et al. (1984) on eumenorrheic and amenorrheic runners: The latter group had a lesser density of the spine whereas that of the radius was the same as the group who were menstruating normally.

Hypokalemic Periodic Paralysis

This is a familial condition characterized by muscle weakness, and episodes of paralysis that are accompanied by hypokalemia. Paralysis can be induced by the administration of sodium bicarbonate, glucose and insulin, epinephrine, and cortisone. Muscle samples reveal an increased sodium content and a decrease in potassium, but a normal ratio of ECF to ICF volume, as judged by chloride content. The ratio of Na_e to K_e has been determined to be 1.80, or greatly in excess of that for normal subjects (1.06). Muscle samples taken during episodes of paralysis show a markedly increased ECF/ICF ratio, an increase in K content, and a fall in Na content. Treatment with a low-Na, high-K diet brings about clinical improvement. The pathogenesis of the disease is unknown. There is also a form of periodic paralysis associated with hyperkalemia and a reduction of muscle K content (Talso et al., 1963; Coppen and Reynolds, 1966).

11

Concluding Remarks

Prior to the introduction of modern techniques for estimating body composition one was limited to conclusions drawn from metabolic balance, and analogies from animal experiments. Although the balance technique can assess changes in body composition, it cannot provide information on actual body content of any element. In this respect the techniques described in previous chapters have a distinct advantage. Furthermore, they offer the opportunity of monitoring relatively long-term changes in body composition, a procedure that although theoretically possible is extremely difficult to achieve in practice by the metabolic balance method. As a result we have acquired a host of new information on growth and aging, on sex differences in body composition, and on changes that take place in response to various influences, nutritional and otherwise. We now have the means of monitoring changes in body composition by methods that require relatively little cooperation on the part of the subject and that pose a negligible hazard.

It is now possible to assess the present status of body composition in patients, in addition to determining changes that may occur as the result of treatment or disease. An example is the discovery of the supranormal lean body mass (LBM) in obesity. Abnormal body fluid volumes can be identified, and designated with regard to disproportions among plasma volume, extracellular and intracellular fluid volumes. Total exchangeable electrolyte can be assayed, and it has been shown that the ratio of total exchangeable sodium to total exchangeable potassium is altered in certain disease states. The changes that occur during growth and aging in the ratio of body fluid volumes have been established.

One awaits with interest the application of newer techniques—electrical impedance, CAT scan, nuclear magnetic resonance.

The section to follow provides an overview of the alterations in body composition that have been recorded in various situations. By bringing all of these together the constellation of effects may be better appreciated.

LBM and Fat: Concordance and Discordance

The previous chapters have reviewed the various circumstances leading to a change in body composition. In some instances the induced changes in LBM and fat are in the same direction: an increase in one is accompanied by an increase in the other, the same being true for a decrease. In a sense, these two body components are companions in that significant changes in body weight involve both, albeit in variable proportions. Under other circumstances LBM and fat behave as strangers; an increase in one is associated with a decrease in the other, and vice versa.

Figure 11.1 is a schematic representation of the changes in LBM and fat that have been recorded for adult human subjects as a result of various maneuvers. Information on the hibernating bear and on animals with hy-

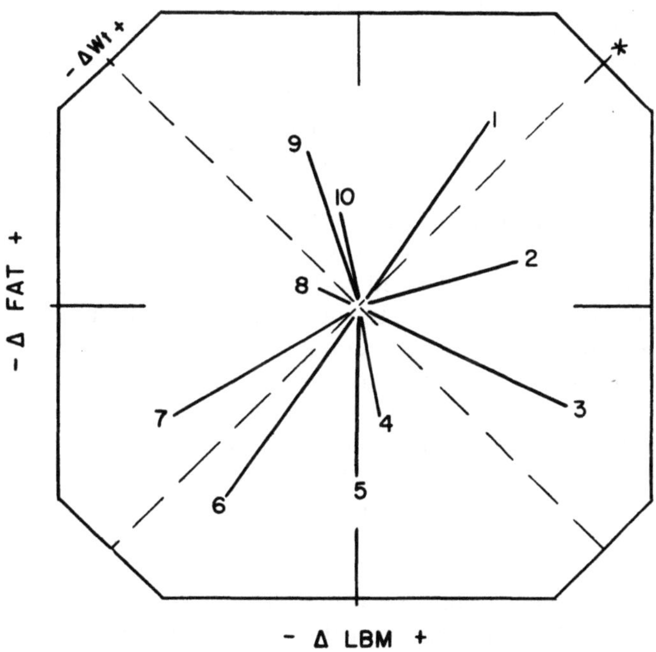

FIGURE 11.1. Observed changes in LBM and fat the result of various maneuvers. The diagonal dotted line headed by ΔW indicates no change in body weight. Dotted line headed by asterisk indicates equivalent changes in LBM and fat. Length of solid lines roughly proportional to the magnitude of the observed changes. Line 1, overfed normal subjects; 2, repletion of undernourished subjects, also normal pregnancy; 3, effect of androgens; 4, effect of exercise; 5, hibernating bear; 6, low-energy diet (obese subjects); 7, low-energy diet (nonobese subjects); 8, zero gravity, bed rest; 9, experimental hypothalamic obesity (animals), Cushing's syndrome; 10, high-energy–low-protein diet.

pothalamic injury are also included since these represent interesting paradoxes. The dotted and interrupted solid lines define eight triangles, and it is to be seen that all eight of these are occupied by at least one solid line radiating from the center.

When the stimulus is nutritional in nature, a change in body weight is seen to involve both LBM and fat, and usually in the same direction. A deficit in energy intake brings about a decrease in both, a surfeit of energy an increase in both. The former circumstance invokes an increase in gluconeogenesis from protein, so as to maintain the supply of glucose, essential to cerebral function. Although the contribution of protein to the total energy expenditure is of the order of only 15% even in fasting, each gram of protein is accompanied by 3.5 ml of water, so the amount of weight lost by the catabolism of protein is appreciable. When there is a surplus of energy, a majority of the excess is stored as fat, yet a portion of the weight gain consists of LBM. The reason for this must lie in the increase in serum insulin, and perhaps in other hormones that have anabolic properties. In this view LBM and fat do not behave as completely independent entities, and so are concordant.

Established states that have evolved from nutritional excesses or deficiencies also exhibit concordance between LBM and fat. The vast majority of obese individuals have supranormal LBM in comparison to their age, sex, and height peers. Patients with anorexia nervosa have a reduced LBM as well as less body fat. Among all of the factors that operate to change body composition nutrition is the most powerful.

Normal pregnancy is associated with an increase in both LBM—partly the result of the growth of the fetus and uterus—and body fat. However, the relative contribution of each component to the total weight gain is different from that resulting from the overfeeding of normal subjects (compare lines 1 and 2 in Figure 11.1). So, too, is the response of undernourished individuals to nutritional rehabilitation, for in the early phase of recovery the restoration of LBM proceeds more quickly than body fat. Later on, fat accumulation may predominate, as shown so clearly in the experiments conducted by Ancel Keys and his associates (1950) (see p. 235).

With regard to dietary-induced weight loss it is important to recognize the difference in behavior between the obese and the nonobese (lines 6 and 7 in Figure 11.1). Although both types of subjects lose both LBM and fat during the course of significant weight loss, the proportions differ. Thin individuals lose proportionally more LBM than fat, obese individuals proportionally more fat than LBM. The obese are thus more tolerant of fasting.

The best example of discordance between changes in LBM and changes in body fat is provided by the effect of androgen administration (line 3 in Figure 11.1). When large doses of anabolic steroids are given over a sufficient period of time, LBM increases and body fat declines, and since the change in the former usually exceeds that of the latter there is an increase in body weight. This implies that energy intake was greater during the experimental period. Since the synthesis of protein, and hence an increase in LBM, requires less energy than is released by the burning of an equal quantity of fat, one could anticipate some increase in body weight when androgens are given in the face of a constant diet. However, an experiment to test such an hypothesis appears not to have been done.

The feeding of an abnormal diet can also lead to discordant results. Low-protein–high-energy diets cause a gain in body fat and a slight reduction in LBM (line 10 in Figure 11.1). Animal experiments have shown that protein-deficient diets fed to the young augments body fat content often at the expense of LBM. Such a circumstance could have been foreseen, for there are limits to the protein-sparing effects of carbohydrate and fat.

Discordant effects have also been observed in a number of studies of individuals who have engaged in programs of exercise and/or physical training, there being a variable rise in LBM and a fall in body fat content. The effect on LBM is actually rather modest—certainly less than that which accrues from large doses of androgens—and some studies have failed to document a change in either body component.

Although diet-induced obesity is usually associated with an increase in LBM, animals that become hyperphagic and obese following injury to the hypothalamus have a subnormal LBM in the face of an increased body fat content. Hence experimental hypothalamic obesity is not an appropriate model for human obesity, nor is Cushing's syndrome.

The hibernating bear provides yet another example of discordance. This remarkable animal loses appreciable quantities of body fat during winter sleep without a concomitant change in LBM (line 5 in Figure 11.1). It accomplishes this feat by reabsorbing water, nitrogen, and electrolytes through the bladder wall. The bear appears to be unique in this respect, for there are no recorded instances of other mammals (including man) being able to conserve LBM in the fact of dietary-induced weight loss of comparable degree. Other hibernators lose both LBM and fat.

The fact that nutrition exerts such a powerful influence on body composition makes for difficulty in evaluating the effects of exercise and/or physical training on body composition. The practical problems encountered in maintaining strict control of diet in such circumstances are formidable, and most investigators who have studied the effect of exercise have not recorded the dietary intake of their subjects. If for example energy expenditure during the exercise period exceeds energy intake by a substantial margin, it is likely that the desired increment in LBM could be lessened. It can be said that there are no extant data to show that LBM can be preserved, much less increased, by exercise in the face of a significant energy deficit. The best example is afforded by the work of Keys et al. (1950) whose adult male subjects lost large amounts of LBM (and fat) on

1600-kcal diets despite a superimposed program of vigorous exercise. As estimated by densitometry over half of their weight loss consisted of LBM.

Attempts at forestalling the loss of LBM on low-energy diets by providing generous amounts of protein have not succeeded in completely sparing protein as body weight is lost. Although the nitrogen losses are less than those that occur during fasting, nonetheless some loss of body N, and LBM, does take place.

Longitudinal vs. Cross-Sectional Data

With regard to the adult, the few longitudinal studies of LBM, of skeletal mass and of stature which have been done have generally recorded a smaller decline with age than is apparent from cross-sectional data. This suggests that the latter were derived from different biological populations, and so may not delineate the true rate of aging. Perhaps the elderly subjects were less well nourished during childhood than were today's young adults, and so failed to achieve their full growth potential. Whatever the reason for this apparent discrepancy, there is a real need for further longitudinal data on body composition of the adult, if for no other reason than to detect—and possibly explain—variations in the rate of aging. Children do not all grow at the same rate, and one can entertain the notion that variation in the rate of aging is in part genetic, just as is the rate of growth. Data on longevity—which is known to be inherited—can be offered in support of this contention.

However, cross-sectional data serve to describe the present status of the population, and so document the differences between the aged and the young as of today. If perchance nutrition is a prime factor in exaggerating the true rate of aging future cross-sectional observations would be expected to show a smaller decline, since present-day young and middle-aged adults were probably better nourished during childhood than were their parents.

It must be admitted that body composition data do not accurately reflect the decline in functional capacity with age. By age 80 years cardiac output, renal function, vital capacity, and maximum oxygen consumption are only 40–70% of their values at age 30. Muscle strength declines by about 20% between age 30 and 65 years, and older adults are much less trainable than young adults. The decrease in strength most closely approximates the decrease in LBM with age. The reasons for functional decline include an increase in the work of breathing, a decrease in pulmonary diffusing capacity, in alveolar capillary blood volume, and in myocardial perfusion; reduction in the number of capillaries per muscle fiber; diminution of maximal arteriovenous oxygen difference; decrease in elastic tissue; and an increase in cross-linkage of collagen (Shephard, 1982). These effects are compounded by a decrease in physical activity which characterizes many elderly people. When they do exercise the recovery period is prolonged.

A question of some interest is whether regular exercise can retard the aging process as concerns both body composition and function. In the study of Dill et al. (1967) world class athletes suffered a decline in maximum oxygen consumption with age regardless of exercise patterns. Modern life styles—the ubiquitous automobile, labor-saving devices, the decline of walking as a pastime—contribute to our sedentary existence. The fact that bed rest is associated with losses of nitrogen and calcium raises the possibility that even less severe restrictions in physical activity could compromise both muscle and skeletal integrity.

Physical training and/or exercise does enhance LBM, albeit to a modest degree, and animal experiments suggest that nutrition is more effective in augmenting LBM in the young animal. If one's goal is a larger LBM, and its corollary greater strength, would it not be proper to suggest that programs of regular exercise combined with adequate nutrition be aimed at the child and adolescent? The likelihood of success would conceivably be greater than in the adult years. If such programs could result in a larger skeletal mass, could this conceivably reduce the likelihood of clinically significant osteoporosis in later years? Others have advocated an increase in dietary calcium intake as a preventive measure. The frequency of osteoporosis with its attendant disability are such as to warrant efforts at prevention.

Other Considerations

Since the lean body mass comprises the active metabolic portion of the body, fat being relatively inert in this respect, it is to be expected that a number of physiological functions would be better related to LBM than to body weight. The sex difference in BMR and in blood volume is rather small when these are related to LBM. Maintenance energy needs for the adult are more closely related to LBM than to body fat. A compilation of data on endogenous urinary nitrogen excretion shows a good correlation ($r = 0.79$) with body cell mass over a range of sizes which includes elderly women and young adult Caucasian and Oriental males (Huang et al., 1972; Rand et al., 1976; Uauy et al., 1978). In considering such matters as nutrient requirements, the age and sex differences in LBM are of importance. Setting the value for young adult males at 100, the illustration in Chapter 5 (Figure 5.1) suggests that LBM in elderly males would be 87, in young adult females 68, and in elderly females it would be 60. This rather substantial variation has been partly taken into account by the National Research Council (1980) in making up the Recommended Dietary Allowance for energy intake. The slope of the regression line describing the relationship between endogenous urinary nitrogen and body cell mass is 0.062

g N/day per kg BCM, which suggests that protein requirement is a function of BCM, and hence LBM, rather than body weight with its variable component of fat. On this basis the need for protein—and perhaps other dietary nutrients as well—should be less in the elderly than in the young adult, and less in women than in men.

Drug doses are another consideration. Some have a volume of distribution which approximates extracellular fluid volume (aminoglycosides); others are bound to muscle (digoxin), or distributed in fat (thiopental). With regard to this last drug, there is one report showing that sleep time in rats following administration is inversely proportional to body fat content (Hermann and Wood, 1952). Distribution volume of diazepam is greater in the obese than in the nonobese (Abernethy and Greenblatt, 1982). The fact that body water volume is less in women than in men makes them less tolerant of alcohol.

The hazard to health from obesity is well known. Recent studies suggest that the distribution of body fat as well as the total amount of fat is important in this regard, the so-called android distribution being a bigger risk factor than the gynecoid distribution. In their body composition studies of a large group of women, many of whom were obese, Weinsier et al. (1985) found, as expected, a relation between blood pressure and body weight. Interestingly enough, they also found that LBM accounted for much more of the total variance in blood pressure than did body fat.

Some may be disappointed in the lack of emphasis on adipocyte size and number in this book. There are two ways for the adipose organ to grow in response to a surfeit of energy: the cells can grow in size or they can multiply; both processes operate in the establishment of the obese state. Adipocytes provide the framework for the deposition and storage of body fat, and it is likely that their role is a permissive one.

Our modern tools have allowed us to look at some of the intricacies of the internal environment of the human body, in a manner that does not disturb its function. Perhaps these peeks at internal form thus provided by body compositionists may stimulate others to draw the curtain of biologic ignorance a bit wider, to relate the individual cell to the body compartments which have been identified. Body composition is a form of chemical anatomy, defining the *milieu* in which mammalian cells are able to perform their function. Without their lifelines—the body fluids—they are helpless.

Das wenige verschwindet leicht dem Blicke
Der vorwärts sieht, wie viel noch übrig bleibt.

J. W. von Goethe in "Iphigenie auf Tauris"

The little that is done seems nothing
when we look forward and see how much
we have yet to do.

M. Arnold's translation

References

Abernethy D.R. and Greenblatt D.J. (1982) Pharmacokinetics of drugs in obesity. *Clinical Pharmacokinetics*, 7:108–124.

Abraham S.F., Beumont P.J.V., Fraser I.S., and Llewellyn D. (1982) Body weight, exercise and menstrual status among ballet dancers in training. *British Journal of Obstetrics and Gynaecology*, 89:507–510.

Adams P., Davies G.T., and Sweetnam P. (1970) Osteoporosis and the effects of ageing on bone mass in elderly men and women. *Quarterly Journal of Medicine*, 39:601–615.

Albanese A.A., Lorenze E.J., and Orto L.A. (1962) Nutritional and metabolic effects of some newer steroids. *New York State Journal of Medicine*, 62:1607–1613.

Alén M., Häkkinen K., and Komi P.V. (1984) Changes in neuromuscular performance and muscle fiber characteristics of elite power athletes self-administering androgenic and anabolic steroids. *Acta Physiologica Scandinavica*, 122:535–544.

Allen D., Adinoff M.D., and Hollister R. (1983) Steroid-induced fractures and bone loss in patients with asthma. *New England Journal of Medicine*, 309:265–268.

Allen T.H., Anderson E.C., and Langham W.H. (1960) Total body potassium and gross body composition in relation to age. *Journal of Gerontology* 15:348–357.

Allen T.H., Peng M.T., Chen K.P., Huang T.F., Chang C., and Fang H.S. (1956) Prediction of blood volume and adiposity in man from body weight and cube of height. *Metabolism*, 5:328–345.

Alleyne G.A.O. (1975) Mineral metabolism in protein-calorie malnutrition. In R.E. Olsen (Ed.), *Protein-Calorie Malnutrition* New York: Academic Press, pp. 201–228.

Alleyne G.A.O., Viteri F., and Alvarado J. (1970) Indices of body composition in infantile malnutrition: total body potassium and urinary creatinine. *American Journal of Clinical Nutrition*, 23:875–878.

Almond D.J., King R.F.G.J., Burkinshaw L., Oxby C.B., and McMahon M.J. (1984) Measurement of short-term changes in the fat content of the body: a comparison of three methods in patients receiving intravenous nutrition. *British Journal of Nutrition*, 52:215–222.

Aloia J.F., Cohn S.H., Babu T., Abesamis C., Kalici N., and Ellis K. (1978) Skeletal mass and body composition in marathon runners. *Metabolism*, 27:1793–1796.

Aloia J.F., Roginsky M., Ellis K., Shukla K., and Cohn S. (1974) Skeletal metabolism and body composition in Cushing's syndrome. *Journal of Clinical Endocrinology and Metabolism*, 39:981–985.

Aloia J.F., Roginsky M.S., Jowsey J., Dombrowski C.S., Shukla K.K., and Cohn S.H. (1972) Skeletal metabolism and body composition in acromegaly. *Journal of Clinical Endocrinology and Metabolism,* 35:543–551.

Aloia J.F., Vaswani A., Atkins H., Zanzi I., Ellis K., and Cohn S.H. (1977) Radiographic morphometry and osteopenia in spinal osteoporosis. *Journal of Nuclear Medicine,* 18:425–431.

Amberson W.R., Nash T.P., Mulder A.G., and Binns, D. (1938) The relationship between tissue chloride and plasma chloride. *American Journal of Physiology,* 122:224–235.

Anasuya, A. and Narasinga Rao B.S. (1970) Relationship between body collagen and urinary hydroxyproline excretion in young rats fed on a low-protein or low-calorie diet. *British Journal of Nutrition,* 24:97–107.

Andersch B., Hahn L., Andersson M., and Isaksson B. (1978) Body water and weight in patients with premenstrual tension. *British Journal of Obstetrics and Gynaecology,* 85:546–550.

Anderson J., Osborn S.B., Tomlinson R.W.S., Newton D., Rundo J., Salmon L. and Smith J.W. (1964) Neutron-activation analysis in man in vivo: a new technique in medical investigation. *Lancet,* ii:1201–1205.

Anderson T.L., Muttart C.R., Bieber M.A., Nicholson J.F., and Heird W.C. (1979). A controlled trial of glucose versus glucose and amino acids in premature infants. *Journal of Pediatrics,* 94:947–951.

Angeleli A.Y.O., Burini R.G., and Campana A.O. (1978). Body collagen in protein-deficient rats. *Journal of Nutrition,* 108:1147–1154.

Annegers J. (1954) A study of total body water in rats and in mice. *Proceedings of the Society for Experimental Biology and Medicine,* 87:454–456.

Applegate E.A., Upton D.E., and Stern J.S. (1984) Exercise and detraining: effect on food intake, adiposity and lipogenesis in Osborne-Mendel rats made obese by a high fat diet. *Journal of Nutrition,* 114:447–459.

Apte S.V. and Iyengar L. (1970) Absorption of dietary iron in pregnancy. *American Journal of Clinical Nutrition,* 23:73–77.

Apte S.V. and Iyengar L. (1972) Composition of the human foetus. *British Journal of Nutrition,* 27:305–312.

Arara N.S. and Rochester D.F. (1982) Effect of body weight and muscularity on human diaphragm muscle mass, thickness and area. *Journal of Applied Physiology,* 52:64–70.

Archibald E.H., Harrison J.E., and Pencharz P.B. (1983) Effect of a weight-reducing high-protein diet on the body composition of obese adolescents. *American Journal of Diseases of Children,* 137:658–662.

Atchley D.W., Loeb R.F., Richards D.W., Jr., Benedict E.M., and Driscoll M.E. (1933) On diabetic acidosis. *Journal of Clinical Investigation,* 12:297–326.

Auld P.A.M., Bhanganandа P., and Mehta S. (1966) The influence of an early caloric intake with I-V glucose on catabolism of premature infants. *Pediatrics,* 37:592–596.

Ausman L.M., Rasmussen L.M., and Gallina D.L. (1981) Spontaneous obesity in maturing squirrel monkeys fed semipurified diets. *American Journal of Physiology,* 241:R316–R321.

Barac-Nieto M., Spurr G.B., Lotero H., and Maksud M.G. (1978) Body composition in chronic undernutrition. *American Journal of Clinical Nutrition,* 31:23–40.

Barac-Nieto M., Spurr G.B., Lotero H., Maksud M.G., and Dahners H.W. (1979) Body composition during nutritional repletion of severely undernourished men. *American Journal of Clinical Nutrition,* 32:981–991.

Barlow J.S. and Manery J.F. (1954) The changes in electrolytes, particularly chloride, which accompany growth in chick muscle. *Journal of Cellular and Comparative Physiology,* 43:165–191.

Barnard D.L., Ford, J., Garnett E.S., Mardell R.J., and Whyman A.E. (1969) Changes in body composition produced by prolonged total starvation and refeeding. *Metabolism*, 18:564–569.

Barnett E. and Nordin B.E.C. (1960) The radiological diagnosis of osteoporosis. *Clinical Radiology*, 11:166–174.

Barnett H.L. and Fellers F.X. (1947) A simple quantitative method for intravenous injection of small volumes of fluid. *Science*, 106:401–402.

Bauer J.H. and Brooks C.S. (1982) Body-fluid composition in normal and hypertensive man. *Clinical Science*, 62:43–49.

Beddoe A.H. and Hill G.L. (1985) Clinical measurement of body composition using *in vivo* activation analysis. *Journal of Parenteral and Enteral Nutrition*, 9:504–520.

Beddoe A.H., Streat S.J., and Hill G.L. (1984) Evaluation of an in vivo prompt gamma neutron activation facility for body composition studies in critically ill intensive care patients: results in 41 normals. *Metabolism*, 33:270–280.

Beddoe A.H., Streat S.J., and Hill G.L. (1985) Hydration of fat-free body in protein-depleted patients. *American Journal of Physiology*, 249:E227–E233.

Beddoe A.H., Zuidmeer H., and Hill G.L. (1984) A prompt gamma in vivo neutron activation analysis facility for measurement of total body nitrogen in the critically ill. *Physics in Medicine and Biology*, 29:371–383.

Behnke A.R., Jr., Feen B.G., and Welham W.C. (1942) The specific gravity of healthy men. *Journal of the American Medical Association*, 118:495–498.

Behnke A.R. and Wilmore J.H. (1974) *Evaluation and Regulation of Body Build and Composition*. Englewood Cliffs, NJ: Prentice-Hall.

Beisel W.R. (1976) The influence of infection or injury on nutritional requirements. In J.I. McKigney and H.N. Munro (Eds.), *Nutrient Requirements in Adolescence*. Cambridge, MA: Massachusetts Institute of Technology Press, pp. 257–275.

Beisel W.R., Sawyer W.D., Ryll E.D., and Crozier D. (1967) Metabolic effects of intracellular infections in man. *Annals of Internal Medicine*, 67:744–779.

Belko A.Z., Meredith M.P., Kalkwarf M.N.S., Obarzanek E., Weinberg S., Roach R., McKeon G., and Roe D.A. (1985) Effects of exercise on riboflavin requirements: biological validation in weight reducing women. *American Journal of Clinical Nutrition*, 41:270–277.

Bell E.F. (1984) Body water in infancy. Sixth Ross Conference on Medical Research, December 17, 1984.

Benedict F.G. (1915) *A Study of Prolonged Fasting*. Washington: Carnegie Institute.

Benedict F.G., Miles W.R., Roth P., and Smith H.M. (1919) *Human Vitality and Efficiency under Prolonged Restricted Diet*. Washington: Carnegie Institute.

Benedict F.G. and Ritzman E.G. (1923) *Undernutrition in Steers: Its Relation to Metabolism, Digestion, and Subsequent Realimentation*. Washington: Carnegie Institute.

Benjamin E. (1914) Der Eiweissnährschaden des Säuglings. *Zeitschrift für Kinderheilkunde*, 10:185–302.

Beretta-Piccoli C., Davies D.L., Boddy K., Brown J.J., Cumming A.M.M., East B.W., Fraser R., Lever A.F., Padfield P.L., Semple P.F., Robertson J.I.S., Weidmann P., and Williams E.D. (1982) Relation of arterial pressure with body sodium, body potassium and plasma potassium in essential hypertension. *Clinical Science*, 63:257–270.

Bergen W.G., Kaplan M.L., Merkel R.A., and Leveille G.A. (1975) Growth of adipose and lean tissue mass in hind limbs of genetically obese mice during preobese and obese phases of development. *American Journal of Clinical Nutrition*, 28:157–161.

Bergström J. (1962) Muscle electrolytes in man. *Scandinavian Journal of Clinical and Laboratory Investigation,* 14:Suppl. 68.

Berman B., Krieger A., and Naiman J.L. (1979) A new method for calculating volumes of blood required for partial exchange transfusion. *Journal of Pediatrics,* 94:86–89.

Bernays E.A. (1986) Diet-induced head allometry among foliage-chewing insects and its importance for graminivores. *Science,* 231:495–496.

Bernstein R.E. (1959) Alterations in metabolic energetics and cation transport during aging of red cells. *Journal of Clinical Investigation,* 38:1572.

Bessard T., Schutz Y., and Jéquier E. (1983) Energy expenditure and postprandial thermogenesis in obese women before and after weight loss. *American Journal of Clinical Nutrition,* 38:680–693.

Bessey P.Q., Walters J.M., Aoki T.T., and Wilmore D.W. (1984) Combined hormone infusion simulates the metabolic response to injury. *Annals of Surgery,* 200:264–281.

Bischoff E. (1863) Einige Gewichts—und Trocken—Bestimmungen der Organe des menschlichen Körpers. *Zeitschrift für Rationelle Medizin,* 20:75–118.

Bistrian B.R. and Blackburn G.L. (1983) Assessment of protein-calorie malnutrition in the hospitalized patient. In H.A. Schneider, C.E. Anderson and D.B. Coursin (Eds.), *Nutrition Support of Medical Practice,* second edition. Philadelphia: Harper & Row.

Bistrian B.R., Winterer J., Blackburn G.L., Young V., and Sherman M. (1977) Effect of a protein-sparing diet and brief fast on nitrogen metabolism in mildly obese subjects. *Journal of Laboratory and Clinical Medicine,* 89:1030–1035.

Björntorp P., Carlgren G., Isaksson B., Krotkiewski M., Larsson B., and Sjöström L. (1975) Effect of an energy-reduced dietary regimen in relation to adipose tissue cellularity in obese women. *American Journal of Clinical Nutrition,* 28:445–452.

Blaxter K.L. (1971) Methods of measuring the energy metabolism of animals and interpretation of results obtained. *Federation Proceedings,* 30:1436–1443.

Blyth B., McRae C.U., Espiner E.A., Nicholls M.G., Conaglen J.V., and Gilchrist N. (1985) Effect of stilboestrol on sodium balance, cardiac state, and renin-angiotensin-aldosterone activity in prostatic carcinoma. *British Medical Journal,* 291:1461–1464.

Boddy K., Hume R., King P.C., Weyers E., and Rowan T. (1974) Total body, plasma and erythrocyte potassium and leucocyte ascorbic acid in "ultra-fit" subjects. *Clinical Science,* 46:449–456.

Boddy K., King P.C., and Davies D.L. (1973) The relationship between total body potassium and exchangeable body potassium measured at 24 and 44 hr after administration of ^{43}K. *European Journal of Clinical Investigation,* 3:188–192.

Bogardus C., La Grange B.M., Horton E.S., and Sims E.A.H. (1981) Comparison of carbohydrate-containing and carbohydrate-restricted hypocaloric diets during the treatment of obesity. *Journal of Clinical Investigation,* 68:399–404.

Bogle S., Burkinshaw L., and Kent J.T. (1985) Estimating the composition of tissue gained or lost from measurements of elementary body composition. *Physics in Medicine and Biology,* 30:369–384.

Bonnard G.D. (1968) Cortical thickness and diaphysial diameter of the metacarpal bones from the age of three months to eleven years. *Helvetica Paediatrica Acta,* 5:445–463.

Booth R.A.D., Goddard B.A., and Paton A. (1966) Measurement of fat thickness in man: a comparison of ultrasound, Harpenden calipers, and electrical conductivity. *British Journal of Nutrition,* 20:719–725.

Borgstedt A., Forbes G.B., and Reina J.C. (1970) Total body potassium and lean body mass in patients with Duchenne dystrophy and their female relatives. *Neuropaediatrie,* 1:447–451.

Borisov B.K. and Marei A.N. (1974) Weight parameters of adult human skeleton. *Health Physics,* 27:224–229.

Borkan G.A., Gerzof S.G., Robbins A.H., Hults D.E., Silbert C.K., and Silbert J.E. (1982) Assessment of abdominal fat by computed tomography. *American Journal of Clinical Nutrition,* 36:172–177.

Borkan G.A., Hults D.E., and Glynn R.J. (1983) Role of longitudinal change and secular trend in age differences in male body dimensions. *Human Biology,* 55:629–641.

Borkan G.A. and Norris A.H. (1977) Fat redistribution and the changing body dimensions of the adult male. *Human Biology,* 49:495–514.

Bortz W.M. (1969) A 500 pound weight loss. *American Journal of Medicine,* 47:325–331.

Bortz W.M., Wroldson A., and Morris P. (1967) Fat, carbohydrate, salt and weight loss. *American Journal of Clinical Nutrition,* 20:1104–1112.

Bouchard C., Savard R., Després J.-P., Tremblay A., and Leblanc C. (1985) Body composition in adopted and biological siblings. *Human Biology,* 57:61–75.

Boyden T.W., Pamenter R.W., Gross D., Stanforth P., Rotkis T., and Wilmore J.H. (1982) Prolactin responses, menstrual cycles, and body composition of women runners. *Journal of Clinical Endocrinology and Metabolism,* 54:711–714.

Bracco E.F., Yang M.-U., Segal K., Hashim S.A., and van Itallie T.B. (1984) A new method for estimation of body composition in the live rat. *Proceedings of the Society for Experimental Biology and Medicine,* 174:143–146.

Brans Y.W., Summers J.E., Dweck H.S., and Cassady G. (1974) A noninvasive approach to body composition in the neonate: dynamic skinfold measurements. *Pediatric Research,* 8:215–222.

Bratteby L.-E. (1968) Studies on erythro-kinetics in infancy. *Acta Paediatrica Scandinavica,* 57:132–136.

Bray G.A. and Gallagher T.F., Jr. (1975) Manifestations of hypothalamic obesity in man. *Medicine,* 54:301–330.

Breibart S., Lee J.S., McCoord A., and Forbes G.B. (1960) Relation of age to radiomagnesium exchange in bone. *Proceedings of the Society for Experimental Biology and Medicine,* 105:361–363.

Brittenham G.M., Farrell D.E., Harris J.W., Feldman E.S., Danish E.H., Muir W.A., Tripp J.H., and Bellon E.M. (1982) Magnetic-susceptibility measurement of human iron stores. *New England Journal of Medicine,* 307:1671–1675.

Brøchner-Mortensen I., Rickers H., and Balslev I. (1980) Renal function and body composition before and after intestinal bypass operation in obese patients. *Scandinavian Journal of Clinical and Laboratory Investigation,* 40:695–702.

Brody S. (1945) *Bioenergetics and Growth.* New York: Reinhold.

Brook C.G.D. (1971) Determination of body composition of children from skinfold measurements. *Archives of Disease in Childhood,* 46:182–184.

Brook C.G.D., Huntley R.M.C., and Slack J. (1975) Influence of heredity and environment in determination of skinfold thickness in children. *British Medical Journal,* 2:719–721.

Brooke M.H. and Engel W.K. (1969) The histographic analysis of human muscle biopsies with regard to fiber types. 4. Children's biopsies. *Neurology,* 19:591–605.

Brooke O.G. and Cocks T. (1974) Resting metabolic rate in malnourished babies in relation to total body potassium. *Acta Paediatrica Scandinavica,* 63:817–825.

Brown M.R., Klish W.J., Hollander J., Campbell M.A., and Forbes G.B. (1983) A high protein, low calorie liquid diet in the treatment of very obese adolescents: long term effect on lean body mass. *American Journal of Clinical Nutrition,* 38:20–31.

Brožek J. (1952) Changes of body composition in man during maturity and their nutritional implications. *Federation Proceedings,* 11:784–793.
Brožek J. (1954) Physical activity and body composition. *Archives of Industrial Hygiene,* 5:193–212.
Brožek J., Grande F., Anderson T., and Keys A. (1963) Densitometric analysis of body composition: revisions of some quantitative assumptions. *Annals of the New York Academy of Science,* 110:113–140.
Brožek J. and Henschel A. (Eds) (1961) *Techniques for Measuring Body Composition.* Washington: National Academy of Sciences.
Brožek J. and Keys A. (1951) The evaluation of leanness-fatness in man: norms and interrelationships. *British Journal of Nutrition,* 5:194–206.
Bruce Å., Andersson M., Arvidsson B., and Isaksson B. (1980) Body composition: Prediction of normal body potassium, body water and body fat in adults with basis of body height, body weight and age. *Scandinavian Journal of Clinical and Laboratory Investigation,* 40:461–474.
Bullen B.A., Shriner G.S., Beitins I.Z., von Mering G., Turnbull B.A., and McArthur J.W. (1985) Induction of menstrual disorders by strenuous exercise in untrained women. *New England Journal of Medicine,* 312:1349–1353.
Burch G., Reaser P., and Cronvich J. (1947) Rates of sodium turnover in normal subjects and in patients with congestive heart failure. *Journal of Laboratory and Clinical Medicine,* 32:1169–1191.
Burch G.E., Threefoot S.A., and Ray C.T. (1950) Rates of turnover and biologic decay of chloride space in the dog determined with the long-life isotope, Cl^{36}. *Journal of Laboratory and Clinical Medicine,* 35:331–347.
Burkinshaw L., Hill G.L., and Morgan D.B. (1979) Assessment of the distribution of protein in the human body by in-vivo neutron activation analysis. In *Nuclear Activation Techniques in the Life Sciences 1978.* Vienna: International Atomic Energy Agency.
Burkinshaw L. and Morgan D.B. (1985) Mass and composition of the fat-free tissues of patients with weight-loss. *Clinical Science,* 68:455–462.
Burmeister W. and Bingert A. (1967) Die quantitativen Veränderungen der menschlichen Zellmasse zwischen dem 8. und 90. Lebensjahr. *Klinische Wochenschrift,* 45:409–416.
Buskirk E.R. (1961) Underwater weighing and body density: a review of procedures. In J. Brožek and A. Henschel (Eds.), *Techniques for Measuring Body Composition* (pp. 90–107). Washington: National Academy of Sciences.
Buskirk E.R., Anderson K.L., and Brožek J. (1956) Unilateral activity and bone and muscle development in the forearm. *Research Quarterly,* 27:127–131.
Butenandt O., Eder R., Wohlfarth K., Bidlingmaier F., and Knorr D. (1976) Mean 24-hour growth hormone and testosterone concentrations in relation to pubertal growth spurt in boys with normal or delayed puberty. *European Journal of Pediatrics,* 122:85–92.
Butte N.F., Wills C., Smith E.O., and Garza C. (1985) Prediction of body density from skinfold measurements in lactating women. *British Journal of Nutrition,* 53:485–489.
Butterfield G.E. and Calloway D.H. (1984) Physical activity improves protein utilization in young men. *British Journal of Nutrition,* 51:171–184.
Cabak V., Dickerson J.W.T., and Widdowson E.M. (1963) Response of young rats to deprivation of protein or of calories. *British Journal of Nutrition,* 17:601–626.
Calloway D.H. and Kurzer M.S. (1982) Menstrual cycle and protein requirements of women. *Journal of Nutrition,* 112:356–366.
Calloway D.H. and Margen S. (1971) Variation in endogenous nitrogen excretion and dietary nitrogen utilization as determinants of human protein requirements. *Journal of Nutrition,* 101:205–216.

Calloway D.H. and Spector H. (1954) Nitrogen balance as related to caloric and protein intake in active young men. *American Journal of Clinical Nutrition,* 2:405–412.

Calloway D.H. and Zanni E. (1980) Energy requirements and energy expenditure of elderly men. *American Journal of Clinical Nutrition,* 33:2088–2092.

Camerer W. and Söldner. (1900) Die chemische Zusammensetzung des Neugeborenen. *Zeitschrift für Biologie,* 39:173–192.

Carlberg K.A., Buckman M.T., Peake G.T., and Riedesel M.L. (1983) Body composition of oligo/amenorrheic athletes. *Medical Science in Sports and Exercise,* 15:215–217.

Cassady G. and Milstead R.R. (1971) Antipyrine space studies and cell water as estimates in infants of low birth weight. *Pediatric Research,* 5:673–682.

Cathcart E.P. (1907) Über die Zusammensetzung des Hungerharns. *Biochemische Zeitschrift,* 6:109–148.

Catzeflis C., Schutz Y., Micheli J.-L., Welsch C., Arnaud M.J., and Jéquier E. (1985) Whole body protein synthesis and energy expenditure in very low birth weight infants. *Pediatric Research,* 19:679–687.

Century T.J., Fenichel I.R., and Horowitz S.B. (1970) The concentration of water, sodium and potassium in the nucleus and cytoplasm of amphibian oocytes. *Journal of Cell Science,* 7:5–13.

Cheek D.B. (1968) *Human Growth.* Philadelphia: Lea & Febiger.

Cheek D.B. (1975) *Fetal and Postnatal Cellular Growth.* New York: John Wiley & Sons.

Cheek D.B., Parra A., and White J. (1975) Overnutrition. In D.B. Cheek (Ed.), *Fetal and Postnatal Cellular Growth.* New York: Wiley & Sons, pp. 462–465.

Cheek D.B., Petrucco O.M., Gillespie A., Ness D., and Green R.C. (1985) Muscle cell growth and the distribution of water and electrolyte in human pregnancy. *Early Human Development,* 11:293–305.

Cheek D.B., Schultz R.B., Parra A., and Reba R.C. (1970) Overgrowth of lean and adipose tissues in adolescent obesity. *Pediatric Research,* 4:268–279.

Cheek D.B. and West C.D. (1955) An appraisal of methods of tissue chloride analysis: the total carcass chloride, exchangeable chloride, potassium and water of the rat. *Journal of Clinical Investigation,* 34:1744–1755.

Cheek D.B., Wishart J., MacLennan A.H., and Haslam R. (1984) Cell hydration in the normally grown, the premature and low weight for gestational age infant. *Early Human Development,* 10:75–84.

Cheek D.B., Wishart J., MacLennan A.H., Haslam R., and Fitzgerald A. (1982) Hydration in the first 24 hours of postnatal life in normal infants born vaginally or by Caesarean section. *Early Human Development,* 7:323–330.

Chen K.P., Damon A., and Elliot O. (1963) Body form, composition, and some physiological functions of Chinese on Taiwan. *Annals of the New York Academy of Science,* 110:760–777.

Chien S., Peng M.T., Chen K.P., Huang T.F., Chang C., and Fang H.S. (1975) Longitudinal measurements of blood volume and essential body mass in human subjects. *Journal of Applied Physiology,* 39:818–824.

Chobanian A.V., Burrows B.A., and Hollander W. (1961) Body fluid and electrolyte composition in arterial hypertension. II. Studies in mineralocorticoid hypertension. *Journal of Clinical Investigation,* 40:416–422.

Christiansen C. and Rödbro P. (1975) Bone mineral content and estimated total body calcium in normal adults. *Scandinavian Journal of Clinical and Laboratory Investigation,* 35:433–439.

Christiansen C., Rödbro P., and Jensen H. (1975a) Bone mineral content in the forearm measured by photon absorptiometry. *Scandinavian Journal of Clinical and Laboratory Investigation,* 35:323–330.

Christiansen C., Rödbro P., and Nielsen C.T. (1975b) Bone mineral content and estimated total body calcium in normal children and adolescents. *Scandinavian Journal of Clinical and Laboratory Investigation,* 35:507–510.

Clark D.A., Kay T.D., Tatsch R.F., and Theis C.F. (1978) Estimations of body composition by various methods. *Aviation, Space, and Environmental Medicine,* 48:701–704.

Clark L.C., Jr., Thompson H.L., Beck E.I., and Jacobson W. (1951) Excretion of creatine and creatinine by children. *American Journal of Diseases of Children,* 81:774–783.

Clarys J.P., Martin A.D., and Drinkwater D.T. (1985) Gross tissue weights in the human body by cadaver dissection. *Human Biology,* 56:459–473.

Claus-Walker J., Singh J., Leach C.S., Hatton D.V., Hubert C.W., and DiFerrante N. (1977) The urinary excretion of collagen degradation products by quadriplegic patients and during weightlessness. *Journal of Bone and Joint Surgery,* 59-A:209–212.

Clemente G., Ferro-Luzzi A., Mariani A., Santaroni G., and Tranquilli G.B. (1973) Evaluation of analytical models based on anthropometry and age for the prediction of body fat. *Nutrition Reports International,* 7:157–168.

Clemmons D.R., Klibanski A., Underwood L.E., McArthur J.W., Ridgway E.C., Beitins I.Z., and Van Wyk J.J. (1981) Reduction of plasma immunoactive somatomedin-C during fasting in humans. *Journal of Clinical Endocrinology and Metabolism,* 53:1247–1250.

Cohn C., Joseph D., and Shrago E. (1957) Effect of diet on body composition. I. Production of increased body fat without overweight (nonobese obesity) by force feeding normal rat. *Metabolism,* 6:381–387.

Cohn S.H. (1981a) *In vivo* neutron activation analysis: state of the art and future prospects. *Medical Physics,* 8:145–154.

Cohn S.H. (Ed.) (1981b) *Non-invasive Measurements of Bone Mass and Their Clinical Application.* Boca Raton, FL: CRC Press.

Cohn S.H., Abesamis C., Zanzi I., Aloia J.F., Yasumura S., and Ellis K.J. (1977) Body elemental composition: comparison between black and white adults. *American Journal of Physiology,* 232:E419–E422.

Cohn S.H. and Dombrowski C.S. (1971) Measurement of total-body calcium, sodium, chlorine, nitrogen and phosphorus in man by in vivo neutron activation analysis. *Journal of Nuclear Medicine,* 12:499–505.

Cohn S.H., Vartsky D., Yasumura S., Sawitsky A., Zanzi I., Vaswani A., and Ellis K.J. (1980) Compartmental body composition based on total-body nitrogen, potassium, and calcium. *American Journal of Physiology,* 239:E524–E530.

Cohn S.H., Vartsky D., Yasumura S., Vaswani A.N., and Ellis K.J. (1983) Indexes of body cell mass: nitrogen versus potassium. *American Journal of Physiology,* 244:E305–E310.

Cohn S.H., Vaswani A.N., Yasumura S., Yuen K., and Ellis K.J. (1984) Improved methods for determination of body fat by in vivo neutron activation. *American Journal of Clinical Nutrition,* 40:255–259.

Cohn S.H., Vaswani A., Zanzi I., and Ellis U.J. (1976) Effect of aging on bone mass in adult women. *American Journal of Physiology*, 230:143–148.

Cole D.F. (1952) The effects of oestradiol on the skeletal muscle and liver of the rat. *Journal of Endocrinology,* 8:179–186.

Cole D.F. (1953) The action of desoxycorticosterone acetate and of dietary sodium and potassium on skeletal muscle electrolyte in adrenalectomized rats. *Acta Endocrinologica,* 14:245–253.

Coleman T.G., Manning R.D., Jr., Norman R.A., Jr., and Guyton A.C. (1972) Dynamics of water isotope distribution. *American Journal of Physiology,* 223:1371–1375.

Collins J.P., McCarthy I.D., and Hill G.L. (1979) Assessment of protein nutrition

in surgical patients—the value of anthropometrics. *American Journal of Clinical Nutrition,* 32:1527–1530.
Collipp P.J., Curti V., Thomas J., Sharma R.K., Maddaiah V.T., and Cohn S.H. (1973) Body composition changes in children receiving human growth hormone. *Metabolism,* 22:589–595.
Comstock G.W. and Livesay V.T. (1963) Subcutaneous fat determinations from a community-wide chest x-ray survey in Muscogee County, Georgia. *Annals of the New York Academy of Science,* 110:475–491.
Conway J.M., Norris K.H., and Bodwell C.E. (1984) A new approach for the estimation of body composition: infrared interactance. *American Journal of Clinical Nutrition,* 40:1123–1130.
Cook D.R., Gualtier W.S., and Galla S.J. (1969) Body fluid volumes of college athletes and non-athletes. *Medicine and Science in Sports,* 1:217–220.
Coppen A.J. and Reynolds E.H. (1966) Electrolyte and water distribution in familial hypokaemic periodic paralysis. *Journal of Neurology, Neurosurgery and Psychiatry,* 29:107–112.
Corsa L., Jr., Olney J.M., Jr., Steenberg R.W., Bell M.R., and Moore F.D. (1950) The measurement of exchangeable potassium in man by isotope dilution. *Journal of Clinical Investigation,* 29:1280–1295.
Cowan M.M. and Gregory L.W. (1985) Responses of pre- and postmenopausal females to aerobic conditioning. *Medicine and Science in Sports and Exercise,* 17:138–143.
Cox J.E., Laughton W.B., and Powley T.L. (1985) Precise estimation of carcass fat from total body water in rats and mice. *Physiology and Behavior,* 35:905–910.
Craig A.B. and Ware D.E. (1967) Effect of immersion in water on vital capacity and residual volume of lungs. *Journal of Applied Physiology,* 23:329–343.
Crews E.L. III, Fuge K.W., Oscai L.B., Holloszy J.O., and Shank R.E. (1969) Weight, food intake, and body composition: effects of exercise and of protein deficiency. *American Journal of Physiology,* 216:359–363.
Crist D.M., Stackpole P.J., and Peake G.T. (1983) Effects of androgenic-anabolic steroids on neuromuscular power and body composition. *Journal of Applied Physiology,* 54:366–370.
Cronk C.E. and Roche A.F. (1982) Race- and sex-specific reference data for triceps and subscapular skinfolds and weight/stature^2. *American Journal of Clinical Nutrition,* 35:351–354.
Culebras J.M. and Moore F.D. (1977) Total body water and the exchangeable hydrogen. Theoretical calculations of nonaqueous exchangeable hydrogen in man. *American Journal of Physiology,* 232:R54–R59.
Cuthbertson A. (1978) Carcass evaluation of cattle, sheep and pigs. *World Review of Nutrition and Dietetics,* 28:210–235.
Cuthbertson D.P. (1929) The influence of prolonged muscular rest on metabolism. *Biochemical Journal,* 23:1328–1345.
Cuthbertson D.P. (1932) Observations on the disturbances of metabolism produced by injury to the limbs. *Quarterly Journal of Medicine,* N.S. 1:233–246.
Cuthbertson D.P., Fell G.S., Smith C.M., and Tilstone W.J. (1972) Metabolism after injury: effects of severity, nutrition, and environmental temperature on protein, potassium, zinc and creatine. *British Journal of Surgery,* 59:925–931.
Cuthbertson D.P., McCutcheon A., and Munro H.N. (1937) A study of the effect of overfeeding in man. *Biochemical Journal,* 31:681–705.
Dalén N. and Olsson K.E. (1974) Bone mineral content and physical activity. *Acta Orthopaedica Scandinavica,* 45:170–174.
Dalén N., Hallberg D., and Lamke B. (1975) Bone mass in obese subjects. *Acta Medica Scandinavica,* 197:353–355.

Darrow D.C., DaSilva M.M., and Stevenson S.S. (1945) Production of acidosis in premature infants by protein milk. *Journal of Pediatrics,* 27:43–58.

Darrow D.C. and Hellerstein S. (1958) Interpretation of certain changes in body water and electrolytes. *Physiological Reviews,* 38:114–137.

Darrow D.C., Pratt E.L., Flett J., Jr., Gamble A.H., and Wiese H.F. (1949) Disturbances of water and electrolytes in infantile diarrhea. *Pediatrics,* 3:129–156.

Datlof S., Colman P.D., Forbes G.B., and Kreipe R.E. (1986) Ventricular dilation on CAT scans of patients with anorexia nervosa. *American Journal of Psychiatry,* 143:96–98.

Davies D.L. and Robertson J.W.K. (1973) Simultaneous measurement of total exchangeable potassium and sodium using ^{43}K and ^{24}Na. *Metabolism,* 22:133–137.

Davis Y.E. and Brewer N. (1935) Effect of physical training on blood volume, hemoglobin, alkali reserve, and osmotic resistance of erythrocytes. *American Journal of Physiology,* 113:586–591.

Dawson N.J., Stephenson S.K., and Fredline D.K. (1972) Body composition of mice subjected to genetic selection for different body proportions. *Comparative Biochemistry and Physiology,* 42B:679–691.

Day T.D. (1952) The permeability of the interstitial connective tissue and the nature of the interfibrillary substance. *Journal of Physiology,* 117:1–8.

Deb S., Martin R.J., and Hershberger T.V. (1976) Maintenance requirement and energetic efficiency of lean and obese Zucker rats. *Journal of Nutrition,* 106:191–197.

Deitrick J.E., Whedon G.D., and Shorr E. (1948) Effects of immobilization upon various metabolic and physiologic functions of normal men. *American Journal of Medicine,* 4:3–36.

Delwaide P.A. (1973) Body potassium measurements of whole-body counting: screening of patient populations. *Journal of Nuclear Medicine,* 14:40–48.

Dequeker J. and Johnston C.C., Jr. (Eds.) (1982) *Non-Invasive Bone Measurements: Methodological Problems.* Oxford: IRL Press.

Deskins W.J., Winter D.C., Sheng H.P., and Garza C. (1985) Use of a resonating cavity to measure body volume. *Journal of the Acoustical Society of America,* 17:756–758.

Després J.P., Bouchard C., Tremblay A., Savard R., and Marcotte M. (1985) Effects of aerobic training on fat distribution in male subjects. *Medicine and Science in Sports,* 17:113–118.

Detenbeck L.C. and Jowsey J. (1969) Normal aging in the bone of the adult dog. *Clinical Orthopaedics and Related Research,* 65:76–80.

Dewasmes G., LeMaho Y., Cornet A., and Groscolas R. (1980) Resting metabolic rate and cost of locomotion in long-term fasting emperor penguins. *Journal of Applied Physiology,* 49:888–896.

Dewey J.R., Bartley M.H., Jr., and Armelagos G.J. (1969) Rates of femoral cortical bone loss in two Nubian populations. *Clinical Orthopaedics and Related Research,* 65:61–66.

Dickerson J.W.T. and Widdowson E.M. (1960) Chemical changes in skeletal muscle during development. *Biochemical Journal,* 74:247–257.

Dill D.B., Robinson S., and Ross J.C. (1967) A longitudinal study of 16 champion runners. *Journal of Sports Medicine,* 7:4–27.

Ditschuneit H., Wechsler J.G., and Ditschuneit H.H. (1985) Clinical experience with a very low calorie diet. In G.L. Blackburn and G.A. Bray (Eds.), *Management of Obesity by Severe Calorie Restriction.* Littleton, MA: PSG Publishing.

Doré C., Hesp R., Wilkins D., and Garrow J.S. (1982) Prediction of energy requirements of obese patients after massive weight loss. *Human Nutrition: Clinical Nutrition,* 36C:41–48.

Drenick E.J. (1967) Weight reduction with low-calorie diets. *Journal of the American Medical Association,* 202:136–138.
Drenick E.J., Swendseid M.E., Blahd W.H., and Tuttle S.G. (1964) Prolonged starvation as treatment for severe obesity. *Journal of the American Medical Association,* 187:100–105.
Drinkwater B.L., Nilson K., Chestnut C.H. III, Bremner W.J., Shainhaltz S., and Southworth M.B. (1984) Bone mineral content of amenorrheic and eumenorrheic athletes. *New England Journal of Medicine,* 311:277–281.
Du Bois D. and Du Bois E.F. (1916) Clinical calorimeter. A formula to estimate the approximate surface if height and weight be known. *Archives of Internal Medicine,* 17:863–871.
Duncan D.L. (1958) The interpretation of studies of calcium and phosphorus balance in ruminants. *Nutrition Abstracts and Reviews,* 28:695–715.
Durnin J.V.G.A. and Taylor A. (1960) Replicability of measurements of density of the human body as determined by underwater weighing. *Journal of Applied Physiology,* 15:142–144.
Durnin J.V.G.A. and Womersley J. (1974) Body fat assessed from total body density and its estimation from skinfold thickness: Measurements on 481 men and women aged from 16 to 72 years. *British Journal of Nutrition,* 32:77–97.
Durrant M.L., Garrow J.S., Royston P., Stalley S.F., Sunkin S., and Warwick P.M. (1980) Factors influencing the composition of the weight lost by obese patients on a reducing diet. *British Journal of Nutrition,* 44:275–286.
East B.W., Boddy K., and Price W.H. (1976) Total body potassium content in males with X and Y chromosome abnormalities. *Clinical Endocrinology,* 5:43–52.
Edelman I.S. (1961) Body water and electrolytes. In J. Brožek and A. Henschel (Eds.), *Techniques for Measuring Body Composition.* Washington: National Academy of Sciences, pp. 140–154.
Edelman I.S., Leibman J., O'Meara M.P., and Birkenfeld L.W. (1958) Interrelationship between serum sodium concentration, serum osmolarity, and total exchangeable sodium, total exchangeable potassium and total body water. *Journal of Clinical Investigation,* 37:1236–1256.
Edmonds C.J. and Smith T. (1981) Total body potassium in relation to thyroid hormones and hyperthyroidism. *Clinical Science,* 60:311–318.
Edmonds C.J., Smith T., Griffiths R.D., MacKenzie J., and Edwards R.H.T. (1985). Total body potassium and water, and exchangeable sodium, in muscular dystrophy. *Clinical Science,* 68:379–385.
Edwards D.A.W. (1950) Observations on the distribution of subcutaneous fat. *Clinical Science,* 9:259–270.
Egusa G., Beltz W.F., Grundy S.M., and Howard B.V. (1985) Influence of obesity on the metabolism of apolipoprotein B in humans. *Journal of Clinical Investigation,* 76:596–603.
Elia M., Carter A., Bacon S., Winearb C.G., and Smith R. (1981) Clinical usefulness of urinary 3-methylhistidine excretion in indicating muscle protein breakdown. *British Medical Journal,* 282:351–355.
Elia M., Carter A., and Smith R. (1979) The 3-methylhistidine content of human tissues. *British Journal of Nutrition,* 42:567–570.
Elkinton J.R. and Danowski T.S. (1955) *The Body Fluids.* Baltimore: Williams & Wilkins, p. 78.
Ellis K.J. and Cohn S.H. (1975a) Correlation between skeletal calcium mass and muscle mass in man. *Journal of Applied Physiology,* 38:455–460.
Ellis K.J. and Cohn S.H. (1975b) Predicting radial bone mineral content in normal subjects. *International Journal of Nuclear Medicine and Biology,* 2:53–58.
Ellis K.J., Vaswani A., Zanzi I., and Cohn S.H. (1976) Total body sodium and chlorine in normal adults. *Metabolism,* 25:645–654.

Ellis K.J., Yasumura S., Vartsky D., Vaswani A.N., and Cohn S.H. (1982) Total body nitrogen in health and disease: effects of age, weight, height, and sex. *Journal of Laboratory and Clinical Medicine*, 99:917–926.

Ernest I. (1967) Changes in body composition after therapeutically induced remission in 12 cases of Cushing's syndrome. *Acta Endocrinologica*, 54:411–427.

Exner G.U., Prader A., Elsasser U., Rüegregger P., and Anliker M. (1979) Bone densitometry using computed tomography. *British Journal of Radiology*, 52:14–23.

Exton-Smith A.N., Millard P.H., Payne P.R., and Wheeler E.F. (1969) Pattern of development and loss of bone with age. *Lancet*, ii:1154–1157.

Fahey T.D. and Brown C.H. (1973) The effects of an anabolic steroid on the strength, body composition, and endurance of college males when accompanied by a weight training program. *Medical Science in Sports*, 5:272–276.

Falkner F. (Ed.) (1966) *Human Development*. Philadelphia: W. B. Saunders.

Fanelli M.T. and Kuczmarski R.J. (1984) Ultrasound as an approach to assessing body composition. *American Journal of Clinical Nutrition*, 39:703–709.

Farr V. (1966) Skinfold thickness as an indication of maturity of the newborn. *Archives of Disease in Childhood*, 41:301–308.

Fehling H. (1876) Beitrage zur Physiologie des placentaren Stoffverkehrs. *Archiv für Gynaekologie*, 11:523.

Fenn W.O. (1939) The deposition of potassium and phosphate with glycogen in rat liver. *Journal of Biological Chemistry*, 128:297–307.

Fenn W.O., Bale W.F., and Mullins L.J. (1942) The radioactivity of potassium from human sources. *Journal of General Physiology*, 25:345–353.

Ferrell C.L., Laster D.B., and Prior R.L. (1982) Mineral accretion during prenatal growth of cattle. *Journal of Animal Science*, 54:618–624.

Fidanza F., Keys A., and Anderson J.T. (1953) The density of body fat in man and other mammals. *Journal of Applied Physiology*, 6:252–256.

Filer L.J., Jr., Baur L.S., and Rezabeck H. (1960) Influence of protein and fat content of the diet on the body composition of piglets. *Pediatrics*, 25:242–247.

Finkelstein J.W., Roffwarg H.P., Boyer R.M., Kream J., and Hellman L. (1972) Age-related change in the 24-hour spontaneous secretion of growth hormone. *Journal of Clinical Endocrinology and Metabolism*, 35:665–670.

Fisler J.S., Drenick E.J., Blumfield D.E., and Swendseid M.E. (1982) Nitrogen economy during very low calorie reducing diets: quality and quantity of dietary protein. *American Journal of Clinical Nutrition*, 35:471–486.

Flanagan J.B., Davis A.K., and Overman R.R. (1950) Mechanism of extracellular sodium and chloride depletion in the adrenalectomized dog. *American Journal of Physiology*, 160:89–102.

Flynn M.A., Hanna F.M., and Lutz R.N. (1967a) Estimation of body water compartments of preschool children. I. Normal children. *American Journal of Clinical Nutrition*, 20:1125–1128.

Flynn M.A., Hanna F.M., Asfour R.Y., and Lutz R.N. (1967b) Estimation of body water compartments of preschool children. II. Undernourished children. *American Journal of Clinical Nutrition*, 20:1129–1133.

Flynn M.A., Murthy Y., Clark J., Comfort G., Chase G., and Bentley A.E.T. (1970) Body composition of Negro and White children. *Archives of Environmental Health*, 20:604–607.

Flynn M.A., Woodruff C., and Chase G. (1972) Total body potassium in normal children. *Pediatric Research*, 6:239–245.

Fohlin L. (1977) Body composition, cardiovascular and renal function in adolescent patients with anorexia nervosa. *Acta Paediatrica Scandinavica, Suppl.* 268:1–20.

Fomon S.J., Haschke F., Ziegler E.E., and Nelson S.E. (1982) Body composition of reference children from birth to age 10 years. *American Journal of Clinical Nutrition*, 35:1169–1175.

Forbes G.B. (1955) Inorganic chemical heterogony in man and animals. *Growth,* 19:75–87.

Forbes G.B. (1960) Studies on sodium in bone. *Journal of Pediatrics,* 56:180–190.

Forbes G.B. (1962) Methods for determining composition of the human body. *Pediatrics,* 29:477–494.

Forbes G.B. (1964) Lean body mass and fat in obese children. *Pediatrics,* 34:308–314.

Forbes G.B. (1970) Weight loss during fasting: implications for the obese. *American Journal of Clinical Nutrition,* 23:1212–1219.

Forbes G.B. (1974) Stature and lean body mass. *American Journal of Clinical Nutrition,* 27:595–602.

Forbes G.B. (1976a) The adult decline in lean body mass. *Human Biology,* 48:161–171.

Forbes G.B. (1976b) Calcium accumulation by the human fetus. *Pediatrics,* 57:976–977.

Forbes G.B. (1977) Nutrition and growth. *Journal of Pediatrics,* 91:40–42.

Forbes G.B. (1978) Body composition in adolescence. In F. Falkner and J. Tanner (Eds.), *Human Growth: An Advanced Treatise, Vol. II.* New York: Plenum Press, pp. 239–272.

Forbes G.B. (1981a) Pregnancy in the teenager: biologic aspects. In E.R. McAnarney and G. Stickle (Eds.), *Pregnancy and Childbearing during Adolescence.* New York: Alan R. Liss, pp. 85–90.

Forbes G.B. (1981b) Nutritional requirements in adolescence. In R.M. Suskind (Ed.), *Textbook of Pediatric Nutrition.* New York: Raven Press, pp. 381–391.

Forbes G.B. (1981c) Nutritional adequacy of human breast milk for premature infants. In E. Lebenthal (Ed.), *Textbook of Gastroenterology and Nutrition in Infancy.* New York: Raven Press, pp. 321–329.

Forbes G.B. (1982) Some influences on lean body mass: exercise, androgens, pregnancy, and food. In P.L. White and T. Mondeika (Eds.), *Diet and Exercise: Synergism in Health Maintenance.* Chicago: American Medical Association, pp. 75–90.

Forbes G.B. (1983) Unmeasured losses of potassium in balance studies. *American Journal of Clinical Nutrition,* 38:347–348.

Forbes G.B. (1984) Energy intake and body weight: a reexamination of two "classic" studies. *American Journal of Clinical Nutrition,* 39:349–350.

Forbes G.B. (1985a) Body composition as affected by physical activity and nutrition. *Federation Proceedings,* 44:343–347.

Forbes G.B. (1985b) The effect of anabolic steroids on lean body mass: the dose response curve. *Metabolism,* 34:571–573.

Forbes G.B., Brown M.R., Welle S.L., and Lipinski B.A. (1986) Deliberate overfeeding in women and men: energy cost and composition of the weight gain. *British Journal of Nutrition* 56:1–9.

Forbes G.B. and Bruining G.J. (1976) Urinary creatinine excretion and lean body mass. *American Journal of Clinical Nutrition,* 29:1359–1366.

Forbes G.B., Deisher R.W., Perley A.M., and Hartmann A.F. (1950) Effect of hyaluronidase on the subcutaneous absorption of electrolytes in humans. *Science,* 111:177–179.

Forbes G.B. and Drenick E.J. (1979) Loss of body nitrogen on fasting. *American Journal of Clinical Nutrition,* 32:1570–1574.

Forbes G.B., Hursh J., and Gallup J. (1961) Estimation of total body fat from potassium-40 content. *Science,* 133:101–102.

Forbes G.B., Kreipe R.E., Lipinski B.A., and Hodgman C.H. (1984) Body composition changes during recovery from anorexia nervosa: comparison of two dietary regimes. *American Journal of Clinical Nutrition,* 40:1137–1145.

Forbes G.B. and Lewis A. (1956) Total sodium, potassium, and chloride in adult man. *Journal of Clinical Investigation,* 35:596–600.

Forbes G.B. and McCoord A. (1963) Changes in bone sodium during growth in the rat. *Growth,* 27:285–294.

Forbes G.B. and McCoord A.B. (1969) Long-term behavior of radiosodium in bone: Comparison with radiocalcium and effects of various procedures. *Calcified Tissue Research,* 4:113–128.

Forbes G.B., Mizner G.L., and Lewis A. (1959) Effect of age on radiosodium exchange in bone (rat). *American Journal of Physiology,* 190:152–156.

Forbes G.B. and Perley A.M. (1951) Estimation of total body sodium by isotopic dilution. *Journal of Clinical Investigation,* 30:558–565, 566–574.

Forbes G.B., Schultz F., Cafarelli C., and Amirhakimi G.H. (1968) Effects of body size on potassium-40 measurement in the whole body counter (tilt-chair technique). *Health Physics,* 15:435–442.

Forbes G.B. and Welle S.L. (1983) Lean body mass in obesity. *International Journal of Obesity,* 7:99–108.

Forbes R.M., Cooper A.R., and Mitchell H.H. (1953) The composition of the human body as determined by chemical analysis. *Journal of Biological Chemistry,* 203:359–366.

Forse R.A. and Shizgal H.M. (1980) The assessment of malnutrition. *Surgery,* 88:17–24.

Fowler W.A. (1984) The quest for the origin of the elements. *Science,* 226:922–935.

Franklin B., Buskirk E., Hodgson J., Gahagan H., Kollias J., and Mendez J. (1979) Effects of physical conditioning on cardiorespiratory function, body composition and serum lipids in relatively normal weight and obese middle-aged women. *International Journal of Obesity,* 3:97–109.

Frisancho A.R. (1981) New norms of upper limb fat and muscle areas for assessment of nutritional status. *American Journal of Clinical Nutrition,* 34:2540–2545.

Frisancho A.R., Klayman J.E., and Matos J. (1977) Influence of maternal nutritional status on prenatal growth in a Peruvian urban population. *American Journal of Physical Anthropology,* 46:265–274.

Frisch R.E. and McArthur J.W. (1974) Menstrual cycles: fatness as a determinant of minimum weight for height necessary for their maintenance or onset. *Science,* 185:949–951.

Frisch R.E. and Revelle R. (1970) Height and weight at menarche and a hypothesis of critical body weight and adolescent events. *Science,* 169:397–399.

Frisch R.E., Revelle R., and Cook S. (1973) Components of weight at menarche and the initiation of the adolescent growth spurt in girls: estimated total water, lean body weight and fat. *Human Biology,* 45:469–483.

Frohman L.A. and Bernardis L.L. (1968) Growth hormone and insulin levels in weanling rats with ventromedial hypothalamic lesions. *Endocrinology,* 82:1125–1132.

Fuchs R.J., Theis C.F., and Lancaster M.C. (1978) A nomogram to predict lean body mass in men. *American Journal of Clinical Nutrition,* 31:673–678.

Fuller M.F., Houseman R.A., and Cadenhead A. (1971) The measurement of exchangeable potassium in living pigs and its relation to body composition. *British Journal of Nutrition,* 26:203–214.

Gamble J.L., Jr., Robertson J.S., Hannigan C.A., Foster C.G., and Farr L.E. (1953) Chloride, bromide, sodium, and sucrose spaces in man. *Journal of Clinical Investigation,* 32:483–489.

Garn S.M. (1961) Radiographic analysis of body composition. In J. Brožek, & A. Henschel (Eds.), *Techniques for Measuring Body Composition.* Washington: National Academy of Science, pp. 36–58.

Garn S.M. (1970) *The Earlier Gain and the Later Loss of Cortical Bone in Nutritional Perspective*. Springfield, IL: Charles C Thomas.
Garn S.M. and LaVelle M. (1983) Reproductive histories of low-weight girls and women. *American Journal of Clinical Nutrition,* 37:862–866.
Garn S.M. and Petzhold A.S. (1983) Characteristics of the mother and child in teenage pregnancy. *American Journal of Diseases of Children,* 137:365–368.
Garn S.M. and Ryan A.S. (1982) The effect of fatness on hemoglobin levels. *American Journal of Clinical Nutrition,* 36:189–191.
Garn S.M., Poznanski A.K., and Larson K. (1976) Metacarpal lengths, cortical diameters and areas from the 10-state nutrition survey. In C.F.G. Jaworski (Ed.), *Proceedings of the First Workshop on Bone Morphometry*. Ottawa: University of Ottawa Press, pp. 367–391.
Garn S.M. and Wagner B. (1969) The adolescent growth of the skeletal mass and its implications to mineral requirements. In F.P. Heald (Ed.), *Adolescent Nutrition and Growth*. New York: Appleton-Century-Crofts, pp. 139–161.
Garrel D.R., Todd K.S., and Calloway D.H. (1984) Effects of marginally negative energy balance on insulin binding to erythrocytes of normal men. *American Journal of Clinical Nutrition,* 39:716–721.
Garrow J.S. (1981) *Treat Obesity Seriously*. London: Churchill Livingstone.
Garrow J.S., Stalley S., Diethelm R., Pittet Ph., Hesp R., and Halliday D. (1979) A new method for measuring body density of obese adults. *British Journal of Nutrition,* 42:173–183.
Genant H.K., Cann C.E., and Faul D.D. (1982) Quantitative computed tomography for assessing vertebral bone mineral. In J. Dequeker and C.C. Johnston, Jr., (Eds.), *Non-Invasive Bone Measurements: Methodological Problems* Oxford: IRL Press, pp. 215–249.
Gibson J.G., II, Peacock W.C., Seligman A.M., and Sack T. (1946a) Circulating red cell volumes measured simultaneously by the radioactive iron and dye method. *Journal of Clinical Investigation,* 25:838–847.
Gibson J.G., II, Seligman A.M., Peacock W.C., Aub J.C., Fine J., and Evans R.D. (1946b) The distribution of red cells and plasma in large and minute vessels of the normal dog, determined by radioactive isotopes of iron and iodine. *Journal of Clinical Investigation,* 25:848–857.
Gnaedinger R.H., Pearson A.M., Reineke E.P., and Hix V.M. (1963) Body composition of market weight pigs. *Journal of Animal Science,* 22:495–500.
Goldman J.K., Bernardis L.L., and Frohman L.A. (1974) Food intake in hypothalamic obesity. *American Journal of Physiology,* 227:88–91.
Goldman R.F., Haisman M.F., Bynum G., Horton E.S., and Sims E.A.H. (1975) Experimental obesity in man: metabolic rate in relation to dietary intake. In G.A. Bray (Ed.), *Obesity in Perspective*. Washington: Department of Health, Education and Welfare, Pub. No. (NIH)75–708, pp. 165–186.
Goldspink G. (1964) The combined effect of exercise and reduced food intake on skeletal muscle fibers. *Journal of Cellular and Comparative Physiology,* 63:209–216.
Gollnick P.D., Timson B.F., Moore R.L., and Reidy M. (1981) Muscular enlargement and number of fibers in skeletal muscles of rats. *Journal of Applied Physiology,* 50:936–943.
Goodford P.J. and Leach E.H. (1961) The extracellular space of the smooth muscle of the guinea-pig taenia coli. *Journal of Physiology (London),* 186:1–10.
Göranzon H. and Forsum E. (1985) Effect of reduced energy intake versus increased physical activity on the outcome of nitrogen balance experiments in man. *American Journal of Clinical Nutrition,* 41:919–928.
Gotfredsen A., Jensen J., Borg J., and Christiansen C. (1986) Measurement of lean body mass and body fat using dual photon absorptiometry. *Metabolism,* 35:88–93.

Graham E.R. (1969) Small animal and infant liquid scintillation counter. *International Journal of Applied Radiation Isotopes,* 20:249–254.

Graybiel A.L. and Sode J. (1971) Diuretics, potassium depletion, and carbohydrate intolerance. *Lancet, July 31, 1971,* p. 265.

Greenleaf J.E., Bernauer E.M., Juhos L.T., Young H.L., Morse J.T., and Staley R.W. (1977) Effects of exercise on fluid balance and body composition of man during 14-day bed rest. *Journal of Applied Physiology,* 43:126–132.

Greenway R.M., Houser H.B., Lindau O., and Weir D.R. (1970) Long-term changes in gross body composition of paraplegic and quadriplegic patients. *Paraplegia,* 7:301–318.

Greenwood M.R.C., Maggio C.A., Koopmans H.S., and Sclafani A. (1982) Zucker fa fa rats maintain their obese body composition ten months after jejunoileal bypass surgery. *International Journal of Obesity,* 6:513–525.

Greer F.R., Lane J., Weiner S., and Mazess R.B. (1983) An accurate and reproducible absorptiometric technique for determining bone mineral content in newborn infants. *Pediatric Research,* 17:259–262.

Greer F.R., Searcy J.E., Levin R.S., Steichen J.J., Asch P.S., and Tsang R.C. (1981) Bone mineral content and serum 25-hydroxyvitamin D concentration in breast-fed infants with and without supplemental vitamin D. *Journal of Pediatrics,* 98:696–701.

Gregerson M.I. and Rawson R.A. (1959) Blood volume. *Physiological Reviews,* 39:307–342.

Gresson C.R., Bird D.L., and Simpson F.O. (1973) Plasma volume, extracellular fluid volume, and exchangeable sodium concentrations in the New Zealand strain of genetically hypertensive rat. *Clinical Science,* 44:349–358.

Griggs R.C., Forbes G.B., Moxley R.T. III, and Herr B.E. (1983) The assessment of muscle mass in progressive neuromuscular disease. *Neurology,* 33:158–165.

Griggs R.C., Kingston W., Herr B.E., Forbes G.B., and Moxley R.T., III (1985) Myotonic dystrophy: Effect of testosterone on total body potassium and on creatinine excretion. *Neurology,* 35:1035–1040.

Grüner O. and Salmen A. (1961) Vergleichende Körperwasserbestimmungen mit Hilfe von N-acetyl-4-aminoantipyrin und alkohol. *Klinische Wochenschrift,* 39:92–97.

Gryfe C.I., Exton-Smith A.N., Payne P.R., and Wheeler E.F. (1971) Pattern of development of bone in childhood and adolescence. *Lancet,* i:523–526.

Gulick A. (1922) A study of weight regulation in the adult human body during overnutrition. *American Journal of Physiology,* 60:371–395.

Gundlach B.L., Nijikrake H.G.M., and Hautvast J.G.A.J. (1980) A rapid and simplified plethysmometric method for measuring body volume. *Human Biology,* 52:23–33.

Gurney J.M. and Jelliffe D.B. (1973) Arm anthropometry in nutritional assessment: nomogram for rapid calculation of muscle circumference and cross-sectional muscle and fat areas. *American Journal of Clinical Nutrition,* 26:912–915.

Guyton A.C. (1981) *Textbook of Medical Physiology,* sixth edition. Philadelphia: W.B. Saunders.

Gwinup G., Chelvam R., and Steinberg T. (1971) Thickness of subcutaneous fat and activity of underlying muscles. *Annals of Internal Medicine,* 74:408–411.

Hagan R.D., Upton S.J., Wong L., and Whittam J. (1986) The effects of aerobic conditioning and/or caloric restriction in overweight men and women. *Medicine and Science in Sports and Exercise,* 18:87–94.

Häggmark T., Jansson E., and Svane B. (1978) Cross-sectional area of the thigh muscle in man measured by computed tomography. *Scandinavian Journal of Clinical and Laboratory Investigation,* 38:355–360.

Hahn P., Ross J., Bale W., Balfour W.M., and Whipple G.H. (1942) Red cell and plasma volumes (circulating and total) as determined by radioiron and by dye. *Journal of Experimental Medicine,* 75:221–232.

Halliday D. (1971) An attempt to estimate total body fat and protein in malnourished children. *British Journal of Nutrition,* 26:147–153.

Han P.W. (1967) Hypothalamic obesity in rats without hyperphagia. *Transactions of the New York Academy of Science,* 30:229–243.

Hankin M.E., Theile H.M., and Steinbeck A.W. (1972) Cortisol and aldosterone excretion and plasma cortisol concentrations in normal and obese female subjects. *Clinical Science,* 43:289–298.

Hansen-Smith F.M., Picou D., and Golden M.N.H. (1978) Quantitative analysis of nuclear population in muscle from malnourished and recovered children. *Pediatric Research,* 12:167–170.

Harris L.H. (1961) The protein anabolic action of mestanolone. *Journal of Clinical Endocrinology and Metabolism,* 21:1099–1105.

Harris P.M. (1980) Changes in adipose tissue of the rat due to early undernutrition followed by rehabilitation. I. Body composition and adipose tissue cellularity. *British Journal of Nutrition,* 43:15–31.

Harvey T.C., James H.M., and Chettle D.R. (1979) Birmingham Medical Research Expeditionary Society 1977 Expedition: effect of a Himalayan trek on whole-body composition, nitrogen and potassium. *Postgraduate Medical Journal,* 55:475–477.

Haschke F. (1983) Body composition of adolescent males. *Acta Paediatrica Supplement,* 307:1–23.

Hastings A.B. and Eichelberger L. (1937) The exchange of salt and water between muscle and blood. *Journal of Biological Chemistry,* 117:73–93.

Hausberger F.X. and Hausberger B.C. (1966) Castration-induced obesity in mice. *Acta Endocrinologica,* 53:571–583.

Hawkins W.W. (1964) Iron, copper, and cobalt. In G.H. Beaton and E.W. McHenry (Eds), *Nutrition: A Comprehensive Treatise,* Vol. I New York: Academic Press, pp. 309–372.

Haxhe J.J. (1963) *La Composition Corporelle Normale.* Paris: Librairie Maloine S.A.

Heaney R.P., Gallagher J.C., Johnston C.C., Neer R., Parfitt A.M., and Whedon G.D. (1982) Calcium nutrition and bone health in the elderly. *American Journal of Clinical Nutrition,* 36:986–1013.

Heaney R.P. and Skillman T.G. (1971) Calcium metabolism in normal human pregnancy. *Journal of Clinical Endocrinology,* 33:661–670.

Heaton J.M. (1972) The distribution of brown adipose tissue in the human. *Journal of Anatomy,* 112:35–39.

Heggeness F.W. (1962) Nutritional status and physiological development of the preweanling rat. *American Journal of Physiology,* 203:545–549.

Heim T. (1982) Homeothermy and its metabolic cost. In J.A. Davis and J. Dobbing (Eds), *Scientific Foundations of Paediatrics,* second edition. Baltimore: University Park Press, pp. 91–128.

Helander E.A.S. (1961) Influence of exercise and restricted activity on the protein composition of skeletal muscle. *Biochemical Journal,* 78:478–482.

Hempelmann L.H., Lisco H., and Hoffman J.G. (1952) The acute radiation syndrome: a study of nine cases and a review of the problem. *Annals of Internal Medicine,* 36:279–510.

Hermann G. and Wood H.C. IV (1952) Influence of body fat on duration of thiopental anesthesia. *Proceedings of the Society for Experimental Biology and Medicine,* 80:318–319.

Hervey E. and Hervey G.R. (1967) The effects of progesterone on body weight and composition in the rat. *Journal of Endocrinology,* 37:361–384.

Hervey G.R., Hutchinson I., Knibbs A.V., Burkinshaw L., Jones P.R.M., Norgan N.G., and Levell M.J. (1976) "Anabolic" effects of methandienone in men undergoing athletic training. *Lancet*, ii:699–701.

Hervey G.R., Knibbs A.V., Burkinshaw L., Morgan D.B., Jones P.R.M., Chettle D.R., and Vartsky D. (1981) Effects of methandienone on the performance and body composition of men undergoing athletic training. *Clinical Science*, 60:457–461.

Heymsfield S.B. and McManus C.B. (1985) Tissue components of weight loss in cancer patients. *Cancer*, 55 (Suppl.):238–249.

Heymsfield S.B., Olafson R.P., Kutner M.H., and Nixon D.W. (1979) A radiographic method of quantifying protein-calorie undernutrition. *American Journal of Clinical Nutrition*, 32:693–702.

Hill D.E. (1975) Cellular growth of the Rhesus monkey placenta. In D.B. Cheek (Ed.), *Fetal and Postnatal Cellular Growth*. New York: John Wiley & Sons, pp. 283–288.

Hill G.L., McCarthy I.D., Collins J.P., and Smith A.H. (1978) A new method for the rapid measurement of body composition in critically ill surgical patients. *British Journal of Surgery*, 65:732–735.

Hines H.M. and Knowlton G.C. (1937) Electrolyte and water changes in muscle during atrophy. *American Journal of Physiology*, 120:719–723.

Hoffer E.C., Meador C.K., and Simpson D.C. (1969) Correlation of whole-body impedance with total body water volume. *Journal of Applied Physiology*, 27:531–534.

Hohenauer L. and Oh W. (1969) Body composition in experimental intrauterine growth retardation in the rat. *Journal of Nutrition*, 99:23–26.

Hollander W., Chobanian A.V., and Burrows B.A. (1961) Body fluid and electrolyte composition in arterial hypertension. I. Studies in essential, renal, and malignant hypertension. *Journal of Clinical Investigation*, 40:408–415.

Hoppeler H., Howald H., Conley K., Lindstedt S.L., Claasen H., Vock P., and Weibel E.R. (1985) Endurance training in humans: aerobic capacity and structure of skeletal muscle. *Journal of Applied Physiology*, 59:320–327.

Horber F.F., Scheidegger J.R., Grünig B.E., and Frey F.J. (1985) Evidence that prednisone-induced myopathy is reversed by physical training. *Journal of Clinical Endocrinology and Metabolism*, 61:83–88.

Horsman A. and Kirby P.A. (1972) Geometric properties of the second metacarpal. *Calcified Tissue Research*, 10:289–301.

Horton R. and Tait J.F. (1966) Androstendione production and interconversion rates measured in peripheral blood and studies on the possible sites of conversion to testosterone. *Journal of Clinical Investigation*, 45:301–313.

Howard J.E., Bigham R.S., Jr., and Mason R.E. (1946) Studies on convalescence. V. Observations on the altered protein metabolism during induced malarial infections. *Transactions of the Association of American Physicians*, 59:242–247.

Howell D.S., Delchamps E., Riemer W., and Kiem I. (1960) A profile of electrolytes in the cartilaginous plate of growing ribs. *Journal of Clinical Investigation*, 39:919–929.

Huang P.C., Chong H.E., and Rand W.M. (1972) Obligatory urinary and fecal nitrogen losses in young Chinese men. *Journal of Nutrition*, 102:1605–1614.

Huenemann R.L., Hampton M.C., Behnke A.R., Shapiro L.R., and Mitchell B.W. (1974) *Teenage Nutrition and Physique*. Springfield, IL: Charles Thomas.

Hume R. (1966) Prediction of lean body mass from height and weight. *Journal of Clinical Pathology*, 19:389–391.

Huston T.P., Puffer J.C., and Rodney W.M. (1985) The athletic heart syndrome. *New England Journal of Medicine*, 313:24–32.

Huxley J.S. (1932) *Problems of Relative Growth*. New York: The Dial Press.

Hytten F.E. and Chamberlain G. (Eds) (1980) *Clinical Physiology in Obstetrics*. Oxford: Blackwell Scientific Publications.
Hytten F.E. and Leitch I. (1971) *The Physiology of Human Pregnancy*, 2nd edition. Oxford: Blackwell Scientific Publications.
Hytten F.E., Taylor K., and Taggart N. (1966) Measurement of total body fat in man by absorption of ^{85}Kr. *Clinical Science*, 31:111–119.
Ikkos D., Ljunggren H., and Luft R. (1956) The relation between extracellular and intracellular water in acromegaly. *Acta Endocrinologica*, 21:211–225.
Ikkos D., Luft R., and Sjörgren B. (1954) Body water and sodium in patients with acromegaly. *Journal of Clinical Investigation*, 33:989–994.
Inokuchi S., Ishikawa H., Iwamoto S., and Kimura T. (1975) Age related changes in the histological composition of the rectus abdominis muscle of the adult human. *Human Biology*, 47:231–249.
International Atomic Energy Agency (1970) *Directory of Whole Body Radioactivity Monitors*. Vienna.
International Commission on Radiological Protection (1975) *Report of the Task Group on Reference Man*, No. 23. Oxford: Pergamon Press.
Iob V. and Swanson W.W. (1938) Mineral growth. *Growth*, 2:252–256.
Irsigler K., Veitl V., Sigmund A., Tschegg E., and Kunz K. (1979) Calorimetric results in man: energy output in normal and overweight subjects. *Metabolism*, 28:1127–1132.
Itoh S. and Schwartz I.L. (1956) Sodium and potassium content of isolated nuclei. *Nature*, 178:494.
Jackson A.A., Picou D., and Reeds P.J. (1977) The energy cost of repleting tissue deficits during recovery from protein-energy malnutrition. *American Journal of Clinical Nutrition*, 30:1514–1517.
Jacobson G., Seltzer C.C., Bondy P.K., and Mayer J. (1964) Importance of body characteristics in the excretion of 17-ketosteroids and 17-ketogenic steroids in obesity. *New England Journal of Medicine*, 271:651–656.
James A.H., Brooks L., Edelman I.S., Olney J.M., and Moore F.D. (1954) Body sodium and potassium. I. Simultaneous measurement of exchangeable sodium and potassium in man by isotope dilution. *Metabolism*, 3:313–323.
James H.M., Dabek J.T., Chettle D.K., Dykes P.W., Fremlin J.H., Hardwicke J., Thomas B.J., and Vartsky D. (1984) Whole body cellular and collagen nitrogen in healthy and wasted men. *Clinical Science*, 67:73–82.
James W.P.T., Bailes J., Davies H.L., and Dauncey M.J. (1978) Elevated metabolic rates in obesity. *Lancet*, i:1122–1125.
Jasani B.M. and Edmonds C.J. (1971) Kinetics of potassium distribution in man using isotope dilution and whole-body counting. *Metabolism*, 20:1099–1106.
Jen K.-L.C. and Hansen B.C. (1983) Changes in body composition during experimental induction of obesity in monkeys. *Abstracts of the Fourth Int. Congress on Obesity*, p. 14A.
Jen K.-L.C., Hansen B.C., and Metzger B.L. (1985) Adiposity, anthropometric measures, and plasma insulin levels of rhesus monkeys. *International Journal of Obesity*, 9:213–224.
Jéquier E. and Schutz Y. (1983) Long-term measurements of energy expenditure in humans using a respiration chamber. *American Journal of Clinical Nutrition*, 38:989–998.
Johanson D.C. and Edey M.A. (1981) *Lucy: The Beginnings of Humankind*. New York: Simon and Schuster.
Johnston D.W. (1968) Body characteristics of palm warblers following an overwater flight. *The Auk*, 85:13–18.
Johnston F.A. and McMillan T.J. (1952) The amount of nitrogen retained by 6 young women on an intake of approximately 70 gm of protein a day. *Journal of Nutrition*, 47:425–435.

Johnston F.E. (1982) Relationships between body composition and anthropometry. *Human Biology,* 54:221–245.

Johnston F.E. (1985) Systematic errors in the use of the Mellits-Cheek equation to predict body fat in lean females. *New England Journal of Medicine,* 312:588–589.

Johnstone F.D., Campbell D.M., and MacGillivray I. (1981) Nitrogen balance studies in human pregnancy. *Journal of Nutrition,* 111:1884–1893.

Johnstone F.D., MacGillivray I., and Dennis K.J. (1972) Nitrogen retention in pregnancy. *Journal of Obstetrics and Gynaecology of the British Commonwealth,* 79:777–781.

Jones H.H., Priest J.D., Hayes W.C., and Nagel D.A. (1977) Humeral hypertrophy in response to exercise. *Journal of Bone and Joint Surgery,* 59A:204–208.

Jones R.W.A., El-Bishti M.M., Bloom S.R., Burke J., Carter J.E., Counahan R., Dalton R.N., Morris M.C., and Chantler C. (1980) The effects of anabolic steroids on growth, body composition, and metabolism in boys with chronic renal failure on regular dialysis. *Journal of Pediatrics,* 97:559–566.

Kagan B.M., Stanincova V., Felix N.S., Hodgman J., and Kalman D. (1972) Body composition of premature infants: relation to nutrition. *American Journal of Clinical Nutrition,* 25:1153–1164.

Kata F.H. and Kappas A. (1967) The effects of estradiol and estriol on plasma levels of cortisol and thyroid hormone-binding globulins and on aldosterone and cortisol secretion rates in man. *Journal of Clinical Investigation,* 46:1768–1777.

Katz J. (1896) Die mineralischen Bestandtheile des Muskelfleisches. *Archiv für die Gessamte Physiologie des Menschen und der Thiere,* 63:1–85.

Kaufman L. and Wilson C.J. (1973) Determination of extracellular fluid volume by fluorescent excitation analysis of bromine. *Journal of Nuclear Medicine,* 14:812–815.

Keith N.M., Rowntree L.G., and Gerachty J.T. (1915) A method for the determination of plasma and blood volume. *Archives of Internal Medicine,* 16:547–576.

Kennedy N.S.J., Eastell R., Smith M.A., and Tothill P. (1983) Normal levels of total body sodium and chlorine by neutron activation analysis. *Physics in Medicine and Biology,* 28:215–221.

Kenyon A.T., Knowlton K., Sandiford I., Koch F.C., and Lotwin G. (1940) A comparative study of the metabolic effects of testosterone propionate in normal men and women and in eunuchoidism. *Endocrinology,* 26:26–45.

Kenyon A.T., Sandiford I., Bryan A.H., Knowlton K., and Koch F.C. (1938) Effect of testosterone propionate on nitrogen, electrolyte, water and energy metabolism in eunuchoidism. *Endocrinology,* 23:135–153.

Kerpel-Fronius E. and Kovach S. (1948) The volume of extracellular body fluids in malnutrition. *Pediatrics,* 2:21–23.

Keutmann E.H., Bassett S.H., and Warren S.L. (1939) Electrolyte balances during artificial fever with special reference to loss through the skin. *Journal of Clinical Investigation,* 18:239–250.

Keys A., Anderson J.T., and Brožek J. (1955) Weight gain from simple overeating. *Metabolism,* 4:427–432.

Keys A. and Brožek J. (1953) Body fat in adult man. *Physiological Reviews,* 33: 245–345.

Keys A., Brožek J., Henschel A., Mickelsen O., and Taylor H.L. (1950) *The Biology of Human Starvation.* Minneapolis: University of Minnesota Press.

Keys A., Fidanza F., Karvonen M.J., Kimura N., and Taylor H.L. (1972) Indices of relative weight and obesity. *Journal of Chronic Disease,* 25:329–343.

Keys A., Taylor H.L., and Grande F. (1973) Basal metabolism and age of adult man. *Metabolism,* 22:579–587.

Khosla T. (1968) Unfairness of certain events in the Olympic Games. *British Medical Journal,* 4:111–113.

King J.C. (1981) Assessment of nutritional status in pregnancy. *American Journal of Clinical Nutrition,* 34(Suppl.):685–690.

King J.C., Calloway D.H., and Margen S. (1973) Nitrogen retention, total body ^{40}K and weight gain in teenage pregnant girls. *Journal of Nutrition,* 103:772–785.

Kinney J.M., Long C.L., Gump F.E., and Duke J.H. (1968) Tissue composition of weight loss in surgical patients. I. Elective operation. *Annals of Surgery,* 168:459–474.

Kirsch K.A., Johnson R.F., and Gorten R.J. (1971) The significance of total-body hematocrit in measurements of blood compartments. *Journal of Nuclear Medicine,* 12:17–21.

Kirton A.H. and Pearson A.M. (1963) Comparison of methods of measuring potassium in pork and lamb and prediction of their composition from sodium and potassium. *Journal of Animal Science,* 22:125–131.

Kitagawa K. and Miyashita M. (1978) Muscle strengths in relation to fat storage in young men. *European Journal of Applied Physiology,* 38:189–196.

Kleiber M. (1975) *The Fire of Life: An introduction to animal energetics.* Huntington, NY: Robert E. Krieger.

Kley H.K., Deselaers T., Peerenboom H., and Krüskemper H.L. (1980) Enhanced conversion of androstenedione to estrogens in obese males. *Journal of Clinical Endocrinology and Metabolism,* 51:1128–1132.

Klish W.J., Forbes G.B., Gordon A., and Cochran W.G. (1984) A new method for the estimation of lean body mass in infants (EMME instrument): its validation in non-human models. *Journal of Pediatric Gastroenterology and Nutrition,* 3:199–204.

Knittle J.L., Timmers K., Ginsberg-Fellner F., Brown R.E., and Katz D.P. (1979) The growth of adipose tissue in children and adolescents: cross-sectional and longitudinal studies of adipose cell number and size. *Journal of Clinical Investigation,* 63:239–246.

Knowlton A.I. and Loeb E.N. (1957) Depletion of carcass potassium in rats made hypertensive with desoxycorticosterone acetate (DCA) and with cortisone. *Journal of Clinical Investigation,* 36:1295–1300.

Knowlton K., Kenyon A.T., Sandiford I., Lotwin G., and Fricker R. (1942) Comparative study of metabolic effects of estradiol benzoate and testosterone propionate in man. *Journal of Clinical Endocrinology,* 2:671–684.

Kochakian C.D. (1960) Summation of protein anabolic effects of testosterone propionate and growth hormone. *Proceedings of the Society of Experimental Biology and Medicine,* 103:196–197.

Kochakian C.D. (Ed.) (1976) *Anabolic-Androgenic Steroids.* New York: Springer-Verlag.

Korubin V., Maisey M.N. and McIntyre P.A. (1972) Evaluation of technetium-labelled red cells for determination of red cell volume in man. *Journal of Nuclear Medicine,* 13:760–762.

Kotler D.P., Wang J., and Pierson R.N. Jr. (1985) Body composition studies in patients with the acquired immunodeficiency syndrome. *American Journal of Clinical Nutrition,* 42:1255–1265.

Krabbe S., Christensen T., Worm J., Christiansen C., and Transbøl I. (1978) Relationship between haemoglobin and serum testosterone in normal children and adolescents and in boys with delayed puberty. *Acta Paediatrica Scandinavica,* 67:655–658.

Krabbe S., Hummer L., and Christiansen C. (1984a) Longitudinal study of calcium metabolism in male puberty. II. Relationship between mineralization and serum testosterone. *Acta Paediatrica Scandinavica,* 73:750–755.

Krabbe S., Kastrup K.W., and Hummer L. (1984b) Somatomedin A in male puberty: variation with age, maturity, growth and androgens. *Acta Endocrinologica,* 107:312–316.

Kral J.G., Björntorp P., Schersten T., and Sjöström L. (1977) Body composition and adipose tissue cellularity before and after jejuno-ileostomy in severely obese subjects. *European Journal of Clinical Investigation,* 7:413–419.

Kreisberg R.A., Bowdoin B., and Meador C.K. (1970) Measurement of muscle mass in humans by isotopic dilution of creatine-^{14}C. *Journal of Applied Physiology,* 28:264–267.

Kriegel W. and Discherl W. (1964) Zur Wirkung von Aldosteron und Corticosteron auf den Elektrolyt- und Wassergehalt bindegewebiger Organe der Ratte. *Acta Endocrinologica,* 46:47–64.

Kritzinger E.E., Kanengoni E., and Jones J.J. (1972) Effective renin activity in the plasma of children with kwashiorkor. *Lancet,* i:412–413.

Krølner B. and Toft B. (1983) Vertebral bone loss: an unheeded side effect of therapeutic bed rest. *Clinical Science,* 64:537–540.

Krølner B., Toft B., Nielsen S.P., and Tøndevald E. (1983) Physical exercise as prophylaxis against involutional vertebral bone loss: a controlled trial. *Clinical Science,* 64:541–546.

Krotkiewski M., Björntorp P., Sjöström L., and Smith U. (1983) Impact of obesity on metabolism in men and women. *Journal of Clinical Investigation,* 72:1150–1162.

Krotkiewski M., Sjöström L., Björntorp P., Carlgren G., Garellick G., and Smith U. (1977) Adipose tissue cellularity in relation to prognosis for weight reduction. *International Journal of Obesity,* 1:395–416.

Krzywicki H.J. and Chinn K.S.K. (1967) Body composition of a military population, Fort Carson, 1963 I. Body density, fat, and potassium 40. *American Journal of Clinical Nutrition,* 20:708–715.

Krzywicki H.J., Ward G.M., Rahman D.P., Nelson R.A., and Consolazio C.F. (1974) A comparison of methods for estimating human body composition. *American Journal of Clinical Nutrition,* 27:1380–1385.

Kuhlbäck B. (1957) Creatine and creatinine metabolism in thyrotoxicosis and hypothyroidism. *Acta Medica Scandinavica,* 139 (Suppl. 331):5–70.

Kulwich R., Feinstein L., Golumbic C., Hiner R.L., Seymour W.R., and Kauffman W.R. (1961) Relationship of gamma-ray measurements to the lean content of hams. *Journal of Animal Science,* 20:497–502.

Kvist H., Sjöström L., and Tylén U. (1986) Adipose tissue volume determinations in women by computed tomography: technical considerations. *International Journal of Obesity,* 10:53–67.

Lamki L., Ezrin C., Koven I., and Steiner G. (1973) L-Thyroxine in the treatment of obesity without increase in loss of lean body mass. *Metabolism,* 22:617–622.

Landau R.L. (1973) The metabolic influence of progesterone. In R.O. Greep (ed.), *Handbook of Physiology, Endocrinology II, Part I.* Washington: American Physiological Society, pp. 573–589.

Lantigua R.A., Amatruda J.M., Biddle T.L., Forbes G.B., and Lockwood D.H. (1980) Cardiac arrhythmias associated with a liquid protein diet for the treatment of obesity. *New England Journal of Medicine,* 303:735–738.

Lapidus L., Bengtsson C., Larsson B., Pennert K., Rybo E., and Sjöström L. (1984) Distribution of adipose tissue and risk of cardiovascular disease and death: a 12 year follow up of participants in the population study of women in Gothenburg, Sweden. *British Medical Journal,* 289:1257–1261.

Laragh J.H. (1985) Atrial natriuretic hormone, the renin-aldosterone axis, and blood pressure-electrolyte homeostasis. *New England Journal of Medicine,* 313:1330–1340.

Laramore D.C. and Grollman A. (1950) Water and electrolyte content of tissues in normal and hypertensive rats. *American Journal of Physiology,* 161:278–282.

Larsson B., Svärdsudd K., Welin L., Wilhelmsen L., Björntorp L., and Tibblin G. (1984) Abdominal adipose tissue distribution, obesity, and risk of cardiovascular disease and death: 13 year follow up of participants in the study of men born in 1913. *British Medical Journal,* 288:1401–1404.

Lawes J.B. and Gilbert J.H. (1859) Experimental inquiry into the composition of some of the animals fed and slaughtered as human food. *Philosophical Transactions of the Royal Society of London,* 149:493–680.

Laws R.M. (1985) The ecology of the southern ocean. *American Scientist,* 73:26–40.

Lechtig A., Delgado H., Lasky R., Yarbrough C., Klein R.E., Habicht J.-P., and Béhar M. (1975) Maternal nutrition and fetal growth in developing countries. *American Journal of Diseases of Children,* 129:553–556.

LeMaho Y., van Kha H.V., Koubi H., Dewasmes G., Girard J., Ferré P., and Cagnard M. (1981) Body composition, energy expenditure, and plasma metabolites in long-term fasting geese. *American Journal of Physiology,* 241:E342–354.

Leonard J.I., Leach C.S., and Rambaut P.C. (1983) Quantitation of tissue loss during prolonged space flight. *American Journal of Clinical Nutrition,* 38:667–679.

Lesser G.T., Deutsch S., and Markofsky J. (1971) Use of independent measurement of body fat to evaluate overweight and underweight. *Metabolism,* 20:792–804.

Lesser G.T., Deutsch S., and Markofsky J. (1973) Aging in the rat: longitudinal and cross-sectional studies of body composition. *American Journal of Physiology,* 225:1472–1478.

Lesser G.T. and Markofsky J. (1979) Body water compartments with human aging using fat-free mass as the reference standard. *American Journal of Physiology,* 236:R215–220.

Lesser G.T., Perl W., and Steele J.M. (1960) Determination of total body fat by absorption of an inert gas: measurements and results in normal human subjects. *Journal of Clinical Investigation,* 39:1791–1806.

Lesser G.T. and Zak G. (1963) Measurement of total body fat by the simultaneous absorption of two inert gases. *Annals of the New York Academy of Science,* 110:40–54.

Letsky E. (1980) The haematological system. In F. Hytten and G. Chamberlain (Eds.), *Clinical Physiology in Obstetrics.* Oxford: Blackwell Scientific Publications.

Levin M.E., Boisseau V.C., and Avioli L.V. (1976) Effects of diabetes mellitus on bone mass in juvenile and adult-onset diabetes. *New England Journal of Medicine,* 294:241–245.

Lewis D.S., Bertrand, H.A., Masoro E.J., McGill H.C., Jr., Carey K.D., and McMahan C.A. (1983) Preweaning nutrition and fat development in baboons. *Journal of Nutrition,* 113:2253–2259.

Lewis D.S., Rollwitz W.L., Bertrand H.A., and Masoro E.J. (1986) Use of NMR for measurement of total body water and estimation of body fat. *Journal of Applied Physiology,* 60:836–840.

Lindholm B. (1957) Body cell mass during long-term cortisone treatment in asthmatic subjects. *Acta Endocrinologica,* 55:202, 222.

Linquette M., Fossati P., Lefebvre J., and Chechan C. (1969) The measurement of total water, exchangeable sodium and potassium in obese persons. In J. Vague (Ed.), *Pathophysiology of Adipose Tissue* (pp. 302–316). Amsterdam: Excerpta Medica Foundation.

Ljunggren H., Ikkos D., and Luft R. (1961) Basal metabolism in women with obesity and anorexia nervosa. *British Journal of Nutrition,* 15:21–34.
Loeppky J.A., Myhre L.G., Venters M.D., and Luft U.C. (1977) Total body water and lean body mass estimated by ethanol dilution. *Journal of Applied Physiology,* 42:803–808.
Lohman T.G. (1981) Skinfolds and body density and their relation to body fatness: a review. *Human Biology,* 53:181–225.
Loucks A.B. and Horvath S.M. (1985) Athletic amenorrhea: a review. *Medicine and Science in Sports and Exercise,* 17:56–72.
Loucks A.B., Horvath S.M., and Freedson P.S. (1984) Menstrual status and validation of body fat prediction in athletes. *Human Biology,* 56:383–392.
Lowry O.H. and Hastings A.B. (1942) Histochemical changes in ageing. In E.V. Cowdry (Ed.), *The Problems of Ageing,* 2nd edition. Baltimore: Williams and Wilkins.
Lukaski H.C., Bolonchuk W.W., Hall C.B., and Siders W.A. (1986) Validation of tetrapolar bioelectrical impedance method to assess human body composition. *Journal of Applied Physiology,* 60:1327–1332.
Lukaski H. and Mendez J. (1980) Relationship between fat-free weight and urinary 3-methylhistidine excretion in man. *Metabolism,* 29:758–761.
Lukaski H.C., Mendez J., Buskirk E.R., and Cohn S.H. (1981a) A comparison of methods of assessment of body composition including neutron activation analysis of total body nitrogen. *Metabolism,* 30:777–782.
Lukaski H.C., Mendez J., Buskirk E.R., and Cohn S.H. (1981b) Relationship between endogenous 3-methylhistidine excretion and body composition. *American Journal of Physiology,* 240:E302–307.
Lye M., May T., Hammick J., and Ackery D. (1976) Whole-body and exchangeable potassium measurements in normal elderly subjects. *European Journal of Nuclear Medicine,* 1:167–171.
Lykken G.I., Lukaski H.C., Bolonchuk W.W., and Sanstead H.H. (1983) Potential errors in body composition as estimated by whole body scintillation counting. *Journal of Laboratory and Clinical Medicine,* 101:651–658.
Lynch T.N., Jensen R.L., Stevens P.M., Johnson R.L., and Lamb L.E. (1967) Metabolic effects of prolonged bed rest: their modification by simulated altitude. *Aerospace Medicine,* 38:10–20.
MacDonald N.S., Hutchinson D.L., Helper M., and Flynn E. (1972) Movement of calcium in both directions across the primate placenta. *Proceedings of the Society for Experimental Biology and Medicine,* 119:476–481.
MacDougall J.D., Ward G.R., Sale D.G., and Sutton J.R. (1977) Biochemical adaptation of human skeletal muscle to heavy resistance training and immobilization. *Journal of Applied Physiology,* 43:700–703.
MacGillivray I. and Buchanan T.J. (1958) Total exchangeable sodium and potassium in non-pregnant women and in normal and pre-eclamptic pregnancy. *Lancet,* ii:1090–1093.
MacLennan A.H., Hocking A., Seamark R.F., Godfrey B., and Haslam R. (1983) Neonatal water metabolism: an objective postnatal index of intrauterine fetal growth. *Early Human Development,* 8:21–31.
MacMahon S.W., Wilcken D.E.L., and MacDonald G.J. (1986) The effect of weight reduction on left ventricular mass. *New England Journal of Medicine,* 314:334–339.
Macy I.G. and Hunscher H.A. (1934) An evaluation of maternal nitrogen and mineral needs during embryonic and fetal development. *American Journal of Obstetrics and Gynecology,* 27:878–888.
Magnus-Levy A. (1906) Physiologie des Stoffwechsels. In C. von Noorden (Ed.), *Handbuch der Pathologie des Stoffwechsels.* Berlin: Hirschwald, p. 446.

Mahalko J.R. and Johnson L.K. (1980) Accuracy of prediction of long term energy needs. *Journal of the American Dietetic Association,* 77:557–561.

Mahon M., Toman A., Willan P.L.T., and Bagnall K.M. (1984) Variability of histochemical and morphometric data from needle biopsy specimens of human quadriceps femoris muscle. *Journal of Neurological Sciences,* 63:85–100.

Manery J.F. (1954) Water and electrolyte metabolism. *Physiological Reviews,* 34:334–417.

Manery J.F. (1961) Minerals in nonosseous connective tissues (including the blood, lens and cornea). In C.L. Comar, and F. Bronner (Eds.), *Mineral Metabolism, Vol. 1B.* New York: Academic Press, pp. 551–608.

Manery J.F., Danielson, I.S., and Hastings A.B. (1938) Connective tissue electrolytes. *Journal of Biological Chemistry,* 124:359–375.

Manzke E., Chestnut C.H. III, Wergedal J.E., Baylink D.J., and Nelp W.B. (1975) Relationship between local and total bone mass in osteoporosis. *Metabolism,* 24:605–615.

Maresh M.M. (1966) Changes in tissue widths during growth. *American Journal of Diseases of Children,* 111:142–155.

Marriott W.McK. (1923) Anhydremia. *Physiological Reviews,* 3:275–294.

Martin A.D., Drinkwater D.T., and Clarys J.P. (1984) Human body surface area: validation of formulae based on a cadaver study. *Human Biology,* 56:475–488.

Mather K., Bowler R.G., Crooke A.C., and Morris C.J.O.R. (1947) The precision of plasma volume determinations by the Evans blue method. *British Journal of Experimental Pathology,* 28:12–24.

Maughan R.J., Watson J.S., and Weir J. (1983) Relationship between muscle strength and muscle cross-sectional area in male sprinters and endurance runners. *European Journal of Applied Physiology,* 50:309–318.

Maughan R.J., Watson J.S., and Weir J. (1984) The relative proportion of fat, muscle and bone in the normal human forearm as determined by computed tomography. *Clinical Science and Molecular Medicine,* 66:683–689.

Mayer J. Zomzely C., and Furth J. (1956) Body composition and energetics in obesity induced in mice by adrenotropic tumors. *Science,* 123:184–185.

Mays C.W., Lloyd R.D., Taysum D.H., Zundel W.S., and Ziter F.A. (1968) *Potassium concentrations and tissue weights in three patients dying with Duchenne muscular dystrophy.* Utah College of Medicine Report C 00-119-239. Salt Lake City: University of Utah Press, p. 30–39.

Mazess R.B. (1971) Estimation of bone and skeletal weight by direct photon absorptiometry. *Investigative Radiology,* 6:52–60.

Mazess R.B. and Mather W. (1974) Bone mineral content of North Alaska Eskimos. *American Journal of Clinical Nutrition,* 27:916–925.

Mazess R.B., Peppler W.W., Chestnut C. III, Nelp W.B., and Cohn S.H. (1981a) Total bone bone mineral and lean body mass by dual photon absorptiometry. II. Comparison with total body calcium by neutron activation analysis. *Calcified Tissue International,* 33:361–363.

Mazess R.B., Peppler W.W., Harrison J.E., and McNeill K.G. (1981b) Total body bone mineral and lean body mass by dual photon absorptiometry. III. Comparison with trunk calcium by neutron activation analysis. *Calcified Tissue International,* 33:365–368.

Mellits E.D. and Cheek D.B. (1970) The assessment of body water and fatness from infancy to adulthood. *Monographs of the Society for Research in Child Development,* 35:12–26.

Merklin R.J. (1974) Growth and distribution of human fetal brown fat. *Anatomical Record,* 178:637–646.

Metcoff J., Frenk S., Antonowicz I., Gardillo G., and Lopez E. (1960) Relations of intracellular ions to metabolite sequences in muscle in Kwashiorkor. *Pediatrics,* 26:960–972.

Metropolitan Life Insurance Company (1983).

Mettau J.W. (1978) *Measurement of Total Body Fat in Low Birth Weight Infants.* Rotterdam: Bronder-Offset B.V.

Mettau J.W., Degenhart H.J., Visser H.K.A., and Holland W.P.S. (1977) Measurement of total body fat in newborns and infants by absorption and desorption of nonradioactive xenon. *Pediatric Research,* 11:1097–1101.

Migeon C.J., Beitins I.Z., Kowarski A., and Graham G.G. (1973) Plasma aldosterone concentration and aldosterone secretion rate in Peruvian infants with marasmus and kwashiorkor. In L.I. Gardner, and P. Amacher (Eds.), *Endocrine Aspects of Malnutrition.* Santa Ynez, CA: Kroc Foundation, pp. 399–424.

Miller D.S. and Mumford P. (1967) Gluttony: I. An experimental study of low- or high-protein diets. II. Thermogenesis in overeating man. *American Journal of Clinical Nutrition,* 20:1212–1222, 1223–1229.

Miller M., Ward L., Thomas B.J., Cooksley W.G.E., and Shepard R.W. (1982) Altered body composition and muscle protein degradation in nutritionally growth-retarded children with cystic fibrosis. *American Journal of Clinical Nutrition,* 36:492–499.

Miller M.E. and Cappon C.J. (1984) Anion-exchange chromatographic determination of bromide in serum. *Clinical Chemistry,* 30:781–783.

Milne D.B., Canfield W.K., Mahalko J.R., and Sandstead H.H. (1983) Effect of dietary zinc on whole body surface loss of zinc: impact on estimation of zinc retention by balance method. *American Journal of Clinical Nutrition,* 38:181–186.

Milne J.S. (1985) *Clinical Effects of Ageing. A Longitudinal Study.* London: Croom Helm, Ltd.

Mitchell H.H. (1962) *Comparative Nutrition of Man and Domestic Animals.* Vols. I and II. New York: Academic Press.

Moerman M.L. (1982) Growth of the birth canal in adolescent girls. *American Journal of Obstetrics and Gynecology,* 143:528–532.

Mollinger L.A., Spurr G.B., El Ghatit A.Z., Barboriak J.J., Rooney C.B., Davidoff D.D., and Bongard R.D. (1985) Daily energy expenditure and basal metabolic rates of patients with spinal cord injury. *Archives of Physical Medicine and Rehabilitation,* 66:420–426.

Mollison P.O., Veall N., and Cutbush M. (1950) Red cell volume and plasma volume in newborn infants. *Archives of Disease in Childhood,* 25:242–253.

Moore F.D. (1946) Determination of total body water and solids with isotopes. *Science,* 104:157–160.

Moore F.D., Hartsuck J.M., Zollinger R.M., and Johnson J.E. (1968) Reference models for clinical studies by isotope dilution. *Annals of Surgery,* 168:679–700.

Moore F.D., Lister J., Boyden C.M., Ball M.R., Sullivan N., and Dagher F.J. (1968) The skeleton as a feature of body composition: values predicted by isotope dilution and observed by cadaver dissection in an adult female. *Human Biology,* 40:135–188.

Moore F.D., Olesen K.H., McMurray J.D., Parker H.V., Ball M.R., and Boyden C.M. (1963) *The Body Cell Mass and Its Supporting Environment.* Philadelphia: W.B. Saunders.

Moore Ede M.C., Brennan M.F., and Ball M.R. (1975) Circadian variation of intercompartmental potassium fluxes in man. *Journal of Applied Physiology,* 38:163–170.

Morey E.R. and Baylink D.J. (1978) Inhibition of bone formation during space flight. *Science,* 201:1138–1141.

Morgan D.B. and Burkinshaw L. (1983) Estimation of non-fat body tissues from measurements of skinfold thickness, total body potassium and total body nitrogen. *Clinical Science,* 65:407–414.

Morgulis S. (1923) *Fasting and Undernutrition.* New York: E.P. Dutton.

Morse W.I. and Soeldner J.S. (1963) Composition of adipose tissue and the non-adipose body of obese and nonobese men. *Metabolism,* 12:99–107.

Motil K.J., Grand R.J., Matthews D.E., Bier D.M., Maletskos C.J., and Young V.R. (1982) Whole body leucine metabolism in adolescents with Crohn's disease and growth failure during nutritional supplementation. *Gastroenterology,* 82:1359–1368.

Moulton C.R. (1923) Age and chemical development in mammals. *Journal of Biological Chemistry,* 57:79–97.

Mulholland J., Tui C., Wright A.M., and Vinci V.J. (1943) Nitrogen metabolism, caloric intake and weight in postoperative convalescence. *Annals of Surgery,* 117:512–534.

Müller A. (1911) Stoffwechsel- und Respirationsversuche zur Frage der Eiweissmast. *Zentralblatt für die gesamte Physiologie und Pathologie des Stoffwechsels,* 6:617–629.

Myhre L.G. and Kessler W.V. (1966) Body density and potassium-40 measurements of body composition as related to age. *Journal of Applied Physiology,* 21:1251–1255.

Myhre L.G., Hartung G.H., Nunneley S.A., and Tucker D.M. (1985) Plasma volume changes in middle-aged male and female subjects during marathon running. *Journal of Applied Physiology,* 59:559–563.

McAreavey D., Cumming A.M., Boddy K., Brown J.J., Fraser R., Leckie B.J., Lever A.F., Morton J.J., Robertson J.I., and Williams E.D. (1983) The renin-angiotensin system and total body sodium and potassium in hypertensive women taking oestrogen-progestagen oral contraceptives. *Clinical Endocrinology,* 18:111–118.

McCance R.A. and Widdowson E.M. (1951) A method of breaking down the body weights of living persons into terms of extracellular fluid, cell mass and fat, and some applications of it to physiology and medicine. *Proceedings of the Royal Society of London, Series B,* 138:115–130.

McCance R.A. and Widdowson E.M. (1954) Normal renal function in the first two days of life. *Archives of Disease in Childhood,* 29:488–494.

McCartney C.P., Pottinger R.E., and Harrod J.P. (1959) Alterations in body composition during pregnancy. *American Journal of Obstetrics and Gynecology,* 77:1038–1053.

McClure W.B. and Aldrich C.A. (1923) Time required for disappearance of intradermally injected salt solution. *Journal of the American Medical Association,* 81:293–294.

McCracken K.J. and McNiven M.A. (1983) Effects of overfeeding by gastric intubation on body composition of adult female rats and on heat production during feeding and fasting. *British Journal of Nutrition,* 49:193–202.

McGowan A., Jordan M., and MacGregor J. (1975) Skinfold thickness in neonates. *Biology of the Neonate,* 25:66–84.

McKeown T. and Record R.G. (1952) Observations on foetal growth in multiple pregnancies in man. *Journal of Endocrinology,* 8:386–401.

McKeown T. and Record R.G. (1953) The influence of placental size on foetal growth in man, with special reference to multiple pregnancy. *Journal of Endocrinology,* 9:418–426.

McMurrey J.D., Boling E.A., Davis J.M., Parker H.V., Magnus I.C., Ball M.R., and Moore F.D. (1958) Body composition: simultaneous determinations of several aspects by the dilution principle. *Metabolism,* 7:651–667.

McNeill K.G. and Harrison J.E. (1981) Partial body neutron activation—truncal. In S.H. Cohn (Ed.), *Non-Invasive Measurements of Bone Mass and Their Clinical Application*. Boca Raton, FL: CRC Press, pp. 165–190.

McNeill K.G., Harrison J.E., Mernagh J.R., Stewart S., and Jeejeebhoy, K.N. (1982) Changes in body protein, body potassium, and lean body mass during total parenteral nutrition. *Journal of Parenteral and Enteral Nutrition*, 6:106–108.

Nachman H.M., James G.W. III, Moore J.W., and Evans E.I. (1950) A comparative study of red cell volumes in human subjects with radioactive phosphorus tagged red cells and T-1824 dye. *Journal of Clinical Investigation*, 29:258–264.

Naeye R.L. and Dixon J.B. (1978) Distortions in fetal growth standards. *Pediatric Research*, 12:987–991.

Naeye R.L. and Roode P. (1970) The sizes and numbers of cells in visceral organs in human obesity. *American Journal of Clinical Pathology*, 54:251–253.

Nagamine S. and Suzuki S. (1964) Anthropometry and body composition of Japanese young men and women. *Human Biology*, 36:8–15.

National Center for Health Statistics (1974) *Skinfold thickness of youths 12–17 years, United States*, Publication No. (HRA)74-614, Series 11, No. 132. Rockville, MD: Department of Health, Education and Welfare.

National Council for Radiation Protection and Measurements (1975) *Natural Background Radiation in the United States* (Report No. 45). Washington.

National Institute of Neurological Diseases and Stroke (1972) *The Women and Their Pregnancies*. Washington: DHEW Publication No. (NIH) 73-379.

National Research Council (Food and Nutrition Board) (1980) *Recommended Dietary Allowances*. Washington: National Academy of Sciences.

Needham J. (1950) *Biochemistry and Morphogenesis*. Cambridge, U.K.: University Press.

Nelp W.B., Denney J.D., Murano R., Hinn G.M., Williams J.L., Rudd T.G., and Palmer H.E. (1972) Absolute measurement of total body calcium (bone mass) in vivo. *Journal of Laboratory and Clinical Medicine*, 79:430–438.

Nelson E.A. and Craig A.B. (1978) Physiologic responses to a transcontinental bicycle trip. *Physician and Sportsmedicine*, 6:83–93.

Nelson R.A., Huse D.M., Holman R.T., Kimbrough B.O., Wahner H.W., Callaway C.W., and Hayles A.B. (1981) In V.A. Holm, S. Sulzbacher, and P.L. Pipes (Eds.), *The Prader-Willi Syndrome*. Baltimore: University Park Press, pp. 105–120.

Nelson R.A., Jones J.D., Wahner H.W., McGill D.B., and Code C.F. (1975) Nitrogen metabolism in bears: urea metabolism in summer starvation and in winter sleep and role of the urinary bladder in water and nitrogen conservation. *Mayo Clinic Proceedings*, 50:141–146.

Nelson R.A., Wahner H.W., Jones J.D., Ellefson R.D., and Zollman P.E. (1973) Metabolism of bears before, during, and after winter sleep. *American Journal of Physiology*, 224:491–496.

Neuman W.F. and Neuman M.W. (1958) *Chemical Dynamics of Bone Mineral*. Chicago: University of Chicago.

Neumann R.O. (1902) Experimentelle Beiträge zur Lehre von dem täglichen Nahrungsbedarf des Menschen unter besonderer Berücksichtigung der notwendigen Eiweissmenge. *Archiv für Hygiene*, 45:1–87.

Nichols B.L., Alleyne G.A.O., Barnes D.J., and Hazelwood C.F. (1969) Relationship between muscle potassium and total body potassium in infants with malnutrition. *Journal of Pediatrics*, 74:49–57.

Nichols B.L., Alvarado J., Kimzey S.L., Hazelwood C.F., and Viteri F. (1973) Anomalies of the regulation of salt and water in protein-calorie malnutrition. In L.I. Gardner, and P. Amacher (Eds.), *Endocrine Aspects of Malnutrition*. Santa Yncz, CA: Kroc Foundation, pp. 363–398.

Nichols B.L., Rudolph A.J., and Hazlewood C.F. (1972) The role of muscle in extrarenal electrolyte metabolism in the newborn. *Johns Hopkins Medical Journal,* 131:212–219.

Nichols G., Jr., Nichols N., Weil W.B., and Wallace W.M. (1953) The direct measurement of the extracellular phase of tissues. *Journal of Clinical Investigation,* 32:1299–1308.

Nicholson J.P. and Zilva J.F. (1964) Body constituents and functions in relation to height and weight. *Clinical Science,* 27:97–109.

Nicolopoulos D.A. and Smith C.A. (1961) Metabolic aspects of idiopathic respiratory distress (hyaline membrane syndrome) in newborn infants. *Pediatrics,* 28:206–222.

Noppa H., Andersson M., Bengtsson C., Bruce Å., and Isaksson B. (1979) Body composition in middle-aged women with special reference to the correlation between body fat mass and anthropometric data. *American Journal of Clinical Nutrition,* 32:1388–1395.

Noppa H., Andersson M., Bengtsson C., Bruce Å., and Isaksson B. (1980) Longitudinal studies of anthropometric data and body composition. The population study of women in Göteborg, Sweden. *American Journal of Clinical Nutrition,* 33:155–162.

Norgan N.G. and Durnin J.V.G.A. (1980) The effect of 6 weeks of overfeeding on the body weight, body composition, and energy metabolism of young men. *American Journal of Clinical Nutrition,* 33:978–988.

Novak L. (1972) Aging, total body potassium, fat-free mass, and cell mass in males and females between ages 18 and 85 years. *Journal of Gerontology,* 27:438–443.

Novak L.P. (1970) Comparative study of body composition of American and Filipino women. *Human Biology,* 42:206–216.

Novak L.P., Hayles A.B., and Cloutier M.D. (1972) Effect of HGH on body composition of hypopituitary dwarfs. *Mayo Clinic Proceedings,* 47:241–246.

Novak L.P., Tauxe W.N., and Orvis A.L. (1973) Estimation of total body potassium in normal adolescents by whole-body counting: age and sex differences. *Medicine and Science in Sports,* 5:147–155.

Nyhlin H., Dyckner T., Ek B., and Wester P.O. (1985) Plasma and skeletal muscle electrolytes in patients with Crohn's disease. *Journal of American College of Nutrition,* 4:531–538.

Oakley J.P., Parsons R.J., and Whitelaw A.G.L. (1977) Standards for skinfold thickness in British newborn infants. *Archives of Disease in Childhood,* 52:287–290.

Oddoye E.A. and Margen S. (1979) Nitrogen balance studies in humans: long-term effect of high nitrogen intake on nitrogen accretion. *Journal of Nutrition,* 109:363–377.

Odum E.P. and Connell C.E. (1956) Lipid levels in migrating birds. *Science,* 123:892–894.

Odum E.P. and Perkinson J.D., Jr. (1951) Relation of lipid metabolism to migration in birds: seasonal variation in body lipids of the migratory white-throated sparrow. *Physiological Zoology,* 24:216–230.

Olsson S.-Å., Petersson B.G., Sörbis R., and Nilsson-Ehle P. (1984) Effect of weight reduction after gastroplasty on glucose and lipid metabolism. *American Journal of Clinical Nutrition,* 40:1273–1280.

O'Meara M.P., Birkenfeld L.W., Gotch F.A., and Edelman I.S. (1957) The equilibration of radiosodium (Na^{24}), radiopotassium (K^{42}), and deuterium oxide (D_2O) in hydropic human subjects. *Journal of Clinical Investigation,* 36:784–792.

Oscai L.B. (1982) Diet-induced severe obesity: a rat model. *American Journal of Physiology,* 242:R212–R215.

Oscai L.B. and Holloszy J.O. (1969) Effect of weight changes produced by exercise, food restriction or overeating on body composition. *Journal of Clinical Investigation,* 48:2124–2128.
Oscai L.B., Mole P., Krusack L.M., and Holloszy J.O. (1973) Detailed body composition analysis on female rats subjected to a program of swimming. *Journal of Nutrition,* 103:412–418.
Owen G.M., Jensen R.L., and Fomon S.J. (1962) Sex-related difference in total body water and exchangeable chloride during infancy. *Journal of Pediatrics,* 60:858–868.
Parizkova J. (1977) *Body Fat and Physical Fitness.* The Hague: Martinus Nijhoff b.v.
Parizkova J. and Eiselt E. (1980) Longitudinal changes in body build and skinfolds in a group of old men over a 16 year period. *Human Biology,* 52:803–809.
Parizkova J., Vaneckova M., Sprynarova S., and Vamberova M. (1971) Body composition and fitness in obese children before and after special treatment. *Acta Paediatrica Scandinavica,* Supplement 217:80–85.
Parker M.W., Johanson A.J., Rogol A.D., Kaiser D.L., and Blizzard R.M. (1984) Effect of testosterone on somatomedin-C concentrations in prepubertal boys. *Journal of Clinical Endocrinology and Metabolism,* 58:87–90.
Parra A., Argote R.M., Garcia G., Cervantes C., Alatorre S., and Pérez-Pastén E. (1979) Body composition in hypopituitary dwarfs before and during human growth hormone therapy. *Metabolism,* 28:851–857.
Passmore R., Meiklejohn A.P., Dewar A.D., and Thow R.K. (1955) Energy utilization in overfed thin young men. *British Journal of Nutrition,* 9:20–27, 27–37.
Passmore R., Strong J.A., and Ritchie F.J. (1958) The chemical composition of the tissue lost by obese patients on a reducing regimen. *British Journal of Nutrition,* 12:113–122.
Passmore R., Strong J.A., Swindells Y.E., and Eldin N. (1963) The effect of overfeeding on two fat young women. *British Journal of Nutrition,* 17:373–383.
Patrick J. and Golden M. (1977) Leucocyte electrolytes and sodium transport in protein energy malnutrition. *American Journal of Clinical Nutrition,* 30:1478–1481.
Peckham S.C., Entenman C., and Carroll H.W. (1962) The influence of a hypercaloric diet on gross body and adipose tissue composition in the rat. *Journal of Nutrition,* 77:187–197.
Peppler W.W. and Mazess R.B. (1981) Total body bone mineral and lean body mass by dual photon absorptiometry. I. Theory and measurement procedure. *Calcified Tissue International,* 33:353–359.
Peters J.P. and Van Slyke D.D. (1946) *Quantitative Clinical Chemistry, Vol. I.* Baltimore: Williams & Wilkins.
Peterson R.E., O'Toole J.J., Kirkendall W.M., and Kempthorn O. (1959) The variability of extracellular fluid space (sucrose) in man during a 24 hour period. *Journal of Clinical Investigation,* 38:1644–1658.
Pfeiffer L. (1887) Über den Fettgehalt des Körpers und verschiedener Theile desselben bei mageren und fetten Thieren. *Zeitschrift für Biologie,* 23:340–380.
Phinney S.D., Horton E.S., Sims E.A.H., Hanson J.S., Danforth E., Jr., and LaGrange B.M. (1980) Capacity for moderate exercise in obese subjects after adaptation to a hypocaloric, ketogenic diet. *Journal of Clinical Investigation,* 66:1152–1161.
Picou D., Halliday D., and Garrow J.S. (1966) Total body protein, collagen and non-collagen protein in infantile protein malnutrition. *Clinical Science,* 30:345–351.
Pierson R.N., Jr., Lin D.H.Y., and Phillips R.A. (1974) Total-body potassium in health: effects of age, sex, height, and fat. *American Journal of Physiology,* 226:206–212.

Pierson R.N., Jr., Wang J., Colt E.W., and Neuman P. (1982) Body composition measurements in normal man: the potassium, sodium, sulfate and tritium spaces in 58 adults. *Journal of Chronic Diseases,* 35:419–428.

Pierson W.R. (1963) A photogrammetric technique for the estimation of surface area and volume. *Annals of the New York Academy of Sciences,* 110:109–112.

Pipe N.G.J., Smith T., Halliday D., Edmonds C.J., Williams C., and Coltart T.M. (1979) Changes in fat, fat-free mass and body water in normal human pregnancy. *British Journal of Obstetrics and Gynaecology,* 86:929–940.

Pitkin R.M. (1977) Components of weight gain during pregnancy. In H.A. Schneider, C.E. Anderson, and D.B.B. Coursin (Eds.), *Nutritional Support of Medical Practice.* Hagerstown, MD: Harper & Row, pp. 407–421.

Pitts G.C. (1962) Density and composition of the lean body compartment and its relationship to fatness. *American Journal of Physiology,* 202:445–452.

Pitts G.C. and Bullard T.R. (1968) Some interspecific aspects of body composition in mammals. In *Body Composition in Animals and Man.* Washington: National Academy of Sciences, Pub. No. 1598, pp. 45–70.

Pitts G.C., Ushakov A.S., Pace N., Smith A.H., Rahlman D.F., and Smirnova T.A. (1983) Effects of weightlessness on body composition in the rat. *American Journal of Physiology,* 244:R332–R337.

Plocher T.A. and Powley T.L. (1976) Effect of hypohysectomy on weight gain and body composition of the genetically obese yellow (A^y/a) mouse. *Metabolism,* 25:593–602.

Pollock M.L., Wilmore J.H., and Fox S.M. III (1984) *Exercise in Health and Disease.* Philadelphia: W.B. Saunders.

Poulos P.P., Pulos J.G., Pifer M., Van Woert W., and Parks M.E. (1956) A constant-change (single injection) method for the estimation of the volume of distribution of substances in body fluid compartments. *Journal of Clinical Investigation,* 35:921–933.

Poznanski A.K., Kuhns L.R., and Guire K.E. (1980) New standards of cortical mass in the humerus of neonates: a means of evaluating bone loss in the premature infant. *Radiology,* 134:639–644.

Preedy J.R.K. and Aitken E.H. (1956) The effect of estrogens on water and electrolyte metabolism. I. The normal. *Journal of Clinical Investigation,* 35:423–429.

Prentice A.M., Black A.E., Coward W.A., Davies H.L., Goldberg G.R., Murgatroyd P.R., Ashford J., Sawyer M., and Whitehead R.G. (1986) High levels of energy expenditure in obese women. *British Medical Journal,* 292:983–987.

Presta E., Segal K.R., Gutin B., Harrison G.G., and van Itallie T.B. (1983) Comparison in man of total body electrical conductivity and lean body mass derived from body density: validation of a new body composition method. *Metabolism,* 32:524–527.

Presta E., Wang J., Harrison G.G., Björntorp P., Harker W.N., and van Itallie T.B. (1983) Measurement of total body electrical conductivity: a new method for estimation of body composition. *American Journal of Clinical Nutrition,* 37:735–739.

Preston R.L. (1975) Biological responses to estrogen additives in meat producing cattle and lambs. *Journal of Animal Science,* 41:1414–1430.

Preston T., Reeds P.J., East B.W., and Holmes P.H. (1985) A comparison of body protein determination in rats by *in vivo* neutron activation and carcass analysis. *Clinical Science,* 68:349–355.

Prezio J.A., Carreon G., Clerkin E., Meloni C.R., Kyle L.H., and Canary J.J. (1964) Influence of body composition on adrenal function in obesity. *Journal of Clinical Endocrinology and Metabolism,* 24:481–485.

Price W.F., Hazelrig J.B., Kreisberg R.A., and Meador C.K. (1969) Reproducibility of body composition measurements in a single individual. *Journal of Laboratory and Clinical Medicine,* 74:557–563.

Pullar J.D. and Webster A.J.F. (1974) Heat loss and energy retention during growth in congenitally obese and lean rates. *British Journal of Nutrition,* 31:377–392.

Rahn H., Fenn W.O., and Otis A.B. (1949) Daily variation of vital capacity, residual air, and expiratory reserve including a study of the residual air methods. *Journal of Applied Physiology,* 1:725–743.

Rajagopalan B., Thomas G.W., Beilin L.J., and Ledingham J.G.G. (1980) Total body potassium falls with age. *Clinical Science,* 59 (Suppl. 6):427s–429s.

Rand W.M., Young V.R., and Scrimshaw N.S. (1976) Change of urinary nitrogen excretion in response to low protein diets in adults. *American Journal of Clinical Nutrition,* 29:639–644.

Ravussin E., Burnand B., Schutz Y., and Jéquier E. (1982) Twenty-four-hour energy expenditure and resting metabolic rate in obese, moderately obese, and control subjects. *American Journal of Clinical Nutrition,* 35:566–573.

Ravussin E., Schutz Y., Acheson K.J., Dusmet M., Bourquin L., and Jéquier E. (1985) Short term, mixed diet overfeeding in man: no evidence for "luxuskonsumption." *American Journal of Physiology,* 249:E470–E477.

Reichman B., Chessex P., Putet G., Verellen G., Smith J.M., Heim T., and Swyer P.R. (1981) Diet, fat accretion, and growth in premature infants. *New England Journal of Medicine,* 305:1495–1500.

Reid J.T. (Ed.) (1968) *Body Composition in Animals and Man* (Publication No. 1598). Washington: National Academy of Sciences.

Remes K., Knoppasalmi K., and Aldercrentz H. (1979) Effect of long-term physical training on plasma testosterone, androstenedione, luteinizing hormone and sex-hormone-binding globulin capacity. *Scandinavian Journal of Clinical and Laboratory Investigation,* 39:743–749.

Revicki D.A. and Israel R.G. (1986) Relationship between body mass indices and measures of body adiposity. *American Journal of Public Health,* 76:992–994.

Rigotti N.A., Nussbaum S.R., Herzog D.B., and Neer R.M. (1984) Osteoporosis in women with anorexia nervosa. *New England Journal of Medicine,* 311:1601–1606.

Ringe J.D., Rehpenning W., and Kuhlencordt F. (1977) Physiologie Änderung des Mineralgehalts von Radius und Ulna in Abhängigkeit von Lebensalter und Geschlecht. *Fortschritte auf dem Gebiete der Röntgenstrahlen und der Nuklearmedezin,* 126:376–380.

Rolland-Cachera M.F., Sempé M., Guilloud-Bataille M., Patois E., Péquignot-Guggenbuhl F., and Fautrad V. (1982) Adiposity indices in children. *American Journal of Clinical Nutrition,* 36:178–184.

Ross E.J., Marshall-Jones P., and Friedman M. (1966) Cushing's syndrome: diagnostic criteria. *Quarterly Journal of Medicine,* 35:149–193.

Rothwell N.J. and Stock M.J. (1979) Regulation of energy balance in two models of reversible obesity in the rat. *Journal of Comparative and Physiological Psychology,* 93:1024–1034.

Roux J.M., Tordet-Caridroit C., and Chanez C. (1970) Studies in experimental hypotrophy in the rat. *Biology of the Neonate,* 15:342–347.

Rowe J.W., Andres R., Tobin J.D., Norris A.H., and Shock N.W. (1976) The effect of age on creatinine clearance in men: a cross-sectional and longitudinal study. *Journal of Gerontology,* 31:155–163.

Rozovski S.J. and Winick M. (1979) Nutrition and cellular growth. In R.B. Alfin-Slater, and D. Kritchevsky (Eds.), *Human Nutrition, Vol. III.* New York: Plenum Press, pp. 61–102.

Rubner M. (1902) *Die Gesetze des Energieverbrauchs bei der Ernährung.* Leipzig and Vienna: Deutsch.

Rudman D., Millikan W.J., Richardson T.J., Bixler T.J. III, Stackhouse J., and McGarrity W.C. (1975) Elemental balances during intravenous hyperalimentation of underweight adult subjects. *Journal of Clinical Investigation,* 35:94–104.

Rundo J. and Sagild U. (1955) Total and "exchangeable" potassium in humans. *Nature,* 175:774.

Rush D., Stein Z., and Susser M. (1980) A randomized controlled trial of prenatal nutritional supplementation in New York City. *Pediatrics,* 65:683–697.

Russell D.McR., Prendergast P.J., Darby P.L., Garfinkel P.E., Whitwell J., and Jeejeebhoy K.N. (1983) A comparison between muscle function and body composition in anorexia nervosa: the effect of refeeding. *American Journal of Clinical Nutrition,* 38:229–237.

Rutledge M.M., Clark J., Woodruff C., Krause G., and Flynn M.A. (1976) A longitudinal study of total body potassium in normal breastfed and bottle-fed infants. *Pediatric Research,* 10:114–117.

Ryan R.J., Williams J.D., Ansell B.M., and Bernstein L.M. (1957) The relationship of body composition to oxygen consumption and creatinine excretion in healthy and wasted men. *Metabolism,* 6:365–377.

Satwanti B., Bharadwaj H., and Singh I.P. (1978) Estimation of body fat and lean body mass from anthropometric measurements in young Indian women. *Human Biology,* 50:515–527.

Saville P.D. and Whyte M.P. (1969) Muscle and bone hypertrophy. Positive effect of running exercise in the rat. *Clinical Orthopedics,* 65:81–88.

Schemmel R., Mickelsen O., and Gill J.L. (1970) Dietary obesity in rats: body weight and body fat accretion in seven strains of rats. *Journal of Nutrition,* 100:1041–1048.

Schmidt-Nielsen K. (1984) *Scaling: Why is Animal Size So Important?* New York: Cambridge University.

Schoeller D.A., Van Sauten D.W., Peterson D.W., Jaspan J., and Klein P.D. (1980) Total body water measurements in humans with ^{18}O and ^{2}H labelled water. *American Journal of Clinical Nutrition,* 33:2686–2693.

Schteingart D.E. and Conn J.W. (1956) Characteristics of the increased adrenocortical function observed in many obese patients. *Annals of the New York Academy of Science,* 131:388–403.

Schutte J.E. (1980) Prediction of total body water in adolescent males. *Human Biology,* 52:381–391.

Schutte J.E., Longhurst J.C., Gaffney F.A., Bastian B.C., and Blomqvist C.G. (1981) Total plasma creatinine: an accurate measure of total striated muscle. *Journal of Applied Physiology,* 51:762–766.

Scott E.C. and Johnston F.E. (1982) Critical fat, menarche, and the maintenance of menstrual cycles: a critical review. *Journal of Adolescent Health Care,* 2:249–260.

Scott H.W., Jr., Brill A.B., and Price R.R. (1975) Body composition in morbidly obese patients before and after jejunoileal bypass. *Annals of Surgery,* 182:395–404.

Sederberg-Olsen P. and Ibsen H. (1972) Plasma volume and extracellular fluid volume during long-term treatment with propanolol in essential hypertension. *Clinical Science,* 43:165–170.

Segal K.R., Gutin B., Presta E., Wang J., and van Itallie T.B. (1985) Estimation of human body composition by electrical impedance methods: a comparative study. *Journal of Applied Physiology,* 58:1565–1571.

Seitchik J. (1967) Body composition and energy expenditure during rest and work in pregnancy. *American Journal of Obstetrics and Gynecology,* 97:701–713.

Seitchik J., Alper C., and Szutka A. (1963) Changes in body composition during pregnancy. *Annals of the New York Academy of Sciences,* 110:821–829.

Shaffer P.A. (1908) The excretion of kreatinin and kreatin in health and disease. *American Journal of Physiology,* 23:1–22.

Shaffer P.A. and Coleman W. (1909) Protein metabolism in typhoid fever. *Archives of Internal Medicine,* 4:538–600.

Shaw J.C.L. (1973) Parenteral nutrition in the management of sick low birth weight infants. *Pediatric Clinics of North America,* 20:333–358.

Shaw J.C.L. (1976) Evidence for defective skeletal mineralization in low birth weight infants: the absorption of calcium and fat. *Pediatrics,* 57:16–25.

Sheng H.P., Garza C., Huggins R.A., and Smith E.O. (1982) Comparison of measured and estimated fat-free weight, fat, potassium and nitrogen of growing guinea pigs. *Growth,* 46:306–321.

Sheng H.-P. and Huggins R.A. (1979) A review of body composition studies with emphasis on total body water and fat. *American Journal of Clinical Nutrition,* 32:630–647.

Sheng H.P., Huggins R.A., Garza C., Evans H.J., LeBlanc A.D., Nichols B.L., and Johnson P.C. (1981) Total body sodium, calcium and chloride measured chemically and by neutron activation in guinea pigs. *American Journal of Physiology,* 241:R419–R422.

Shephard R.J. (1982) *Physiology and Biochemistry of Exercise.* New York: Praeger Scientific.

Shephard R.J., Hatcher J., and Rode A. (1973) On the body composition of the Eskimo. *European Journal of Applied Physiology,* 32:3–15.

Shires T., Williams J., and Brown F. (1961) Acute changes in extracellular fluids associated with major surgical procedures. *Annals of Surgery,* 154:803–810.

Shizgal H.M. (1981) The effect of malnutrition on body composition. *Surgery, Gynecology and Obstetrics,* 152:22–26.

Shizgal H.M. (1985) Body composition of patients with malnutrition and cancer. *Cancer,* 55:250–253.

Shizgal H.M., Forse R.A., Spanier A.H., and MacLean L.D. (1979) Protein malnutrition following intestinal bypass for morbid obesity. *Surgery,* 86:60–68.

Shizgal H.M., Spanier A.H., Humes J., and Wood C.D. (1977) Indirect measurement of total exchangeable potassium. *American Journal of Physiology,* 233:F253–F259.

Shukla K.K., Ellis K.J., Dombrowski C.S., and Cohn S.H. (1973) Physiological variation of total-body potassium in man. *American Journal of Physiology,* 224:271–274.

Sidney K.H., Shepard R.J., and Harrison J.E. (1977) Endurance training and body composition of the elderly. *American Journal of Clinical Nutrition,* 30:326–333.

Sievert R.M. (1951) Measurements of gamma radiation from the human body. *Archiv Fysik,* 3:337–346.

Sippell W.G., Dörr H.G., Bidlingmaier F., and Knorr D. (1980) Plasma levels of aldosterone, corticosterone, 11-deoxycorticosterone, progesterone, 17-hydroxyprogesterone, cortisol, and cortisone during infancy and childhood. *Pediatric Research,* 14:39–46.

Siri W.E. (1961) Body composition from fluid spaces and density. In J. Brožek and A. Henschel (Eds.), *Techniques for Measuring Body Composition.* Washington: National Academy of Sciences, pp. 223–244.

Sisson T.R.C., Whalen L.E., and Telek A. (1959) The blood volume of infants. II. The premature infant during the first year of life. *Journal of Pediatrics,* 55:430–446.

Sizonenko P.C. (1978) Endocrinology in preadolescents and adolescents. *American Journal of Diseases of Children,* 132:704–712.

Sjöström L. (1980) Fat cells and body weight. In A.J. Stunkard (Ed.), *Obesity* Philadelphia: W.B. Saunders, pp. 86–100.

Sjöström L., Kvist H., Cederblad A., and Tylén U. (1986) Determination of total adipose tissue and body fat in women by computed tomography, ^{40}K and tritium. *American Journal of Physiology,* 250:E736–E745.

Sköldborn H., Arvidsson B., and Andersson M.A. (1972) A new whole body monitoring laboratory. *Acta Radiology* (Suppl.) 313:233–241.

Skrabal F., Arnot R.N., Helus F., Glass H.I., and Joplin G.F. (1970) A method for simultaneous electrolyte investigations in man using ^{77}Br, ^{43}K, and ^{24}Na. *International Journal of Applied Radiation and Isotopes,* 21:183–191.

Skrabal F., Arnot R.N., Joplin G.F., and Fraser T.R. (1972) The effect of glucocorticoid withdrawal on body water and electrolytes in hypopituitary patients with adrenocortical insufficiency as investigated with ^{77}Br, ^{43}K, ^{24}Na and $^{3}H_2O$. *Clinical Science,* 43:79–90.

Slaughter M.H., Lohman T.G., and Boileau R.A. (1978) Relationship of anthropometric dimensions to lean body mass in children. *Annals of Human Biology,* 5:469–482.

Smith C.A. and Nelson N.M. (1976) *The Physiology of the Newborn Infant.* Springfield, IL: Charles Thomas, p. 159.

Smith D.M., Khairi M.R.A., Norton J., and Johnston C.C., Jr. (1976) Age and activity effects on the rate of bone mineral loss. *Journal of Clinical Investigation,* 58:716–721.

Smith D.M., Nance W.E., Kang K.W., Christian J.C., and Johnston C.C., Jr. (1973) Genetic factors in determining bone mass. *Journal of Clinical Investigation,* 52:2800–2808.

Soberman R., Brodie B.B., Levy B.B., Axelrod J., Hollander V., and Steele J.M. (1949) The use of antipyrine in the measurement of total body water in man. *Journal of Biological Chemistry,* 179:31–42.

Southgate D.A.T. and Hey E.N. (1976) Chemical and biochemical development of the human fetus. In *The Biology of Human Fetal Growth,* D.F. Roberts and A.M. Thomson (Eds.). London: Taylor & Francis, pp. 195–209.

Spady D.W., Payne P.R., Picou D., and Waterlow J.C. (1976) Energy balance during recovery from malnutrition. *American Journal of Clinical Nutrition,* 29:1073–1088.

Spanier A.H., Kurtz R.S., Shibata H.R., MacLean L.D., and Shizgal H.M. (1976) Alterations in body composition following intestinal bypass for morbid obesity. *Surgery,* 80:171–177.

Spears C.P., Hyatt K.H., Vogel J.M., and Langfitt S.B. (1974) Unified method for serial study of body fluid compartments. *Aerospace Medicine,* 45:274–278.

Spencer S.A., Vinter J., and Hall D. (1985) The effect in newborn rabbits of overfeeding on fat deposition, gross energetic efficiency, and metabolic rate. *Pediatric Research,* 19:127–130.

Spurr G.B., Barac-Nieto M., Lotero H., and Dahners H.W. (1981) Comparisons of body fat estimated from total body water and skinfold thicknesses of undernourished men. *American Journal of Clinical Nutrition,* 34:1944–1953.

Srikantia S.G. and Mohanram M. (1970) Antidiuretic hormone values in plasma and urine of malnourished children. *Journal of Clinical Endocrinology,* 31:312–314.

Stearns G. (1939) Mineral metabolism of normal infants. *Physiological Reviews,* 19:415–430.

Stein Z., Susser M., Saenger G., and Marolla F. (1975) *Famine and Human Development: The Dutch Hunger Winter of 1944–45.* New York: Oxford University Press.

Steinkamp R., Cohen N.L., Gaffey W.R., McKay T., Bron G., Siri W.E., Sargent T.W., and Isaacs E. (1965) Measures of body fat and related factors in normal adults—II:a simple clinical method to estimate body fat and lean body mass. *Journal of Chronic Diseases,* 18:1291–1307.

Stokholm K.H., Brøchner-Mortensen J., and Hoilund-Carlsen P.F. (1980) Increased glomerular filtration rate and adrenocortical function in obese women. *International Journal of Obesity,* 4:57–63.

Stone M.H., Rush M.E., and Lipner H. (1978) Responses to intensive training and methandrostenelone administration: II. hormonal, organ weights, muscle weights and body composition. *Pflügers Archiv,* 375:147–151.

Stouffer J.R. (1963) Relationship of ultrasonic measurements and x-rays to body composition. *Annals of the New York Academy of Sciences,* 110:31–39.

Streeter D.H.P., Stevenson C.T., Dalakos T.G., Nicholas J.J., Dennick L.G., and Fellerman H. (1969) The diagnosis of hypercortisolism. Biochemical criteria differentiating patients from lean and obese normal subjects and from females on oral contraceptives. *Journal of Clinical Endocrinology and Metabolism,* 29:1191–1211.

Stuart H.C., Hill P., and Shaw C. (1940) The growth of bone, muscle and overlying tissue as revealed by studies of roentgenograms of the leg area. *Monographs of the Society for Research in Child Development, V* (No. 3, serial No. 26), 1–190.

Stunkard A.J., Sørensen T.I.A., Hanis C., Teasdale T.W., Chakraborty R., Schull W.J., and Shulsinger F. (1986) An adoption study of human obesity. *New England Journal of Medicine,* 314:193–198.

Surveyor I. and Hughes D. (1968) Discrepancies between whole-body potassium content and exchangeable potassium. *Journal of Laboratory and Clinical Medicine,* 71:464–472.

Swan R.C., Madisso H., and Pitts R.F. (1954) Measurement of extracellular fluid volume in nephrectomized dogs. *Journal of Clinical Investigation,* 33:1447–1456.

Swanson W.W. and Iob V.I. (1933) Loss of minerals through the skin of infants. *American Journal of Diseases of Children,* 45:1036–1039.

Swanson W.W. and Iob V. (1940) Growth and chemical composition of the human skeleton. *American Journal of Diseases of Children,* 59:107–111.

Tager B.N. and Kirsch H.W. (1942) Creatinin excretion in women: clinical significance in obesity. *Journal of Clinical Endocrinology and Metabolism,* 2:696–699.

Talso P.J., Glynn M.F., Oester Y.T., and Fudema J. (1963) Body composition in hypokalemic periodic paralysis. *Annals of the New York Academy of Sciences,* 110:993–1008.

Talso P.J., Miller C.E., Carballo A.J., and Vasquez I. (1960) Exchangeable potassium as a parameter of body composition. *Metabolism,* 9:456–471.

Talso P.J., Spafford N., and Blaw M. (1953) Metabolism of water and electrolytes in congestive heart failure. *Journal of Laboratory and Clinical Medicine,* 41:281–286, 405–415.

Tanaka K., Kato H., Kikuchi K., et al. (1979) Anthropometric and body composition characteristics of Shindeshi Sumo wrestlers. *Journal of Physical Fitness in Japan,* 28:257–264.

Tanner J.M. (1952) The effect of weight-training on physique. *American Journal of Physical Anthropology,* 10:427–461.

Tanner J.M. (1962) *Growth at Adolescence,* second edition. Oxford: Blackwell Scientific.

Tanner J.M. (1974) Sequence and tempo in the somatic changes in puberty. In M.M. Grumbach, G.D. Grave, and F.E. Mayer (Eds.), *Control of the Onset of Puberty.* New York: J. Wiley & Sons, pp. 448–470.

Tanner J.M. and Whitehouse R.H. (1967) The effect of human growth hormone on subcutaneous fat thickness in hyposomatotrophic and panhypopituitary dwarfs. *Journal of Endocrinology,* 39:263–275.

Tanner J.M. and Whitehouse R.H. (1975) Revised standards for triceps and subscapular skinfolds in British children. *Archives of Disease in Childhood,* 50:142–145.

Taverner D., Bing R.F., Swales J.D., and Thurston H. (1985) Fluid volumes and hemodynamics in hypertension produced by chemical renal medullectomy. *American Journal of Physiology,* 249:H415–H420.

Taylor A., Aksoy Y., Scopes J.W., Mont G. du, and Taylor B.A. (1985) Development of an air displacement method for whole body volume measurement of infants. *Journal of Biomedical Engineering*, 7:9–17.

Thomas L.W. (1962) The chemical composition of adipose tissue of man and mice. *Quarterly Journal of Experimental Physiology*, 47:179–188.

Thomas R.D., Silverton N.P., Burkinshaw L., and Morgan D.B. (1979) Potassium depletion and tissue loss in chronic heart disease. *Lancet*, ii:9–11.

Thompson R.C. (1953) Studies of metabolic turnover with tritium as a tracer. II. gross studies on the rat. *Journal of Biological Chemistry*, 200:731–743.

Thurlby P.L. and Trayburn P. (1978) The development of obesity in preweanling *ob/ob* mice. *British Journal of Nutrition*, 39:397–402.

Toal J.N., Millar F.K., Brooks R.H., and White J. (1961) Sodium retention by rats bearing the Walker carcinosarcoma 256. *American Journal of Physiology*, 200:175–181.

Tobian L., Jr. and Binion J. (1954) Artery wall electrolytes in renal and DCA hypertension. *Journal of Clinical Investigation*, 33:1407–1414.

Tokunaga K., Matsuzawa Y., Ishikawa K., and Tarui S. (1983) A novel technique for the determination of body fat by computed tomography. *International Journal of Obesity*, 7:437–445.

Trayhurn P., James W.P.T., and Gurr M.I. (1979) Studies in the body composition fat distribution and fat cell size and number of "Ad," a new obese mutant mouse. *British Journal of Nutrition*, 41:211–221.

Trenkle A. and Willham R.L. (1977) Beef production efficiency. *Science*, 198:1009–1015.

Trotter M., Broman G.F., and Peterson R.R. (1960) Density of bones of white and Negro skeletons. *Journal of Bone and Joint Surgery*, 42A:50–58.

Trotter M. and Peterson R.R. (1969) Weight of bone during the fetal period. *Growth*, 33:167–184.

Trotter M. and Peterson R.R. (1970) Weight of the skeleton during postnatal development. *American Journal of Physical Anthropology*, 33:313–324.

Truax F.L. (1939) Equality of chloride space and extracellular space of rat liver. *American Journal of Physiology*, 126:402–408.

Tumbleson M.E., Tinsley O.W., Corwin L.A., Jr., Flatt R.E., and Flynn M.A. (1969) Undernutrition in young miniature swine. *Journal of Nutrition*, 99:505–518.

Turcotte D.L. and Schubert G. (1982) *Geodynamics*. New York: John Wiley & Sons.

Tzamaloukas A.H., Jackson J.E., Gallegos J.C., Long D.A., and McLane M.M. (1985) Distribution volume of ethanol as a measure of body water. *Mineral and Electrolyte Metabolism*, 11:123–130.

Tzankoff S.P. and Norris A.H. (1977) Effect of muscle mass decrease on age-related BMR changes. *Journal of Applied Physiology*, 43:1001–1006.

Tzankoff S.P. and Norris A.H. (1978) Longitudinal changes in basal metabolism in man. *Journal of Applied Physiology*, 45:536–539.

Uauy R., Scrimshaw N.S., Rand W.M., and Young V.R. (1978) Human protein requirements: obligatory urinary and fecal nitrogen losses and the factorial estimation of protein needs in elderly males. *Journal of Nutrition*, 108:97–103.

Uezu N., Yamamoto S., Rikimaru T., Kishi K., and Inoue G. (1983) Contributions of individual body tissues to nitrogen excretion in adult rats fed protein-deficient diets. *Journal of Nutrition*, 113:105–114.

Usher R., Shepard M., and Lind J. (1963) The blood volume of the newborn infant and placental transfusion. *Acta Paediatrica Scandinavica*, 52:497–512.

Van Es A.J.H., Vogt J.E., Niessen C.H., Veth J., Rodenburg L., Teeuwse V., and Dhuyvetter J. (1984) Human energy metabolism below, near and above energy equilibrium. *British Journal of Nutrition*, 52:429–442.

Van Gaal L.F., Snyders D., De Leeuw I.H., and Bakaert J.L. (1985) Anthropometric and calorimetric evidence for the protein sparing effects of a new protein supplemented low calorie preparation. *American Journal of Clinical Nutrition,* 41:540–544.

Van Itallie T.B. and Yang M.U. (1984) Cardiac dysfunction in obese dieters: a potentially lethal complication of rapid, massive weight loss. *American Journal of Clinical Nutrition,* 39:695–702.

van Seters A.P., Harinck H.I.J., Moolenaar A.J., and Gundlach B.L. (1983) Changes in body composition on Modifast, and electrolyte retention during carbohydrate refeeding. *Abstracts 4th International Congress on Obesity,* 51A.

van Wieringen J.C. (1972) *Secular Changes of Growth.* Leiden: Institute of Preventive Medicine.

Vaswani A.N., Vartsky D., Ellis K.J., Yasumura S., and Cohn S.H. (1983) Effects of caloric restriction on body composition and total body nitrogen as measured by neutron activation. *Metabolism,* 32:185–188.

Veall N., Fisher H.J., Browne J.C.M., and Bradley J.E.S. (1955) An improved method for clinical studies of total exchangeable sodium. *Lancet,* i:419–422.

Vernadakis A. and Woodbury D.M. (1964) Electrolyte and nitrogen changes in skeletal muscle of developing rats. *American Journal of Physiology,* 206:1365–1368.

Vestergaard P. and Leverett R. (1968) Constancy of urinary creatinine output. *Journal of Laboratory and Clinical Medicine,* 51:211–218.

Virtama P. and Helelä T. (1969) Radiographic measurements of cortical bone. Variations in a normal population between 1 and 90 years of age. *Acta Radiologica* (Stockholm), *Supplement 293.*

Viteri F.E. and Alvarado J. (1970) The creatinine height index: its use in the estimation of the degree of protein depletion and repletion in protein calorie malnourished children. *Pediatrics,* 46:696–706.

Voit E. (1901) Die Bedeutung des Körperfettes für die Eiweisszersetzung des hungernden Tieres. *Zeitschrift für Biologie,* 41:502–549.

von Bezold A. (1857) Untersuchungen über die vertheilung von Wasser, organischer Materie und anorganischen Verbindungen im Thierreiche. *Zeitschrift für Wissenschaftliche Zoologie,* 8:487–524.

von Hevesy G. and Hofer E. (1934) Die Verweilzeit des Wassers im menschlichen Körper, untersucht mit Hilfe von "schwerem" Wasser als Indicator. *Klinische Wochenschrift,* 13:1524–1526.

Von Porat B.T.D. (1951) Blood volume determinations with the Evans Blue dye method. *Acta Medica Scandinavica, Suppl. 256.*

Wagen A., Okken A., Zweens J., and Zijlstra W.G. (1985) Composition of postnatal weight loss and subsequent weight gain in small for dates newborn infants. *Acta Paediatrica,* 74:57–61.

Walgren M.C. and Powley T.L. (1985) Effects of intragastric hyperalimentation on pair-fed rats with ventromedial hypothalamic lesions. *American Journal of Physiology,* 248:R172–R180.

Wallace G.B. and Brodie B.B. (1939) The distribution of administered bromide in comparison with chloride and its relation to body fluids. *Journal of Pharmacology and Experimental Therapeutics,* 65:214–219.

Wallace W.M., Weil W.B., and Taylor A. (1958) The effect of variable protein and mineral intake upon the body composition of the growing animal. In G.E.W. Wolstenholme and M. O'Connor (Eds). *Ciba Foundation Colloquium on Ageing, Vol. 4.* Boston: Little Brown.

Walsh C.H., Soler N.G., James H., Harvey T.C., Thomas B.J., Fremlin J.H., Fitzgerald M.G., and Malius J.M. (1976) Studies in whole body potassium and whole body nitrogen in newly diagnosed diabetics. *Quarterly Journal of Medicine,* 45:295–302.

Walsh F.K., Heyward V.H., and Schau C.G. (1984) Estimation of body composition of female intercollegiate basketball players. *Physician and Sports Medicine*, 12:74–89.
Wang J. and Pierson R.N., Jr. (1976) Disparate hydration of adipose and lean tissue require a new model for body water distribution in man. *Journal of Nutrition*, 106:1687–1693.
Ward P. (1973) The effect of an anabolic steroid on strength and lean body mass. *Medicine and Science in Sports*, 5:277–282.
Warnold I., Carlgren G., and Krotkiewski M. (1978) Energy expenditure and body composition during weight reduction in hyperplastic obese women. *American Journal of Clinical Nutrition*, 31:750–763.
Wassner S.J. and Li J.B. (1982) N^{+}-methylhistidine release: Contributions of rat skeletal muscle, GI tract and skin. *American Journal of Physiology*, 243:E293–E297.
Waterhouse C., Terepka A.R., and Sherman C.D., Jr. (1955) The gross electrolyte composition of certain human malignant tissues. *Cancer Research*, 15:544–549.
Waterlow J.C., Cravioto J., and Stephen J.M.L. (1960) Protein malnutrition in man. *Advances in Protein Chemistry*, 15:163–238.
Waxman M. and Stunkard A.J. (1980) Caloric intake and expenditure of obese boys. *Journal of Pediatrics*, 96:187–193.
Webb P. and Annis J.F. (1983) Adaptation to overeating in lean and overweight men and women. *Human Nutrition: Clinical Nutrition*, 37C:117–131.
Webster J.D., Hesp R., and Garrow J.S. (1984) The composition of excess weight in obese women estimated by body density, total body water, and total body potassium. *Human Nutrition: Clinical Nutrition*, 38C:299–306.
Weil W.B., Jr. and Wallace W.M. (1960) The effect of alterations in extracellular fluid on the composition of connective tissue. *Pediatrics*, 26:915–924.
Weinland E. (1925) Über den Gehalt an einigen Stoffen beim Igel im Winterschlaf. *Biochemische Zeitschrift*, 160:66–74.
Weinsier R.L., Norris D.J., Birch R., Bernstein R.S., Wang J., Yang M.-U., Pierson R.N., Jr., and van Itallie T.B. (1985) The relative contribution of body fat and fat pattern to blood pressure level. *Hypertension* 1:578–585.
Weir E.G. and Hastings A.B. (1939) The distribution of bromide and chloride in tissues and body fluids. *Journal of Biological Chemistry*, 129:547–558.
Weisskopf V.F. (1977) The frontiers and limits of science. *American Scientist*, 65:405–411.
Weldon V.V., Kowarski A., and Migeon C.J. (1967) Aldosterone secretion rates in normal subjects from infancy to adulthood. *Pediatrics*, 39:713–723.
Welman A., Matter S., and Stamford B.A. (1980) Caloric restriction and/or mild exercise: effects on serum lipids and body composition. *American Journal of Clinical Nutrition*, 33:1002–1009.
Whedon G.D. (1956) Steroid hormones in osteoporosis. In E.T. Engle and G. Pincus (Eds.), *Hormones and the Aging Process*. New York: Academic Press, pp. 221–239.
Whedon G.D., Lutwak L., Rambaut P.C., Whittle M.W., Reid J., Smith M.C., Leach C., Stadler C.R., and Sanford D.D. (1976) Mineral and nitrogen balance study observations: the second manned skylab mission. *Aviation, Space and Environmental Medicine*, 47:391–396.
Whipple G.H. (1938) Protein production, exchange in the body including hemoglobin, plasma protein and cell protein. *American Journal of Medical Science*, 196:609–621.
Whipple G.H., Sisson T.R.C., and Lund C.J. (1957) Delayed ligation of the umbilical cord: its influence on the blood volume of the newborn. *Obstetrics and Gynecology*, 10:603–610.

Whitbourne S.K. (1985) *The Aging Body: Physiological Changes and Psychological Consequences*. New York: Springer-Verlag.

Whitelaw A.G.L. (1979) Subcutaneous fat measurement as an indication of nutrition of the fetus and newborn. In H.K.A. Visser (Ed.), *Nutrition and Metabolism of the Fetus and Infant*. The Hague: Martinus Nijhoff, pp. 131–143.

Whyte R.K., Haslam R., Vlainic C., Shannon S., Samulski K., Campbell D., Bayley H.S., and Sinclair J.C. (1983) Energy balance and nitrogen balance in growing low birth weight infants fed human milk or formula. *Pediatric Research*, 17:891–898.

Widdowson E.M. (1981) The demands of the fetal and maternal tissues for nutrients, and the bearing of these on the needs of the mother to "eat for two." In J. Dobbing (Ed.), *Maternal Nutrition in Pregnancy—Eating for Two?* New York: Academic Press, pp. 1–17.

Widdowson E.M. and Dickerson J.W.T. (1964) Chemical composition of the body. In C.L. Comar and F. Bronner (Eds.), *Mineral Metabolism*, Vol. 2, Part A. New York/London: Academic Press, pp. 2–247.

Widdowson E.M. and McCance R.A. (1960) Some effects of accelerating growth. I. General somatic development. *Proceedings of the Royal Society (London), Series B*, 152:188–206.

Wilde W.S. (1962) Potassium. In C.L. Comar and F. Bronner (Eds.), *Mineral Metabolism*, Vol. 1, Part B. New York/London: Academic Press, pp. 73–107.

Wiley F.H. and Newburgh L.H. (1931) The doubtful nature of "luxuskonsumption." *Journal of Clinical Investigation*, 10:733–744.

Wilkinson P.R., Issler H., Hesp R., and Raftery E.B. (1975) Total body and serum potassium during prolonged thiazide therapy for essential hypertension. *Lancet*, i:759–762.

Williams E.D., Boddy K., Brown J.J., Cumming A.M.M., Davies D.L., Harvey I.R., Haywood J.K., Lever A.F., and Robertson J.I.S. (1982) Whole body elemental composition in patients with essential hypertension. *European Journal of Clinical Investigation*, 12:321–325.

Williams E.D., Paterson P.J., Scott R., Boddy K., Haywood J.K., and Harvey I.R. (1980) Effects on body elemental composition of prophylactic diuretic treatment of urinary lithiasis. *Urological Research*, 8:49–52.

Wilmore D.W., Moylan J.A., Jr., Bristow B.F., Mason A.D., Jr., and Pruitt B.A., Jr. (1974) Anabolic effects of human growth hormone and high caloric feedings following thermal injury. *Surgery, Gynecology and Obstetrics*, 138:875–884.

Wilmore J.H. (1969) A simplified method for determination of residual lung volume. *Journal of Applied Physiology*, 27:96–100.

Wilmore J.H. (1983) Body composition in sport and exercise: directions for future research. *Medical Science in Sports and Exercise*, 15:21–31.

Wilson J.D. and Foster D.W. (Eds.) (1985) *Williams Textbook of Endocrinology*, seventh edition. Philadelphia: W.B. Saunders.

Winick M. and Noble A. (1967) Cellular response with increased feeding in neonatal rats. *Journal of Nutrition*, 91:179–182.

Winter J.S.D. (1978) Prepubertal and pubertal endocrinology. In F. Falkner & J.M. Tanner (Eds.), *Human Growth, Vol. II*. New York: Plenum Press, pp. 183–213.

Womersley J., Boddy K., King P.C., and Durnin J.V.G.A. (1972) A comparison of the fat-free mass of young adults estimated by anthropometry, body density and total body potassium content. *Clinical Science*, 43:469–475.

Womersley J., Durnin J.V.G.A., Boddy K., and Mahaffy M. (1976) Influence of muscular development, obesity, and age on the fat-free mass of adults. *Journal of Applied Physiology*, 41:223–229.

Woo R., Garrow J.S., and Pi-Sunyer F.X. (1982) Voluntary food intake during

prolonged exercise in obese women. *American Journal of Clinical Nutrition,* 36:478–484.
Woodward K.T., Randolph C.L., Jr., van Hoek R., Hartgering J.B., Claypool H.A., Manskey A.A., Jr., and Noble J.J. (1960) *The Walter Reed Whole Body Counting Facility.* Washington: Walter Reed Institute of Research.
Wynn V., Abraham R.R., and Densem J.W. (1985) Method for estimating rate of fat loss during treatment of obesity by calorie restriction. *Lancet,* i:482–486.
Yang M.-U., Barbosa-Saldivar J.L., Pi-Sunyer F.X., and Van Itallie T.B. (1981) Metabolic effects of substituting carbohydrate for protein in a low-calorie diet: a prolonged study in obese patients. *International Journal of Obesity,* 5:231–236.
Yao A.C., Lind J., Tüsala R., and Michelsson K. (1969) Placental transfusion in the premature infant with observation on clinical course and outcome. *Acta Paediatrica Scandinavica,* 58:561–566.
Young C.M., Blondin J., Tensuan R., and Fryer J.H. (1963) Body composition studies of "older" women, thirty to seventy years of age. *Annals of the New York Academy of Sciences,* 110:589–607.
Young C.M. and Di Giacomo M.M. (1965) Protein utilization and changes in body composition during weight reduction. *Metabolism,* 14:1084–1094.
Young C.M., Scanlon S.S., Im H.S., and Lutwak L. (1971) Effect on body composition and other parameters in obese young men of carbohydrate level of reduction diet. *American Journal of Clinical Nutrition,* 24:290–296.
Young C.M., Sipin S.S., and Roe D.A. (1968) Body composition of pre-adolescent and adolescent girls. *Journal of the American Dietetic Association,* 53:25, 357, 469, 579.
Young V.R., Garza C., Steinke F.H., Murray E., Rand W.M., and Scrimshaw N.S. (1984) A long-term metabolic balance study in young men to assess the nutritional quality of an isolated soy protein and beef proteins. *American Journal of Clinical Nutrition,* 39:8–15.
Zacharias L., Rand W.M., and Wurtman R.J. (1976) A prospective study of sexual development and growth in American girls: the statistics of menarche. *Obstetrical and Gynecological Survey,* 31:325–337.
Ziegler E.E., O'Donnell A.M., Nelson S.E., and Fomon S.J. (1976) Body composition of the reference fetus. *Growth,* 40:329–341.
Zlatnik F.J. and Burmeister L.F. (1983) Dietary protein in pregnancy: effect on anthropometric indices of the newborn infant. *American Journal of Obstetrics and Gynecology,* 146:199–203.

Index

MIX
Papier aus verantwortungsvollen Quellen
Paper from responsible sources
FSC® C105338

If you have any concerns about our products,
you can contact us on
ProductSafety@springernature.com

In case Publisher is established outside the EU,
the EU authorized representative is:
Springer Nature Customer Service Center GmbH
Europaplatz 3, 69115 Heidelberg, Germany

Printed by Libri Plureos GmbH
in Hamburg, Germany